Water and Climate
in the Western United States

Water and Climate
in the Western United States

William M. Lewis Jr., Editor

University Press of Colorado

Published by the University Press of Colorado
5589 Arapahoe Avenue, Suite 206C
Boulder, Colorado 80303

First paperback edition 2006
Printed in the United States of America

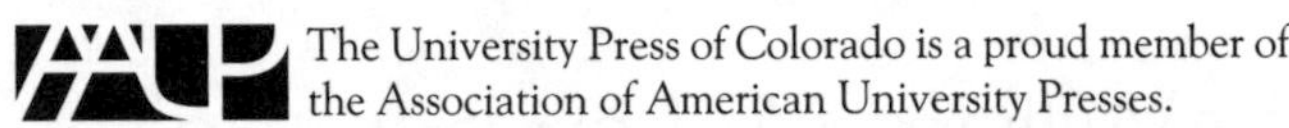

The University Press of Colorado is a proud member of the Association of American University Presses.

The University Press of Colorado is a cooperative publishing enterprise supported, in part, by Adams State College, Colorado State University, Fort Lewis College, Mesa State College, Metropolitan State College of Denver, University of Colorado, University of Northern Colorado, University of Southern Colorado, and Western State College of Colorado.

∞ The paper used in this publication meets the minimum requirements of the American National Standard for Information Sciences—Permanence of Paper for Printed Library Materials. ANSI Z39.48-1992

Library of Congress Cataloging-in-Publication Data

Water and climate in the western United States / William M. Lewis Jr., Editor.
p. cm.
Includes bibliographical references and index.
ISBN-10: 0-87081-727-2 (hardcover : alk. paper)
ISBN-13: 978-0-87081-854-7 (pbk : alk. paper)
ISBN-10: 0-87081-854-6 (pbk : alk. paper)
1. Water-supply—West (U.S.) 2. West (U.S.)—Climate. 3. Hydrometeorology—West (U.S.)
I. Lewis, William M., 1945–
TD223.6 .W34 2003
363.6'1'0978—dc21

2003000200

Design by Daniel Pratt

15 14 13 12 11 10 09 08 07 06 10 9 8 7 6 5 4 3 2 1

Contents

Acronyms and Abbreviations

AOP	Annual Operating Plan
bgd	billion gallons per day
BLM	Bureau of Land Management
C-BT	Colorado–Big Thompson
CC	climate category
CFG	California Fish and Game
CIG	Climate Impacts Group
CIRES	Cooperative Institute for Research in Environmental Sciences (University of Colorado, Boulder)
CPC	Climate Prediction Center
CRN	Climate Reference Network (NOAA)
CV	coefficient of variation
CVP	Central Valley Project
CVPIA	Central Valley Project Improvement Act
CWCD	Colorado Water Conservancy District
DSS	decision support system
DWB	Denver Water Board
DWD	Denver Water Department
DWR	Department of Water Resources (California)
EIR	environmental impact report
ENSO	El Niño Southern Oscillation
EPA	Environmental Protection Agency
ESA	Endangered Species Act
ESP	extended streamflow prediction
ET	evapotranspiration
GASP	Groundwater Appropriators of the South Platte
GCM	general circulation model
GFDL	Geophysical Fluid Dynamics Laboratory
GIS	geographic information system
GISS	Goddard Institute for Space Science
HDB	Hydrologic Database
hPa	hectoPascals
HRU	hydrologic response unit
IPCC	Intergovernmental Panel on Climate Change
km	kilometer
LROC	Long-Range Operating Criteria

m	meter
MLR	multiple linear regression
mm	millimeter
MM	mesoscale model
MMS	Modular Modeling System
MOS	Model Output Statistics
MRF	medium-range forecasting
NAO	North Atlantic Oscillation
NCAR	National Center for Atmospheric Research
NCEP	National Center for Environmental Prediction
NCWCD	Northern Colorado Water Conservancy District
NetCDF	Network Common Data Form
NOAA	National Oceanographic and Atmospheric Administration
NWS	National Weather Service
OGP	Office of Global Programs (NOAA)
PDO	Pacific Decadal Oscillation
PDSI	Palmer Drought Severity Index
PNW	Pacific Northwest
PRISM	Precipitation Regression on Independent Slopes Method
PRMS	Precipitation Runoff Modeling System
RAWS	Remote Automatic Weather Station
RCM	regional climate model
RMSE	root mean square error
SGML	Standard Generalized Markup Language
Snotel	Snowpack Telemetry
SOI	Southern Oscillation Index
SST	sea-surface temperature
SWE	snow-water equivalent
TAO	Tropical Atmosphere Ocean
TOGA	Tropical Ocean Global Atmosphere
UCAN	Unified Climate Access Network
USBR	U.S. Bureau of Reclamation
USDA	U.S. Department of Agriculture
USGS	U.S. Geological Survey
VEMAP	Vegetation Mapping and Analysis Project
VIC	Variable Infiltration Capacity
WGA	Western Governors' Association
WSWC	Western States Water Council
XML	Extensible Markup Language

Preface

Through an accident of nature, this book goes to press as much of the western United States experiences its most severe one-year drought in a number of decades. Many of the points discussed in hypothetical terms in this volume are in fact now well illustrated in Colorado and other states. Supply is inadequate to meet demand, and mechanisms for contingency allocation to uses of highest priority are clumsy at best. Conservation measures in cities have brought the urban population into a concern that in milder droughts primarily affects agriculture. Ancillary effects are numerous. Grass hay, for example, which sold in 2001 for about $3.50 per bale, was selling for $10 per bale as of July 2002. Forced sale of livestock is common because hay is either unavailable or overly expensive. Wildfire is rampant in dry timber, and fires once extinguished threaten mudflows and water-quality problems. Fish mortality has occurred in both reservoirs and flowing water, where warming caused by low flows is inconsistent with the welfare of salmonids. Thus, the chapters of this book dealing with interactions between climate variability and the management and consumption of water for human purposes are now under much more intense consideration than they were just a year ago.

I am grateful for the efforts of authors to prepare their manuscripts in a timely manner and to respond to the external reviewers who provided invaluable constructive criticism. I especially thank Leslie Northcott for preparing the manuscript for publication and Darrin Pratt and the staff of the publisher's office for their pleasant efficiency. Diana Perfect of the Cooperative Institute for Research in Environmental Sciences (CIRES) did all of the early work on solicitation and coordination of reviews before leaving for a new job with the National Oceanographic and Atmospheric Administration (NOAA). Credit also goes to the NOAA Office of Global Programs, which sponsored the book project, and to Susan Avery, director of CIRES, who has been a constant champion of this project. We hope this book can serve a clear need for information on the interaction between natural and human systems as they affect water, the most essential of resources.

—William M. Lewis Jr.
Boulder, Colorado

Introduction

William M. Lewis Jr.

Transplants to the western states from points east of the 100th meridian are often in search of something different. If so, they are gratified immediately by the spectacular openness of the West, which is largely a by-product of aridity, and by the stunning contrast of the mountains—with their snowfields and abundant perennial streams and forests—with the plains, which have none of these. The migrants will have been told since childhood that the West has a water problem, and the landscape indicates as much. At the same time, use of water seems as free and easy as it is back East. If anything, there is a stronger commitment to manage greenery in the West than in many other places with more water. Even corn growing to 8 or 10 feet, a sure sign of water, is commonplace.

What the migrant sees is the by-product of a century and a half of effort by immigrant westerners to manage the water cycle of the West to human ends (Cech, Chapter 11). To an astonishing extent, this attempt has been successful in putting virtually all annual runoff in the West at human disposal. From the viewpoint of John Wesley Powell and many other early observers, harnessing the water cycle in support of development would be a cause for celebration. Westerners take it for granted but also expend much time and effort analyzing, litigating, and pondering the long-term consequences of having such extensive control over a natural resource as essential as water.

The most frustrating aspect of managing water as completely as the West does is that the demands on management are not constant. Availability of water changes quantitatively and qualitatively from year to year in ways that reflect changes in both nature and society. The manager can never rest assured that today's management will be adequate tomorrow (Stakhiv, Chapter 9).

The West has used many of the devices of a technologically oriented society to hedge against water shortages of the most obvious type. Even a casual inspection of hydrologic records shows that interannual variation in water availability at any given point on the hydrologic network has been sufficient over the past twenty-five to fifty years to interfere with any management system that does not incorporate mechanisms for storage and orderly plans for prioritization of water uses. The storage requirement was anticipated very early in the water-development process and is reflected in the thousands of small reservoirs and gigantic federally constructed dams and distribution systems that characterize the West (e.g., Fulp, Chapter 10). American society, through its representatives and regulatory agencies, has decided collectively, however, that little more storage will be built (Getches, Chapter 14) and has left it to the technical experts to find some way of expanding the use of water without increasing the amount of storage.

Prior appropriation has been the system for orderly use of water in times of scarcity. It is thoroughly embedded in law and practice, and yet it has many critics and shows many deficiencies, the essence of which is rigidity and antiquation in the context of present trends in water use (Travis, Chapter 13; Getches, Chapter 14; Loomis et al., Chapter 15; Miller and Gloss, Chapter 16). Even so, no wholesale replacement is yet in view.

Despite constraints that derive from lack of additional storage and persistence of a largely unmodified prior appropriation system for allocation of water, the present system proves adequate to accommodate substantial irregularity in the availability of water from one year to the next. Past performance, however, is not a guarantee of future results. Even the recorded hydrologic record—short as it is—contains the example of the 1930s drought, which was extraordinary compared with conditions that have prevailed since then. A look further into the past, through reconstructions (Woodhouse, Chapter 3), is even more sobering, as such records contain droughts that make the drought of the 1930s seem mild by comparison. Thus, one concern of both the manager and the knowledgeable citizen is the sufficiency, or rather the degree of insufficiency, of the water distribution system in the event of an extreme drought.

The possibility of change in climate also raises persistent anxieties for water management (Stakhiv, Chapter 9). First, one would want to know whether there is likely to be more or less water under the most probable conditions of change. Climate modeling, although in essence hypothetical, can be applied to this question; it shows a surprising diversity of hydrologic responses across the West to synoptic climate change (Strzepek and Yates, Chapter 6).

Not only disaster but also opportunity can be found in the variability of climate. At present, water in the West is managed in the same way that a blackjack player manages bets: through a general knowledge of probability and a certain amount of adaptation to the amount of resource currently on the table. In fact, this is the only possible management strategy in the absence of any means of foretelling the future. Those who study climate, however, recently have shown that climate may be more predictable than it has appeared to be (Dole, Chapter 1). The underlying reason is that some features of climate are not truly stochastic but rather are under the control of master variables (e.g., sea-surface temperature in the Pacific) that have known effects in specific parts of the earth, including the American West. Although these connections are not yet fully exploited analytically, they offer the obvious promise of eliminating the assumption that nothing of next year's climate can be determined in advance. The complex terrain of the West will create problems in prediction, however (Redmond, Chapter 2; Reitsma and Pasteris, Chapter 4).

Prediction of climate, even to a limited degree, would allow more efficient use of water. In effect, it would be the equivalent of creating more storage, but the practical significance of this innovation depends on the reliability of the climatic predictions that can be developed, the degree to which hydrology can be predicted from climate (Clark et al., Chapter 5; Lettenmaier and Hamlet, Chapter 7; Labadie and Garcia, Chapter 8), and the ability of the management infrastructure to use whatever predictive information climatologists can supply. The latter is not a trivial problem, given that managers tend to be conservative—for perfectly understandable reasons. Management, however, will likely show some adaptability, particularly as it comes under pressure to provide water more flexibly and to accommodate changes in growth and allocation. Some intelligent planning and strategizing can probably accelerate the process of assimilation from climatology to hydrology and thence to management and operations.

Another substitute for storage in the face of variation is flexibility of water use. Experiments in flexibility are already visible in the form of temporary reallocations from uses of lower economic value to uses of higher economic value in times of stress (Loomis et al., Chapter 15). In the proper framework, which does not yet exist in most places, these exchanges can occur voluntarily and can be economically beneficial to both sides of the exchange. Alternatively, they could be legally mandated through an overhaul of the prior appropriation system.

Getches (Chapter 14) gives a fascinating glimpse into the dual views of water, one of which holds that water is a public resource and thus may be taken from those who hold it to serve the higher needs of society, whereas the other view holds that water is, under the prior appropriation system, property and thus is no more subject to taking than a piece of land under private ownership. Resolution of tension between these two views is hard to foresee. The view of water as private property still holds in most of the West, although water distributed from federal sources occupies a gray zone within which traditional rights of use could be revoked by the federal government. Treatment of nonfederal water, most of which was either paid for with cash or inherited as property, as something other than property would be disruptive to say the least, especially given that much of the land in the West has little value for agriculture or even for development without some contractually secure source of water.

Water in the West, although under intensive development for more than a century, is far from being a settled matter. It is clear from the chapters of this book that new scientific understanding of the water cycle and new technologies are emerging at a time when demands for water are expanding and changing and old policies and institutions are coming under severe criticism for their inadequacy. No certain conclusions are possible, but the authors of these chapters open many interesting possibilities and provide a convincing case that water watching in the West will be quite exciting over the next decade or two (Limerick, Chapter 17).

PART ONE

Prospects for Understanding and Predicting Variability of Climate

Studies of climatic variability are attracting unprecedented attention from those who manage natural resources. Whereas in the past the primary climatic variables of direct importance to humans (precipitation, temperature) have seemed essentially random variables for any given month or season at any given place, climatologists now have shown convincingly that previously unknown deterministic elements underlie climate variability. These deterministic elements, which are associated with the so-called climate forcing phenomena, can, through statistical analysis, produce some degree of predictability where climate forcing has strong effects. Even more exciting is the prospect that a fundamental understanding of climate forcing mechanisms may produce models that give predictions from first principles on the basis of data that may be available weeks or even months ahead of events that have great practical importance. At the same time, climatologists are showing with increasing spatial and temporal resolution the degree to which climate has varied in the past. Such information carries the obvious implication that we can expect extremes in climate for any given location that commonly exceed the extremes observed since record keeping was initiated.

In combination, these discoveries encourage us to think in terms of extensive use of climatology for advancing wise use of natural resources, especially water.

Chapters 1–4 give a sense of the current state of the art and future prospects for the analysis of climate variability. All four chapters give viewpoints on the potential connections between climate analysis and water management. In addition, they put the peculiarities of the West into context with a broader view of analysis and prediction of climate. The four chapters reveal deficiencies in present capability for analysis and prediction and in most cases propose or foresee remedies for these problems.

In Chapter 1, Dole gives a broad overview of climate prediction, including its methodological basis as well as the climate mechanisms that support prediction. He illustrates with examples the nature of current prediction products and the interface between analysis and prediction for climate and weather. He explains the probabilistic nature of prediction and shows that prediction must be perceived as probabilistic if it is to be useful. Dole makes informed judgments on the limits to predictability of climate and the basis for improvement of predictive skill. Key problems that he analyzes include downscaling as it applies to modeling and the special problems caused by topographic complexity in the West.

Redmond's chapter is an excellent companion for Dole's in that Dole deals primarily with climate, whereas Redmond deals primarily with weather, but both show awareness of continuity in the temporal scale of events that comprise weather and climate. Redmond summarizes the data sources and data types for weather observations and explains the numerous problems of using this information analytically. He underscores the difficulty of doing either analysis or prediction in topographically complex terrain and explains the inherent limitations of dealing with an inadequate observational network. He gives numerous examples of fine-scale variation in weather and explains its causes, as well as some trends that define general expectations for predicting specific kinds of weather events.

In Chapter 3, Woodhouse deals with the climate record as reconstructed through the use of proxy data. She explains the methods that have proven useful, as well as their limitations, and illustrates the possibility for reconstruction with some recently derived data of her own for two watershed segments in Colorado. The possibility of substantial climatic differences in watersheds belonging to the same drainage basin is one striking result of her work.

The final chapter in the set is by Reitsma and Pasteris, who summarize the results of a workshop on storage, retrieval, and use of climatic data. Both the authors and the workshop participants generally were very critical of current data management practices in the field of climatology. Although computing power and electronic storage capabilities have increased greatly, users of data have not always benefited from these technical advances to the degree that might be expected because of incompatibility among data sets and dispersion of data sets in a manner that makes retrieval difficult. The authors and workshop participants recommend a number of changes that would move the U.S. climate community toward a centralized database that is well documented and four-dimensional (the three spatial dimensions plus time), which would be a far better platform for analysis of climate than the current patchwork of databases.

Predicting Climate Variations in the American West: What Are Our Prospects?

ONE

RANDALL M. DOLE

> The West is defined . . . by inadequate rainfall, which means a general deficiency of water. We have water only between the time of its falling as rain or snow and the time when it flows or percolates back into the sea or the deep subsurface reservoirs of the earth. We can't create water, or increase it. We can only hold back and redistribute what there is.
>
> —Wallace Stegner
> *The American West as Living Space* (1994)

Water has always been a precious resource in the American West. Consequently, variations in climate—which play a crucial role in both water supply and demand—have had profound social, economic, and environmental consequences for the region. Extended periods of above-normal precipitation have provided ample water supplies to meet consumptive and agricultural water needs, as well as to sustain and perhaps even promote population growth. Conversely, periods of extended drought have exposed major vulnerabilities in both human and natural systems. With rapid population growth projected to continue for the foreseeable future, the effective management of water resources will become even more vital for the future of the West. Water managers will face increasingly difficult challenges as they attempt to address the often conflicting demands arising from consumptive needs, agriculture, energy production, recreation, environmental and species preservation, and flood protection. Advances in climate forecasts can provide information useful for projecting future water supply and demand and thereby support more effective water-management decisions in this region.

Despite their obvious potential utility, climate forecasts presently are little used in western water-management decisions. One of the principal reasons cited by water managers is the absence of forecasting skill; other important factors include their lack of familiarity with or difficulty in interpreting current forecasts and failure of forecasts to address directly the needs of users. This review first considers the problems and prospects of predicting climate variations in the West, focusing particularly on seasonal and longer timescales. A second section provides an overview of climate predictions, including a discussion of the scientific basis of climate forecasts as well as of the major factors that influence climate variations and forecasts in the western United States. Following this, several examples of climate forecasts are presented, emphasizing especially recent research aimed at developing new forecast products of particular relevance to the water-resources community. The chapter concludes with a summary of major issues that must be addressed if we are to develop new climate forecast capabilities that will better serve water management in this critically water-sensitive region.

OVERVIEW OF CLIMATE PREDICTIONS

General Considerations

Before discussing specific prediction problems related to western water, it is important to consider several basic characteristics of climate prediction. First and perhaps most important is that climate forecasts are intrinsically probabilistic. Whatever scientific advances are made, uncertainties will always be present in climate forecasts. Such uncertainties would occur even if scientists were able to construct a perfect forecasting model. Uncertainties result from the chaotic nature of the climate system. In such a system even small influences—the flap of a butterfly's wings, in the words of Edward Lorenz (1993)—could eventually lead to large differences in outcomes.

For large-scale storms—the predominant producers of wintertime precipitation in the American West—the uncertainty in forecasts is such that beyond a few weeks, skillful daily forecasts are impossible (see Chapter 2). This absence of predictive capability does not, however, imply that all hope for skillful forecasts is lost, as is sometimes stated. For example, it may be possible to predict average temperature or precipitation over given periods. Such forecasts are conventionally referred to as climate forecasts (as contrasted with weather forecasts), of which the most familiar examples are of seasonal mean temperature and precipitation. Climate forecasts are not limited, however, simply to expected mean conditions. For example, climate forecasts in the future could include high or low probability of storms, extremes of heat or cold, or other aspects of climate. Climate forecasts are usually expressed in terms of deviations from long-term (climatological) ex-

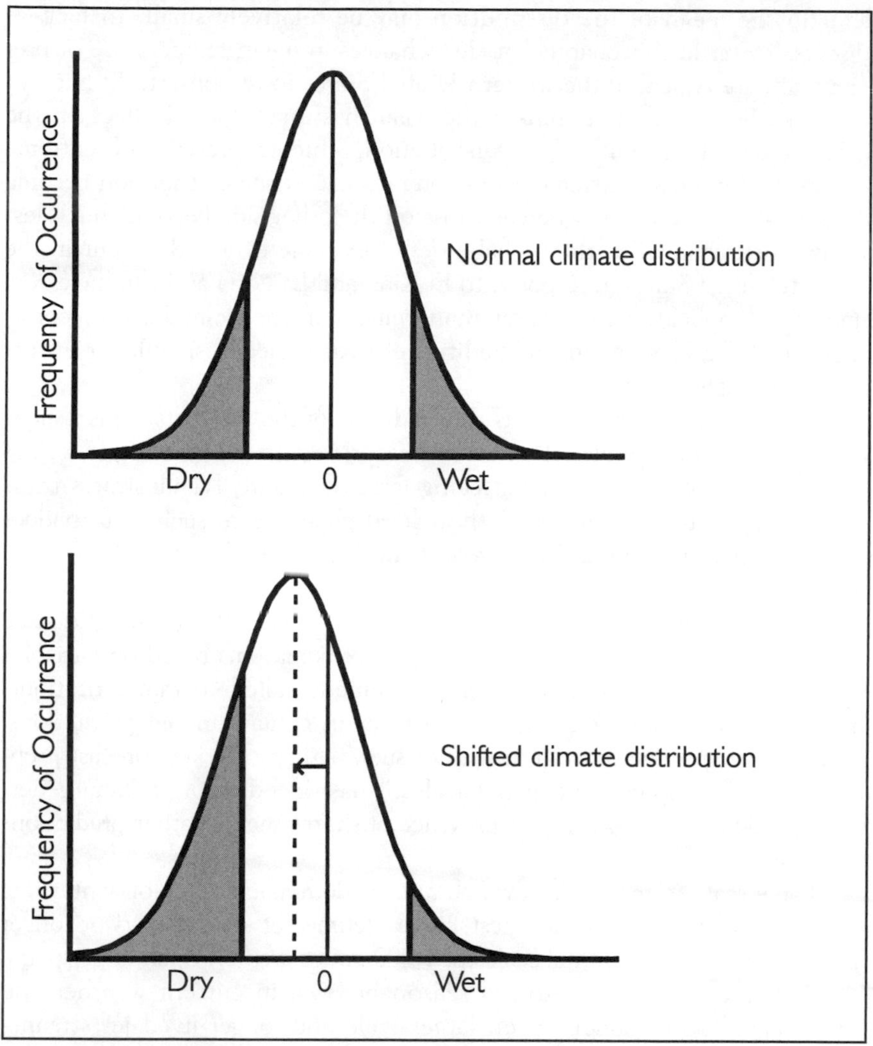

1.1 Illustration of the effect of a shift in the mean of a normal distribution on the frequency of occurrence of extreme events.

pectations, which would, in the absence of any other information, be the best prediction of future conditions.

Figure 1.1 illustrates how a simple shift in the probability distribution for precipitation influences the likelihood of precipitation in a particular region. The shift may be caused by some distant event such as an El Niño or La Niña condition in the tropical Pacific Ocean. As shown in Figure 1.1, a shift in the mean of the distribution may be relatively small. In fact, as discussed later in this chapter, modest changes in mean temperature or precipitation are typical of the western United States in response to El Niño or La Niña. Even a modest shift in the mean has a substantial effect on the probabilities at the tails of the distribution, which represent the extreme events. In particular, extreme events on one side of the distribution become considerably more likely, whereas those on the other side become much less likely. Such changes in the probabilities of extreme events do occur in the western United States in response to El Niño and La Niña and can have very important implications for water management if, for example, they correspond to large changes in probabilities of hydrologically significant events such as droughts or floods.

Another key consideration is the lead time of the forecasts. This issue is closely tied to the spatial scales that can be resolved in the forecast, as well as to the nature of the phenomena being forecast. Figure 1.2 illustrates these relationships schematically. Very short-lived phenomena, such as tornadoes and thunderstorms, are characterized by small spatial scales (1–10 km); these are below the limit of horizontal resolution for most climate models, which typically have resolutions of a few hundred km. Effects of such subgrid-scale processes are included in models through approximations based on variables that the model can resolve through a technique called "parameterization," one of the most important sources of error in climate-model predictions. From the perspective of western water issues, perhaps the key forecast problem on small temporal and spatial scales is flash flooding, a problem generally considered to be within the province of short-range weather predictions (Chapter 2).

Large-scale storm systems, which are the dominant source of wintertime precipitation throughout the West, have lifetimes of several days or longer and characteristic horizontal scales of 1,000 km or greater. Such large spatial and temporal scales are resolved reasonably well in current weather and climate-prediction models. Even larger-scale and longer-lived jet stream–level wave patterns influence storm tracks over weeks and longer and thus change the probability of storms over large regions.

Weather forecasts out to approximately one week are determined largely by the initial state of the atmosphere, although influences of ocean and land-surface conditions also can play a role (e.g., Paegle et al. 1990). As forecasts extend from a week to a season, the effects of slowly evolving boundary conditions, such as of the oceans or land surfaces, become increasingly important in determining the potential for predictability. On timescales of decades and longer, climate varies because of both natural and human-induced forcing; the former includes solar variations, volcanic activity, and slow changes in ocean circulation, and the latter includes greenhouse gases, aero-

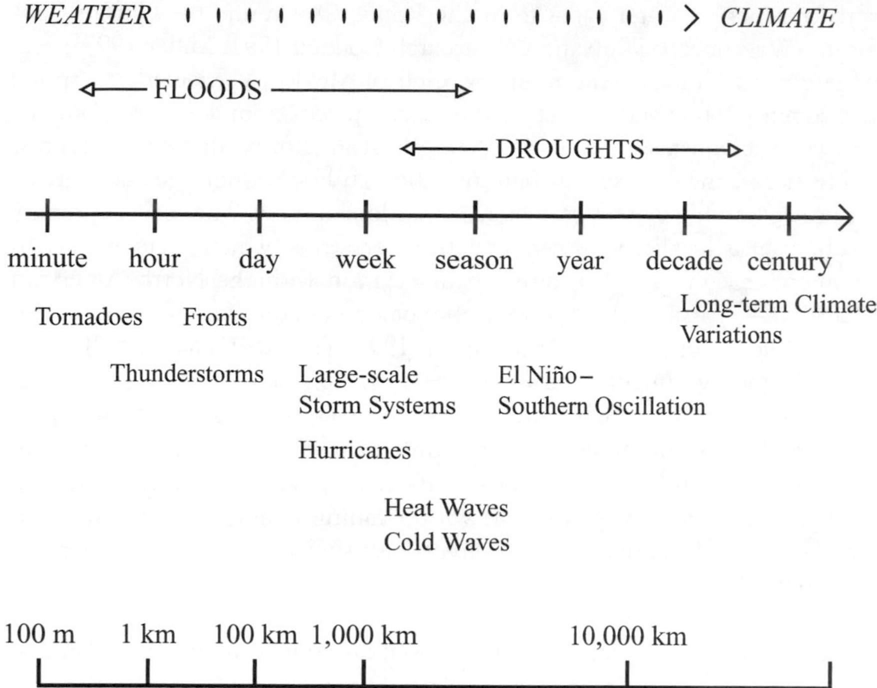

1.2 Representative time and distance scales for some important weather and climate phenomena.

sols, and—at least on local to regional scales—changes in land use (IPCC 2001; Pielke and Avissar 1990; Pielke et al. 1991).

A final major consideration in assessing potential predictability is the phenomenon to be predicted. The general hydrological prediction problem depends on many variables, such as precipitation (quantity and type), surface temperature, evaporation, and—more indirectly—wind, humidity, vegetative cover, and soil type (Chapters 5, 7). All of these variables are associated with different timescales of predictability and are contained either explicitly or implicitly in most climate models. This chapter focuses only on predictions of precipitation and temperature, both of which are fundamental to water supply and demand.

Important Climate Factors for the American West

As Redmond discusses in Chapter 2, topography is especially important in the western United States. Precipitation in the West is largely associated with transport of water vapor from the Pacific Ocean and the Gulf of California (Waggoner 1990; National Research Council 1991; Miller 1997); secondary contributions come from the Gulf of Mexico, particularly in spring and summer. Throughout much of the region, precipitation is caused primarily by two mechanisms. The first is synoptic-scale storms that have greatest intensities in the cold season (late fall through late spring) and show greatest precipitation generally in regions of upslope airflow. The second primary mechanism is localized showers and thunderstorms, which occur mainly in summer through early fall, often in association with the North American Monsoon—a particularly important phenomenon in the southwestern United States (Higgins et al. 1997; Barlow et al. 1998; Yu and Wallace 2000).

The relative importance of the two precipitation mechanisms varies widely over the West. Substantial year-to-year variation is of critical importance to hydrologic budgets throughout the region. As a generalization, however, more northerly regions—particularly western slopes—have an annual maximum precipitation in winter falling mainly as snow, whereas more southern locations at lower elevations usually have summer precipitation that falls as rain.

Mechanisms influencing interannual variability of precipitation in the West are only partially understood. Without question, however, the El Niño–Southern Oscillation (ENSO) phenomenon plays an important role. Major ENSO events typically occur every three to seven years and, because of the relatively slow rate of change in surface heat content of oceans, persist for several seasons once established (Philander 1990). Changes in sea-surface temperatures over the tropical Pacific Ocean shift tropical rainfall patterns in the direction of the warmest water, leading to changes in jet stream patterns that subsequently alter the weather over the American West and other parts of the world (Ropelewski and Halpert 1986, 1996; Kiladis and Diaz 1989; Diaz and Kiladis 1992). During warm ENSO events (El Niño), precipitation typically is anomalously high, especially in the Southwest, whereas during cold ENSO events (La Niña), precipitation in this region usually is anomalously low. As shown in Figure 1.3, El Niño and La Niña correspond to variations in sea-surface temperature (SST). For Arizona and New Mexico, above-normal rainfall corresponds with positive SST departures (El Niño), and La Niña shows the opposite pattern. Note also the apparent trend toward wetter and more frequent El Niño conditions in the past few decades.

Figure 1.4 shows seasonality of ENSO effects on the lower Colorado basin. For Figure 1.4, the top panel shows the distributions for winter, and the bottom panel shows the distributions for spring. The stars show quintiles where differences from the expected frequency (20%) are either significantly greater or less than expected (Wolter et al. 1999 give the method for estimating statistical significance). Note that in both winter and spring the

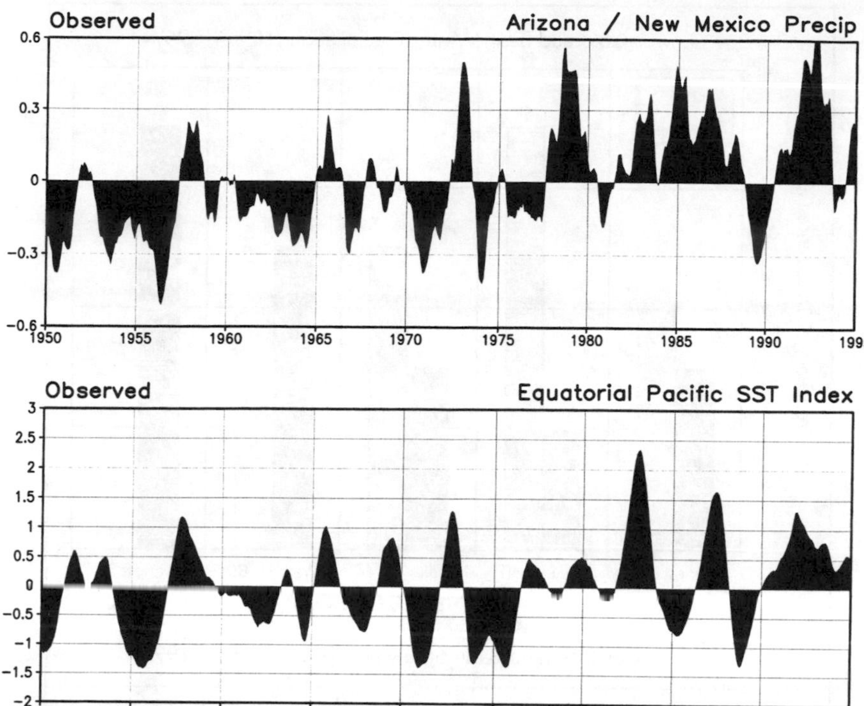

1.3 Time series for the period 1950 to 1995 for an index of annual precipitation over Arizona and New Mexico (TOP PANEL) and an equatorial sea-surface temperature (SST) index (BOTTOM PANEL). Positive values correspond to above-normal rainfall (TOP) or to warmer-than-normal SST index values (BOTTOM).

likelihood of high precipitation is significantly greater than normal in El Niño years, whereas under La Niña conditions the risk of very dry conditions is significantly increased (from K. Wolter, unpublished). Individual ENSO events differ substantially in the timing, location, and intensity of precipitation, with corresponding implications for predictability. Climate responses over the West to such variations in ENSO are not completely understood; such variations may be predictable to some extent in the future.

Longer-term climate variations are also quite evident throughout the West and have major implications for the hydrology of the region. Large, decadal-scale variability appears in part related to changes in the frequency and intensity of ENSO events and to a large-scale pattern of ocean–atmosphere variability called the Pacific Decadal Oscillation (PDO) (Trenberth

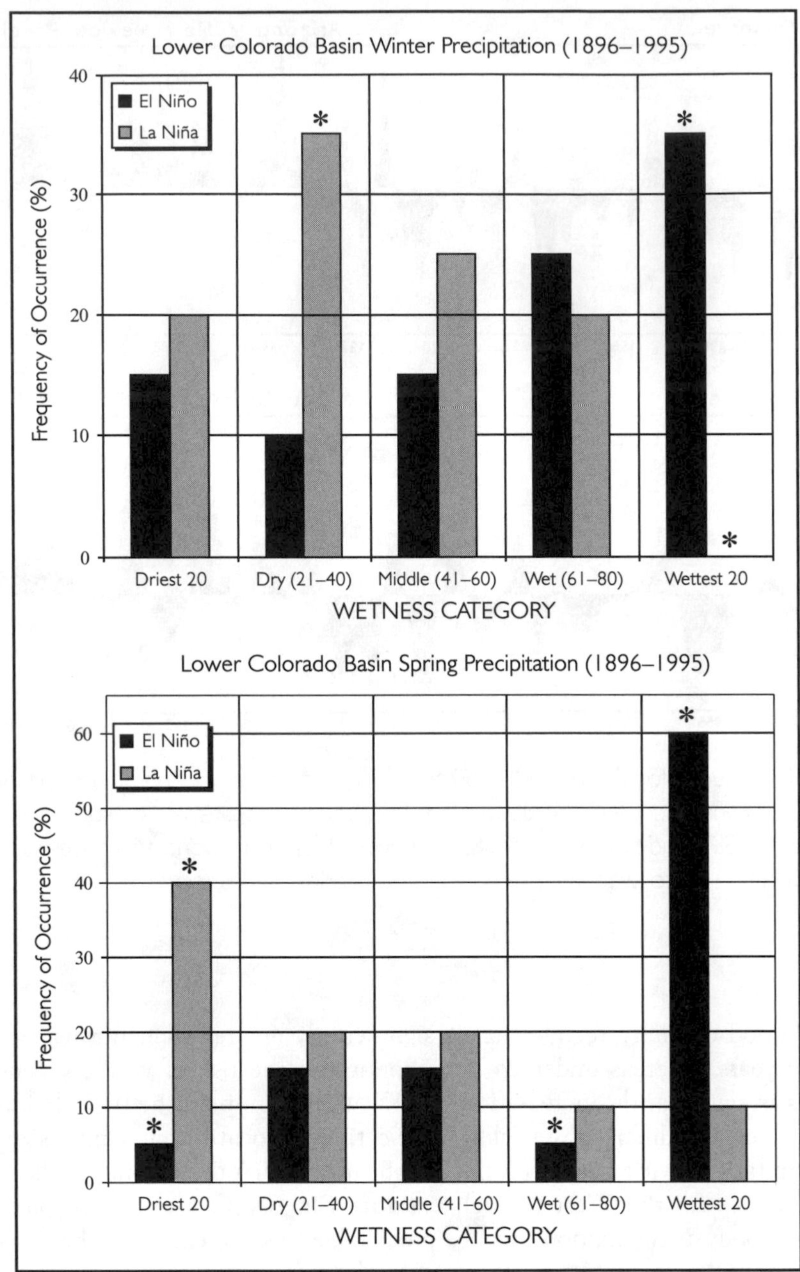

1.4 Distribution of seasonal precipitation for the lower Colorado basin, by quintiles, from driest to wettest, during El Niño or La Niña events for the 100-year period 1896–1995 (stars show statistical significance of departures from expected frequencies).

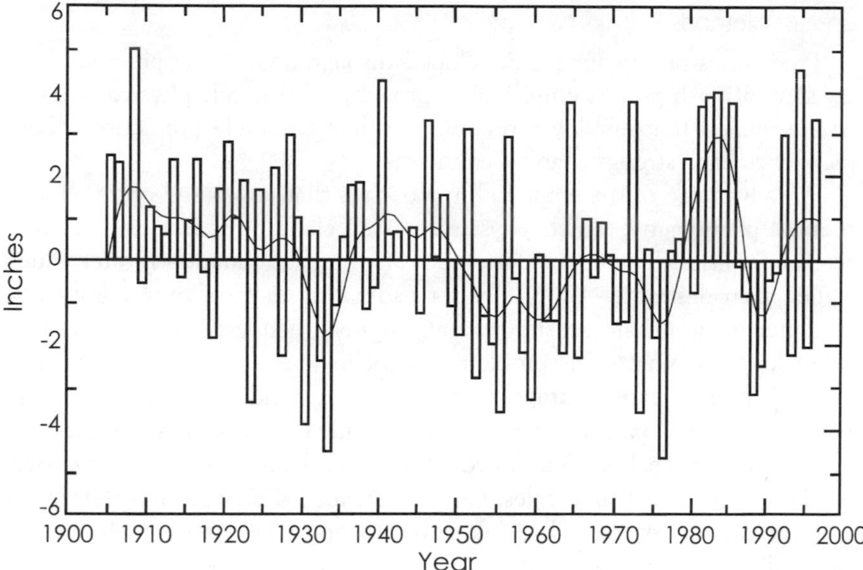

1.5 Annual precipitation for water years in the upper Colorado basin, 1900–2000. Bars denote annual deviations from the long-term mean, and the smooth curve is a running ten-year mean. Positive departures correspond to above-normal precipitation; negative departures correspond to below-normal precipitation.

and Hurrell 1994; Zhang et al. 1997; Gershunov and Barnett 1998). The mechanisms for the PDO and its potential links with ENSO are subjects of current research. To the extent that ENSO and PDO are independent phenomena and the PDO is predictable, significant implications exist for forecasts of longer-term variations in stream flow (Dettinger et al. 2000).

Decadal-scale variability of western climate is correlated with changes in annual stream flow for the Colorado River basin and with lake and groundwater levels in the Great Basin. Tree-ring data suggest that the first two decades of the twentieth century showed anomalously high runoff (Meko et al. 1991). Precipitation analyses also suggest that this twenty-year period was unusually wet (Figure 1.5). Many environmental and legal problems have arisen subsequently because this period of historically high runoff was used in determining water allocations of the Colorado River through the Colorado River Compact of 1922. An improved understanding of interannual and longer-term climate variations is potentially of major importance in the developing future water-management policies for the region (see Chapter 3).

Forecast Methods

Predictions of climate are developed through analyses of previously observed relationships (an empirical approach) or through physically based climate models that employ a technique called ensemble predictions. Both approaches have strengths and weaknesses.

An advantage of the empirical approach is that it is based on previously observed phenomena, which physically based climate models often cannot represent realistically, but the data record for empirical models is often much smaller than desirable—especially for estimating risks of extreme events. Computer models that can simulate atmospheric behavior provide one possibility for overcoming this problem. Paleoclimatic records, often obtained by proxy, also provide an important means of extending the record of climate. Some recent paleoclimatic studies provide evidence that empirical relationships between ENSO and droughts in the United States change over decadal to centennial timescales, perhaps because of changes in other large-scale climatic factors (Cook et al. 2000; Chapter 3). As such changes in statistical relationships may also occur in the future, climate models may offer some advantages over empirical approaches in developing climate forecasts.

Model-based ensemble predictions apply a Monte Carlo strategy—a strategy that uses repeated model runs to estimate probability distributions. Climate-prediction models are run many times with slightly different initial conditions; probability distributions of climate variables are estimated from the results of multiple runs. The choice of different initial states reflects both the incomplete and approximate nature of atmospheric observations and the fact that evolution of weather is very sensitive to these uncertainties.

Modeling has some inherent shortcomings. One is that models must contain many approximations as to how the climate system behaves. Errors in these approximations cause errors in predictions that typically will grow with time. Although in principle climate models can be run many times to overcome deficiencies caused by small numbers of samples, in practice, because of limitations in computing power, operational forecast ensembles typically incorporate only ten to twenty model runs. Computing power also limits the distance scales that can be resolved in most climate models to a few hundred kilometers. Thus the details of topographic features, which are important to western climate, are not properly resolved. In fact, some of the coarser resolutions do not segregate the Great Basin at all but rather merge the Sierras and Rocky Mountains into one larger and substantially flatter range. The problem of applying climate-model forecasts of coarse resolution to the scales of interest for water managers and other decision makers ("downscaling") is a key issue in making climate forecasts useful; prospects for progress in this area will be discussed later in this chapter.

EXAMPLES OF CLIMATE FORECASTS

Most of this section focuses on seasonal to interannual climate forecasts, which are now produced routinely by the National Oceanographic and Atmospheric Administration's (NOAA) Climate Prediction Center (CPC). Some aspects of predictions on other timescales also are discussed because of their relevance to a wide array of water-management issues.

Seasonal to Interannual Climate Forecasts

Plate 1 (top panel) gives an example of an official forecast produced by the NOAA Climate Prediction Center. This forecast was issued in mid-December 1998 for precipitation in January–March 1999. Also shown is the actual precipitation that occurred during the same period (middle panel). The distribution of observed precipitation is given by ranks; the numbers indicate that a given location is in the top five driest or wettest years in the 105-year period since 1895. Prior estimates of the precipitation probabilities for the same period are shown in the bottom panel, which is based on data from ten past events that the CPC considers to be similar (analogous). The forecast distributions are for three equal probability classes (above normal, near normal, below normal). A shift in the forecast distribution toward higher values is indicated by green; brown indicates a shift in the other direction. For example, in the first panel a +20 percent shift would indicate an increase in probability of above-normal precipitation from 33 percent to 53 percent. The numbers indicate the percentage of time in the analogous periods that events occurred in the opposite tercile to the most commonly experienced; for example, over New Mexico, most analogous periods were drier than normal, but up to 10 percent of the years were wetter than normal.

The forecast issued by the CPC was developed subjectively after careful consideration of information derived from both past empirical analyses and climate-model forecasts and thus can be considered a hybrid forecast. The period covered by the forecast showed weak La Niña conditions, and CPC forecasters placed particularly heavy weight on similar past situations, as illustrated in the bottom panel. From analyses of ten past weak La Niña events, forecasters noted that over the southwestern United States most prior cases were in the driest third of the precipitation distributions and, conversely, that very few such years were in the wettest third. In addition to the La Niña conditions, forecasters considered current climate conditions, the state of other modes of climate variability such as the North Atlantic Oscillation (NAO; Hurrell 1996; Hurrell and van Loon 1997), and long-term trends (Huang et al. 1996). Several major aspects of the precipitation variations over the United States were well predicted, particularly the Southwest (dry) and Southeast (dry) and the Pacific Northwest (wet), which are regions where ENSO signals tend to be relatively strong. Some notable deviations were also seen between predictions and observations (e.g., over portions of the northern plains). These differences do not necessarily reflect errors in the forecast because forecasts are probabilistic; rather, they reflect outcomes estimated to be less probable but still possible, much as a biased coin might occasionally land on its less probable side. It is for this reason that analysis of climate forecast skill requires careful assessment of the outcomes of many forecasts.

An area of climate research particularly relevant to hydrological applications involves estimating how particular climate conditions, such as ENSO, alter the probability of extreme events (e.g., Gershunov and Barnett 1998; Wolter et al. 1999; Sardeshmukh et al. 2000). Plate 2 shows an example of such an estimate derived empirically from data over the 100-year period 1896–1995. The figure shows estimated increases in the risk of spring conditions that are very wet (upper 20%) or very dry (lowest 20%), given either El Niño (top panel) or La Niña (bottom panel) conditions the preceding winter. Here risk is defined as the ratio (expressed as percentage) of observed extreme events to the number expected by chance (i.e., the climatological or unconditional risk). Thus 200 percent corresponds to an estimated doubling of the likelihood of extreme events during El Niño (or La Niña) conditions. The results suggest a significantly enhanced likelihood of very wet springs across much of the Southwest following El Niño winters and of very dry springs over the Southwest following winter La Niña conditions. Wolter et al. (1999) present more detailed descriptions of this approach to the analysis of temperature over the continental United States.

ENSO influences seasonal climate variations in the West, particularly in winter and spring, with consequent effects on stream flow and risk of flood (Cayan and Webb 1992; Dettinger et al. 2000). Progress in predicting tropical Pacific sea-surface temperatures also has advanced to the point that forecasts with useful skill are now possible out to at least six months through a variety of statistical or dynamical ocean modeling methods (Barnston et al. 1994). Comparisons of empirical and climate-model forecasts of ENSO effects outside the tropics show that the two methods have approximately the same skill for predictions of seasonal mean climate anomalies (Anderson et al. 1999; Hoerling and Kumar 2000). Some evidence suggests, however, that as computer power increases and climate models continue to be improved, more skillful predictions may be possible with the models—particularly at times when ENSO events have large amplitudes (Hoerling et al. 1997; Hoerling and Kumar 2000).

Because of the short length of the modern data record, climate models provide otherwise unavailable estimates of change in climate probability distributions associated with specific El Niño and La Niña events (Sardeshmukh

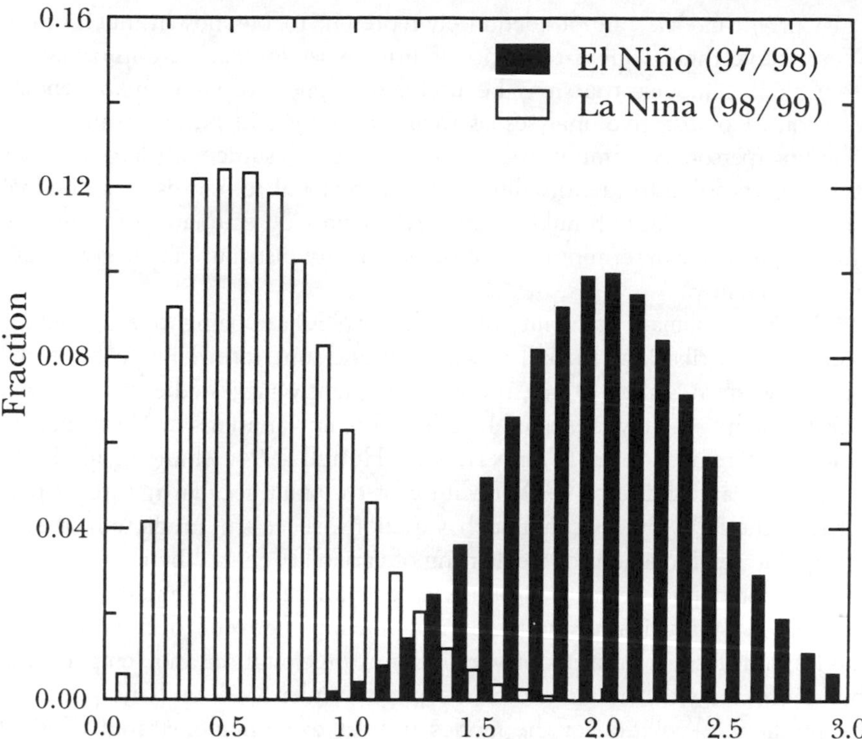

1.6 Estimates of probability distributions obtained from ensemble model runs for November–March precipitation in New Mexico for an El Niño year (1997–1998) and a La Niña year (1998–1999). Precipitation is normalized to average seasonal values (from M. Hoerling, NOAA Climate Diagnostics Center, unpublished).

et al. 2000). Figure 1.6 shows an example of such an estimate for November–March precipitation in 1997–1998 (a major El Niño event) and in 1998–1999 (a weak to moderate La Niña event), as obtained from ensemble analyses with an atmospheric general circulation model that was run many times using observed ocean conditions for the two years but with all other conditions identical, for New Mexico. In addition, a statistical resampling technique was applied to the data used in the model to improve the validity of estimates of the two distributions. The results show only a small overlap in the seasonal precipitation distributions for the El Niño and La Niña years. Modeling shows changes in expected seasonal mean precipitation; extremely wet conditions in the cold season are much more likely with El Niño, whereas drought is more likely during La Niña; none of the El Niño runs suggested drought conditions.

Current models cannot adequately represent topography in much of the West. Nevertheless, near-term possibilities exist for deriving information from climate models that may be useful in support of water-management decisions. Comparative analyses of weak and strong El Niño events by M. Hoerling (personal communication) suggest that with sufficiently large samples it may be possible to identify differences in regional responses to individual El Niño events. Sardeshmukh et al. (2000) provide estimates of ensemble sizes required for determining forecast probability distributions at given levels of reliability.

Modes of climate variability other than ENSO are being investigated for potential contributions to skill in seasonal and long-term climate forecasts. For the western United States, the most frequently suggested example is the PDO, which is usually identified by decadal-scale variations in North Pacific sea-surface temperatures (Trenberth and Hurrell 1994; Zhang et al. 1997; Gershunov and Barnett 1998). Because of the multidecadal nature of patterns of the PDO or any other such variations, successful prediction of evolution and reinforcement or weakening of variability caused by ENSO would have major implications for western water management. Numerous studies have shown statistically significant relationships between decadal climate modes and decadal patterns of variability in precipitation and temperature, but cause-and-effect relationships are poorly understood, and the potential for predicting evolution of the modes has yet to be demonstrated (Enfield and Mestas-Nuñez 2000).

Decadal to Centennial Climate Change

The possibility of future climate change has enormous significance for western water management, particularly for planning and designing infrastructure for future water requirements. Recent developments in climate-change research are discussed extensively in the Third Assessment Report of the Intergovernmental Panel on Climate Change (IPCC 2001). At this point, climate-change projections for regions such as the West are considered unreliable, especially for precipitation. Most models do suggest substantial warming over much of North America, including the West. Some observational analyses (Livezey and Smith 1999) indicate general warming across this region during winter over the past few decades (Plate 3), and evidence from plant phenological records—such as those for flowering of lilac and honeysuckle—suggests an earlier onset of spring throughout much of the western United States (Cayan et al. 2001). Models indicate a likely acceleration of continental warming over the twenty-first century (IPCC 2001). Even without changes in precipitation, such warming would have major implications for western water management. Potential effects would include changes in the annual water cycle (e.g., timing of peak flows), in snowpack and water storage, and in water demand for urban and agricultural needs. Improved climate-change projections on regional scales are urgently needed for the benefit of those who deal with water resources.

PROSPECTS FOR THE FUTURE

Regional climate prediction is still in its early stages. Although the ultimate skill of such predictions is likely to be modest, significant progress has been achieved in recent years, and advances probably will continue over the next decade and beyond. Even modest skill may convey great benefits to society, particularly if the probabilistic nature of climate forecasts is understood and applied effectively in decision-making processes. Future prospects for advances in the use of climate forecasts for western water management will depend on our ability to deal with major challenges that are not only scientific but also technical and institutional (Chapters 9, 16).

One of the most important scientific challenges for the next decade is to downscale climate information and forecasts from global to regional and local scales where most water-management decisions are made. This will require greatly improved regional observational capabilities, as emphasized by Redmond (Chapter 2), as well as the development of refined statistical and dynamical downscaling methods. Current computing capabilities in the United States are inadequate to run global climate models at the resolutions required for water-management purposes (National Research Council 2001). Although major advances in such capabilities are likely over the next decade, near-term improvements in downscaling will be required—most likely through statistical techniques or by nesting high-resolution regional models within global models of coarse resolution.

Although considerable progress has been made in identifying ENSO-climate links in the West, further empirical and model-based research is required to extract maximum predictive information from this known climate signal. Identifying sources of predictability other than ENSO also will be important in advancing forecast capabilities. One focus of current research is on multidecadal modes of variability, such as the PDO and NAO. Potential predictability over decadal periods is unknown, but if such capabilities could be developed, they would have tremendous ramifications for western water management. The role of oceanic conditions is well recognized, but improvements are required in modeling the effects of other boundary processes, such as those involving land surface and vegetation—especially for predicting warm-season temperatures and perhaps also summer precipitation.

Studies of the connections between weather and climate produced rapid advances in research over the past decade; advances are likely to continue, with important practical implications. For example, El Niño and La Niña influence not only the seasonal-mean precipitation but also the frequencies, intensities, and types of storms (e.g., warm subtropical storms with high snowfall levels versus colder storms with low snowfall levels). For the West, a major research priority will be to determine the roles of both remote ocean forcing and local land-surface processes on the potential predictability of warm-season precipitation, particularly that associated with the North American Monsoon.

Projections of long-term climate change must also be refined and focused on regional scales. Critical issues include, in addition to changes in precipitation and temperature, changes in seasonal cycles as well as frequency and magnitude of precipitation types (i.e., rain versus snow). Ultimately, improved predictions of hydrologic budget terms from seasonal to multidecadal timescales in regions such as the Interior West will need to link climate variability on global scales to basin scales and will include critical couplings of atmospheric and regional hydrologic and terrestrial processes. This will require significant advances in computing capabilities and an improved understanding of the wide array of processes that control the regional hydrologic cycle.

Examples of potential future climate products that appear feasible over the next several years include (1) improved drought monitoring and forecasting, (2) downscaling of climate forecasts to finer regional to local scales, (3) quantitative estimates of risk for extreme events, (4) improved early warning for hazards (assessment of threats a week to a season ahead), (5) improved estimates of probability distributions of precipitation and temperature at various lead times, and (6) forecasts of variables more directly linked to applications.

It is possible to envision the development of a strategy for providing forecasts of high-impact events related to weather and climate (Figure 1.7). At lead times of one or more seasons, estimates of seasonal climate risks for events such as major floods or droughts can be based on the state of ENSO or other agents of climate forcing. Such forecasts would provide general information on increased or reduced risks in a given season, but because of limits to predictability discussed earlier, they would not be able to provide detailed information on the timing of such events. The earliest plausible lead time for predicting the timing and regional extent of such events would be approximately one to two weeks. Although forecasts would be probabilistic, decision makers who need time to implement mitigation strategies may find this information particularly useful. At shorter timescales, probabilistic forecasts would eventually merge into conventional weather warnings, such as those now provided by the National Weather Service. This staged approach is consistent with a general strategy being developed by NOAA to establish a seamless suite of products on both climate and weather.

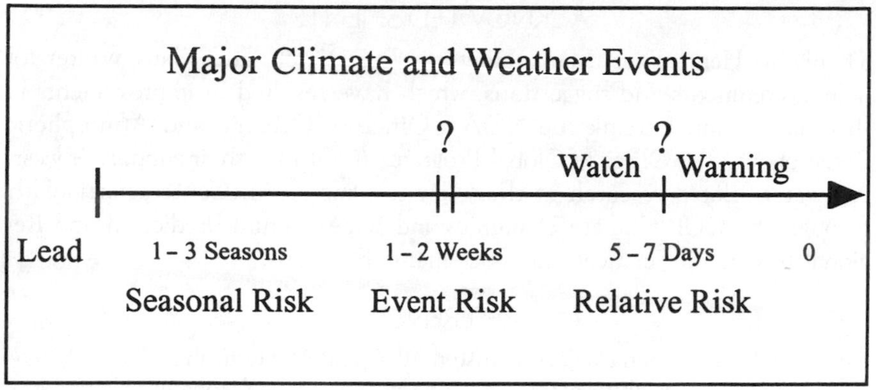

1.7 Diagram of a prediction-staging strategy for major weather and climate events (see text for explanation).

A key issue in all cases will be to translate forecasts or projections into forms that will maximize their usefulness for decision makers. This will require mechanisms that facilitate interaction and communication between scientists and users of climate information. For regional applications, such interactions are likely to be developed best in cooperation with existing institutions that have established regional or local credibility, such as universities or extension services, local weather service forecast offices, river forecast centers, or regional climate centers. The Regional Integrated Science and Assessments program supported through the NOAA Office of Global Programs, in connection with universities throughout the West, provides an innovative model of such an approach.

Perhaps the most difficult challenge for the future will be to overcome legal and institutional barriers that presently limit the most effective use of climate forecasts or that may even discourage the use of innovative new climate products (see Chapter 14). The stakes are great, as changes in demography and land use together with potential future changes in climate present increasing risk of conflict over water at local, regional, and even international levels. Perhaps because of this, there is room for optimism that the West can serve as a model for innovative research that will overcome institutional barriers and develop society's capability to adapt to climate variability and climate change. If this is to occur, effective multidisciplinary partnerships must be developed and sustained. Partnerships are essential because, especially in the American West, as noted often by Bruce Babbitt during his term as Secretary of the Department of the Interior, "water connects us all."

ACKNOWLEDGMENTS

Thanks to Henry Diaz, Marty Hoerling, Don Mock, and Klaus Wolter for helpful comments and suggestions, which have resulted in improvements in this manuscript. I thank the NOAA Office of Oceanic and Atmospheric Research and the Office of Global Programs (OGP) for their support of western water-related research at the Climate Diagnostics Center, particularly through the OGP Climate Dynamics and Experimental Prediction and Regional Integrated Science and Assessments programs.

REFERENCES

Anderson, J., H. van den Dool, A. Barnston, W. Chen, W. Stern, and J. Ploshay. 1999. Present-day capabilities of numerical and statistical models for atmospheric extratropical seasonal simulation and prediction. *Bulletin of the American Meteorological Society* 80: 1349–1361.

Barlow, M., S. Nigam, and E. H. Berbery. 1998. Evolution of the North American Monsoon system. *Journal of Climate* 11: 2238–2257.

Barnston, A. G., H. M. van den Dool, S. E. Zebiak, T. P. Barnett, M. Ji, D. R. Rodenhuis, M. A. Cane, A. Leetmaa, N. E. Graham, C. R. Ropelewski, V. E. Kousky, E. A. O'Lenic, and R. E. Livezey. 1994. Long-lead seasonal forecasts—where do we stand? *Bulletin of the American Meteorological Society* 75: 2097–2114.

Cayan, D. R., S. Kammerdiener, M. D. Dettinger, J. M. Caprio, and D. H. Peterson. 2001. Change in the onset of spring in the western United States. *Bulletin of the American Meteorological Society* 82: 399–415.

Cayan, D. R., and R. H. Webb. 1992. El Niño/Southern Oscillation and Streamflow in the Western United States. In *El Niño: Historical and Paleoclimate Aspects of the Southern Oscillation.* H. F. Diaz and V. Markgraf, eds. Cambridge University Press, New York, pp. 29–68.

Cook, E. R., R. D. D'Arrigo, J. E. Cole, D. W. Stahle, and R. Villalba. 2000. Tree-Ring Records of Past ENSO Variability and Forcing. In *El Niño and the Southern Oscillation: Multiscale Variability and Regional Impacts.* H. F. Diaz and V. Markgraf, eds. Cambridge University Press, New York, pp. 297–324.

Dettinger, M. D., D. R. Cayan, G. J. McCabe, and J. A. Marengo. 2000. Multiscale Streamflow Variability Associated With El Niño/Southern Oscillation. In *El Niño and the Southern Oscillation: Multiscale Variability and Regional Impacts.* H. F. Diaz and V. Markgraf, eds. Cambridge University Press, New York, pp. 113–148.

Diaz, H. F., and G. N. Kiladis. 1992. Atmospheric Teleconnections Associated With the Extreme Phase of the Southern Oscillation. In *El Niño: Historical and Paleoclimate Aspects of the Southern Oscillation.* H. F. Diaz and V. Markgraf, eds. Cambridge University Press, New York, pp. 7–28.

Enfield, D., and A. M. Mestas-Nuñez. 2000. Global Modes of ENSO and Non-ENSO Sea Surface Temperature Variability and Their Associations With Climate. In *El Niño and the Southern Oscillation: Multiscale Variability and Regional Impacts.* H. F. Diaz and V. Markgraf, eds. Cambridge University Press, New York, pp. 89–112.

Gershunov, A., and T. P. Barnett. 1998. Interdecadal modulation of ENSO teleconnections. *Bulletin of the American Meteorological Society* 79: 2715–2725.

Higgins, R. W., Y. Yao, and X. Wang. 1997. Influence of the North American Monsoon system on United States summer precipitation regime. *Journal of Climate* 10: 2600–2622.

Hoerling, M. P., and A. Kumar. 2000. Understanding and Predicting Extratropical Teleconnections Related to ENSO. In *El Niño and the Southern Oscillation: Multiscale Variability and Regional Impacts.* H. F. Diaz and V. Markgraf, eds. Cambridge University Press, New York, pp. 57–88.

Hoerling, M. P., A. Kumar, and M. Zhong. 1997. El Niño, La Niña, and the nonlinearity of their teleconnections. *Journal of Climate* 10: 1769–1786.

Huang, J., H. van den Dool, and A. G. Barnston. 1996. Long-lead seasonal prediction using optimal climate normals. *Journal of Climate* 9: 809–817.

Hurrell, J. W. 1996. Influence of variations in extratropical teleconnections on Northern Hemisphere temperature. *Geophysics Research Letters* 23: 665–668.

Hurrell, J. W., and H. van Loon. 1997. Decadal variations in climate associated with the North Atlantic Oscillation. *Climate Change* 36: 301–326.

IPCC (Intergovernmental Panel on Climate Change). 2001. *Climate Change 2001: The Scientific Basis.* Third Assessment Report of the IPCC. Working Group 1 Technical Summary.

Kiladis, G. N., and H. F. Diaz. 1989. Global climate anomalies associated with extremes in the Southern Oscillation. *Journal of Climate* 2: 1069–1080.

Livezey, R. E., and T. M. Smith. 1999. Covariability of aspects of North American climate with global sea surface temperatures on interannual to interdecadal timescales. *Journal of Climate* 12: 289–302.

Lorenz, E. 1993. *The Essence of Chaos.* University of Washington Press, Seattle.

Meko, D., M. Hughes, and C. Stockton. 1991. Climate Change and Climate Variability: The Paleorecord. In *National Research Council: Managing Water Resources in the West Under Conditions of Climate Uncertainty.* National Academy Press, Washington, DC, pp. 71–100.

Miller, K. A. 1997. *Climate Variability, Climate Change, and Western Water.* Report of the Western Water Policy Review Advisory Commission, Denver, CO. National Technical Information Service, Springfield, VA.

National Research Council. 1991. *Managing Water Resources in the West Under Conditions of Climate Uncertainty.* National Academy Press, Washington, DC.

———. 2001. *Improving the Effectiveness of U.S. Climate Modeling.* National Academy Press, Washington, DC.

Paegle, J., A. Pielke, G. A. Dalu, W. Miller, J. R. Garratt, T. Vukicevic, G. Berri, and M. Nicolini. 1990. Predictability of Flows Over Complex Terrain. In *Atmospheric Processes Over Complex Terrain.* W. Blumen, ed. American Meteorological Society, Boston, pp. 285–299.

Philander, S. G. 1990. *El Niño, La Niña, and the Southern Oscillation.* Academic Press, New York.

Pielke, R. A., and R. A. Avissar. 1990. Influence of landscape structure on local and regional climate. *Landscape Ecology* 4: 133–155.

Pielke, R. A., G. A. Dalu, J. S. Snook, T. J. Lee, and T.G.F. Kittel. 1991. Nonlinear influence of mesoscale land use on weather and climate. *Journal of Climate* 4: 1053–1069.

Ropelewski, C. F., and M. S. Halpert. 1986. North American precipitation and temperature patterns associated with the El Niño/Southern Oscillation (ENSO). *Monthly Weather Review* 114: 2352–2362.

———. 1996. Quantifying Southern Oscillation–precipitation relationships. *Journal of Climate* 9: 1043–1059.

Sardeshmukh, P. D., G. P. Compo, and C. Penland. 2000. Changes of probability associated with El Niño. *Journal of Climate* 13: 4268–4286.

Stegner, W. 1994. *The American West as Living Space.* University of Michigan Press, Ann Arbor.

Trenberth, K. E., and J. W. Hurrell. 1994. Decadal atmospheric-ocean variations in the Pacific. *Climate Dynamics* 9: 303–319.

Waggoner, P. E., ed. 1990. *Climate Change and U.S. Water Resources.* Wiley, New York.

Wolter, K., R. M. Dole, and C. A. Smith. 1999. Short-term climate extremes over the continental United States and ENSO. Part 1: Seasonal temperatures. *Journal of Climate* 12: 3255–3272.

Yu, B., and J. M. Wallace. 2000. The principal mode of interannual variability of the North American Monsoon system. *Journal of Climate* 13: 2794–2800.

Zhang, Y., J. M. Wallace, and D. S. Battisti. 1997. ENSO-like interdecadal variability: 1900–93. *Journal of Climate* 10: 1004–1020.

Climate Variability in the West: Complex Spatial Structure Associated With Topography, and Observational Issues

TWO

Kelly T. Redmond

Variability in a renewable natural resource such as water can manifest itself in many ways, both spatially and temporally. In the American West the use of water resources typically was initiated and often then perpetuated with only cursory consideration or understanding of its variations in space and time. Even now, the spatial field of a climatic variable changing in time has barely been described and is not explained or incorporated into planning or operations involving water management.

EFFECTS OF ELEVATION ON PRECIPITATION

Strong seasonality in precipitation exists throughout the western United States, much of which must be described in context with elevation (Redmond 1994). The winter fractional contribution to the annual total generally increases with elevation and reaches streams seasonally as melted snowpack. Thus the supply of water in the West occurs mostly in the winter and spring at high elevations, whereas the demand for water is highest in the summer (during the growing season) at low elevations. Because the availability of water at any given time is often strongly affected by weather and climate

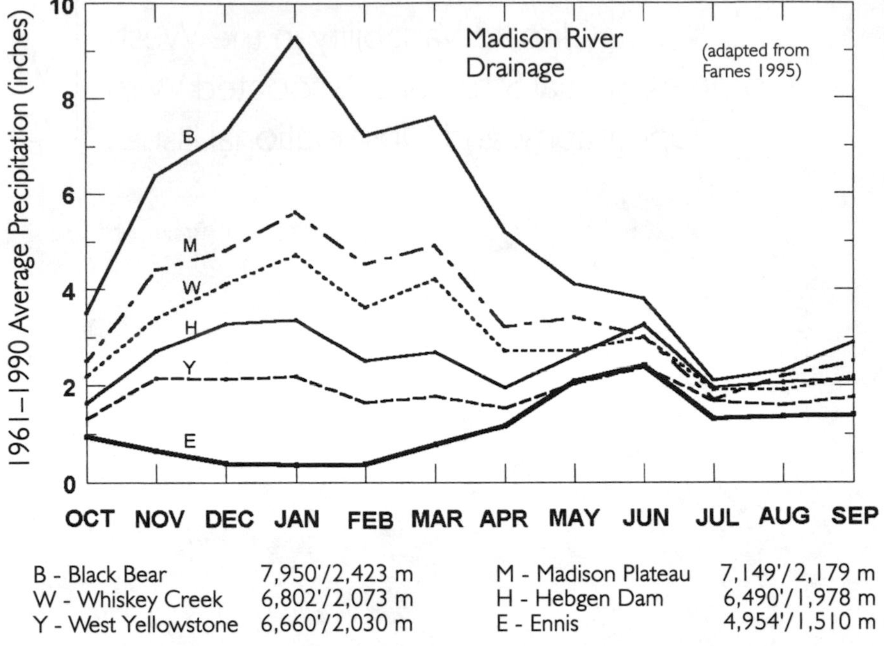

2.1 Monthly precipitation cycle at different elevations in the Madison River drainage in southwest Montana for the water year (October through September, thirty-year average; modified from Farnes 1995).

events that occurred months earlier at a distant location, considerable problems are often found in the perception of water-supply status.

Farnes (1995) presented a number of examples from Montana of elevational differences over short horizontal distances in the percentage of annual precipitation that falls each month. Figure 2.1 shows monthly precipitation climatologies for valley and mountain stations in the Madison River drainage of southwestern Montana. The stations range from 5,000 to 8,000 feet (1,524 to 2,438 m) in elevation. All are within 65 miles (105 km) of each other, and the five highest fall within a 30-mile (48-km) radius. Two prominent patterns are evident. First, as in most of the West, the amount of annual precipitation increases with elevation. In this case the increase is from 13 inches (340 mm) at Ennis to 62 inches (1,560 mm) at Black Bear. Second, the seasonality of precipitation changes with increasing elevation toward a more winter-centered peak; winter is the driest season of the year in the valley and the wettest season of the year in the adjacent mountains. The April-May-June peak that dominates in the valleys is barely visible in the nearby mountains, where it forms only a late spring shoulder on the dominant winter peak.

Other patterns typical of the West also are apparent in Figure 2.1. Hebgen Dam sits in a narrow canyon; the higher surrounding elevations increase local annual precipitation. Thus the precipitation for this site is higher than its elevation would suggest. By contrast, nearby West Yellowstone, which lies at similar elevation, is relatively flat and receives 27 percent less annual precipitation. Such topographic effects are most pronounced in winter (Daly et al. 1994).

Although precipitation generally increases with elevation in the West, the rate of such increase varies spatially. In making comparisons among sites, it is convenient to use annual precipitation as a normalizing factor. For this purpose, each month's precipitation is divided by the annual total for a given site and is expressed as monthly percentage. The rate at which the monthly percentage of average annual precipitation changes with elevation can be either positive or negative, depending on the month (Farnes 1995).

One widely used method for mapping precipitation in topographically complex terrain relies on a technique called PRISM (Precipitation Regression on Independent Slopes Method; Daly and Neilson 1992; Daly et al. 1994, 1997), which explicitly accounts for monthly variation associated with elevation. This technique also accounts for differing precipitation-elevation relationships related to orientation of slopes (e.g., east- versus west-facing slopes), as well as to topographic-scale effects that cause or enhance precipitation.

Plate 4 shows an annual precipitation map for the western contiguous United States for the period 1961–1990. Approximately 2,000 valley and 700 mountain stations served as the primary data source for this map. The important role of topography in producing the rich, fine-scale structure is obvious in that long-term mean annual precipitation can vary by an order of magnitude or more over less than 5 miles (8 km). When temporal variation is taken into account as well, the distribution of precipitation becomes even more complex.

Three to four main natural precipitation seasons are found in the western United States. These can be illustrated by Houghton's (1969) method, which describes the annual precipitation cycle in terms of superposition of contributions from multiple physical processes operating at different times of the year. Wind velocity almost always increases with altitude, and in winter atmospheric circulation it expands southward and strengthens. Thus in winter (October–March) the rate of uplift of moist air increases so that high elevations receive the most precipitation (Plate 5). Late spring is another important season for precipitation on the high plains and westward into parts of the Rockies and the Great Basin (Figure 2.1, Plate 6). Finally, in the southern portion of the West, the southwest monsoon is important during summer (Plate 7).

For October through March and in smaller degree for April through June (Plates 5 and 6), fine spatial structure of precipitation is related to elevation. Winter maxima of precipitation on mountaintops occur immediately adjacent to late spring peaks of precipitation in nearby valley bottoms. Thus climate variability (by definition, the variations through time for a given location or suite of locations) may exhibit patterns with fine spatial structure.

The use of water is adapted to amount and form (frozen or liquid) of precipitation across seasons and elevations. For example, consider a dryland farm adjacent to an irrigated farm. These two farms depend on completely different portions of the climatic spectrum. For the dryland farmer the immediate precipitation and temperature are of paramount importance, but for the irrigating neighbor local circumstances may be of little or no consequence because the water supply originates elsewhere. Throughout the West, many farms and cities are tied to climate in mountains nearby or far away. For example, municipal supply in San Diego relies in part on the previous winter's precipitation in the Never Summer Range of Colorado and the Wind River Mountains of Wyoming.

Local effects cause variation in seasonal precipitation. General patterns of wind within even a single season can interact variably with topography. Within the state of Colorado, for example, one can find precipitation maxima in ten different months (Figure 2.2). Some seasonal patterns have sharp peaks, whereas others are more uniform from month to month.

With different annual precipitation patterns in close juxtaposition, much opportunity exists for misinterpretation of annual precipitation time series. Atmospheric circulation patterns that control regional and local climate in general (as described by Dole, Chapter 1) will respond in variable ways to the influence of factors occurring at continental or global scales. Circulation patterns in the Northern Hemisphere vary slowly. Consequently, a seasonal evolution occurs in the connections between large-scale circulation patterns and the local or regional circulation patterns that bring precipitation. In mountainous regions the interactions between atmospheric movement across a range of size scales can be extraordinarily complex. Thus the annual precipitation time series for a valley bottom can be very different from the annual time series on a mountaintop a short distance away.

EFFECTS OF ELEVATION ON TEMPERATURE

Temperature also can vary greatly over short distances. In the West, for example, temperature inversions can form nearly every night on flat terrain when humidity and wind velocity are low and clouds are absent. Snow cover can facilitate this process. Especially in winter, where local circulation is blocked by topography, or in summer, where marine atmospheric layers are

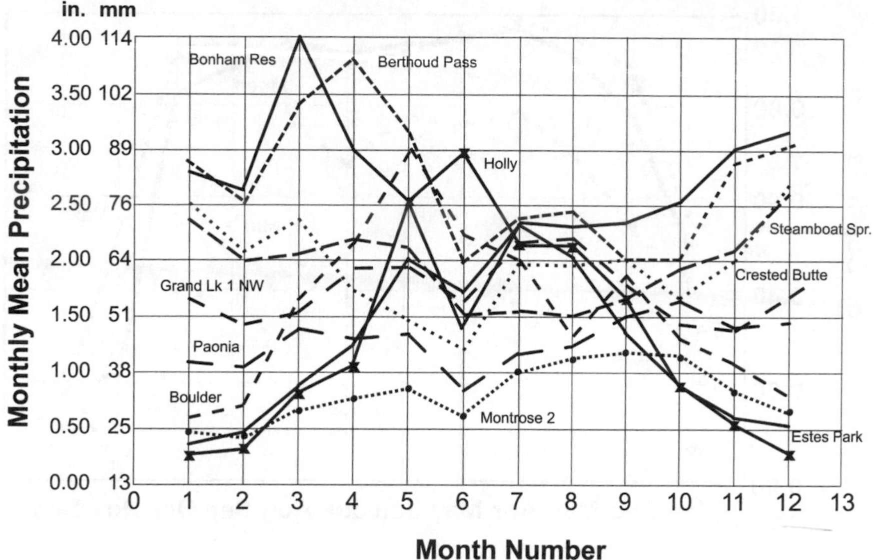

2.2 Precipitation at several Colorado climate stations with different months of peak precipitation (averages; 1961–1990).

present near coastlines, inversions can persist throughout the day. Thus a large-scale weather pattern can produce different temperatures at different elevations or between coastal and interior sites. Persistent weather patterns can lead to monthly departures from average that differ radically within short distances in both magnitude and sign. These local effects, which are most commonly a result of topography, can occur over as little as a few miles or a few thousand feet of elevation. Because the large-scale patterns themselves vary from month to month and from year to year for the same month, temporal variability at closely spaced locations differing in elevation need not be the same everywhere. The reasoning is the same as for precipitation, but the details may be very different.

Figure 2.3 shows a comparison of temperature records at two closely spaced sites in Oregon: Medford at elevation 1,300 ft (396 m) and Sexton Summit, 29 miles (47 km) to the northwest at elevation 3,840 ft (1,170 m). Both are first-order stations that have employed trained observers for several decades and thus have excellent climate records. For the years between 1942 and 1992, the correlation coefficient of the annual mean temperatures at the two sites is only 0.74 ($r^2 = 0.54$). For annual means of daily maximum temperatures, r = 0.76 ($r^2 = 0.58$), and for annual means of daily minimum

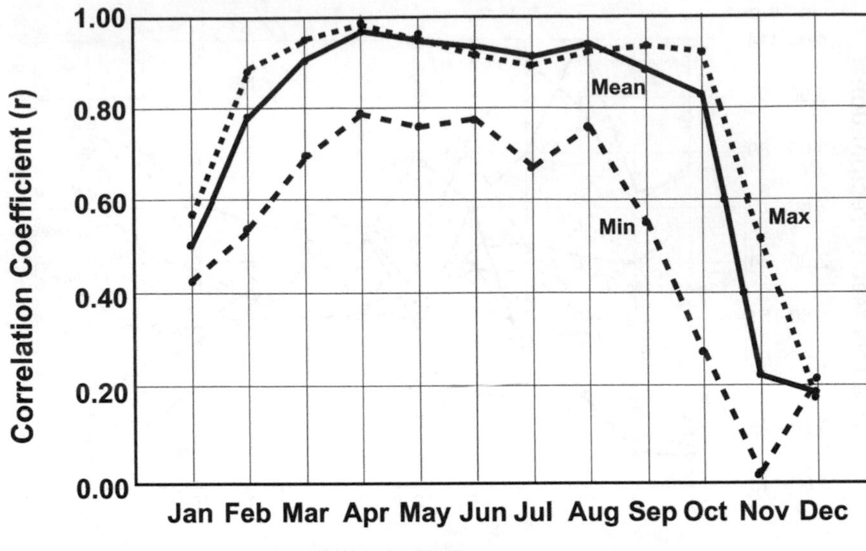

2.3 Correlation coefficients, by month, between Medford and Sexton Summit, Oregon, for mean monthly maximum temperature, mean monthly minimum temperature, and mean monthly temperature, 1942–1992.

temperatures, r = 0.60 (r^2 = 0.36). As expected, the maximum temperatures correlate closely in months when the atmosphere is well mixed and not so well during low-sun seasons when persistent inversions are common. Minimum temperatures correlate most highly in April (r = 0.79, r^2 = 0.62) and reach a minimum correlation in November (r = 0.02, r^2 = 0.00). Other combinations of seasons are shown in Table 2.1. This example shows a mountain-valley difference in temporal climate behavior.

Oregon also furnishes an illustration of a coastal-inland difference in temporal climate behavior. Medford is approximately 80 miles (129 km) inland from the Pacific Ocean. The North Bend Federal Aviation Administration station, 80 miles (129 km) north of this point on the coast, is effectively the same distance from Medford. The annual correlation between the mean temperatures of Medford and North Bend for 53 years within the period 1941–1999 is rather low (r = 0.68, r^2 = 0.46). The correlation is particularly poor in summer, when fog and the marine layer are often more prevalent on warm days immediately inland. In July and August the correlation is essentially zero (Figure 2.4).

Table 2.1—Correlations for temperatures in various seasons, half years, and years at sites in Oregon (Sexton Summit, Medford, North Bend, Newport, Tidewater) and Colorado (Wolf Creek Pass, Alamosa). Missing months affect the number of years available for comparison (N varies between 27 and 66 years; "Avg" refers to mean temperature).

	Sexton Summit versus Medford			*Medford versus North Bend*	*North Bend versus Newport*	*Newport versus Tidewater*	*Wolf Creek Pass versus Alamosa*		
	Avg	*Max*	*Min*	*Avg*	*Avg*	*Avg*	*Avg*	*Max*	*Min*
Dec–Feb	0.49	0.64	0.34	0.81	0.93	0.87	0.29	0.04	0.43
Mar–May	0.95	0.96	0.82	0.76	0.88	0.85	0.85	0.79	0.53
June–Aug	0.90	0.86	0.72	0.12	0.73	0.61	0.74	0.75	0.61
Sep–Nov	0.66	0.79	0.30	0.51	0.81	0.78	0.74	0.70	0.43
Oct–Mar	0.51	0.68	0.26	0.71	0.86	0.80	0.19	0.03	0.28
Apr–Sep	0.88	0.86	0.71	0.42	0.81	0.79	0.88	0.79	0.68
Ann.	0.74	0.76	0.60	0.68	0.83	0.78	0.48	0.23	0.59

To further illustrate the small-scale structure of temperature variability, the record from the National Weather Service cooperative climate station at Tidewater, Oregon, 9 miles inland, was compared with the record from Newport, very near the beach. The correlation of mean annual temperature (37 complete years within the period 1940–1995) at these two Oregon sites is only 0.78 ($r^2 = 0.61$). During August (records available for 53 years), the correlation of mean temperature is just 0.55 ($r^2 = 0.30$). By contrast, the correlation along the coast between Newport and North Bend, 85 miles (137 km) apart, is 0.83 for 55 years (from the period 1931–1999) and ranges from a low of 0.71 ($r^2 = 0.50$) in June to a high of 0.95 ($r^2 = 0.91$) in January (Figure 2.4). Thus variations are more likely to be coherent along the coast than from the coast inland.

In Colorado, Wolf Creek Pass at elevation 10,640 ft (3,243 m) and a nearby valley bottom at Alamosa at 7,540 feet (2,298 m) show a correlation of mean annual temperature for the period 1958–1999 of 0.48 ($r^2 = 0.23$). Figure 2.5 shows the monthly correlations between sites for mean, maximum, and minimum temperatures. Means, maxima, and minima correlate poorly in winter and better in summer. As expected, maxima correlate rea-

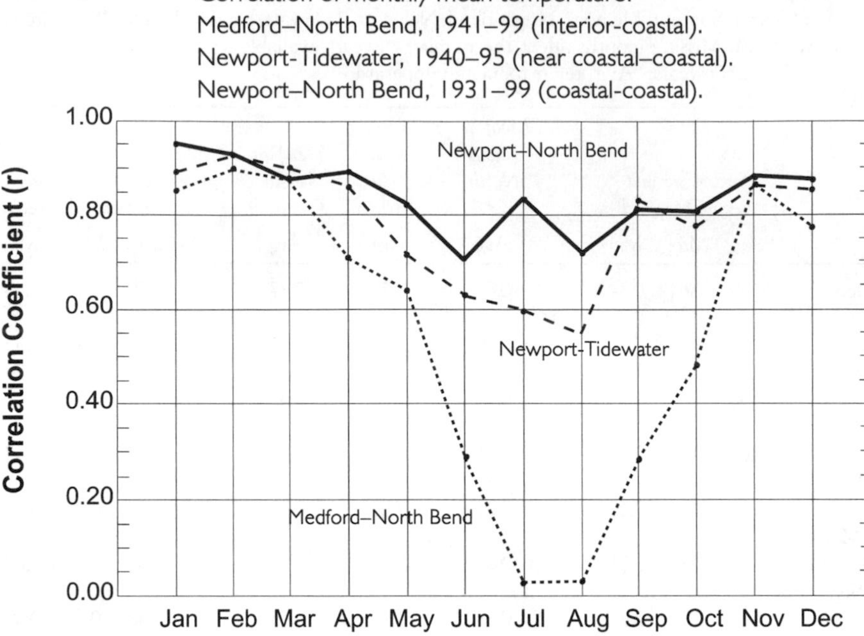

2.4 Correlation coefficients for temperature, by month, among combinations of Medford, North Bend, Newport, and Tidewater, Oregon.

sonably well in some months but have only modest correlations throughout the year. The January correlation between maximum temperatures is slightly negative (–0.12), reflecting a strong tendency for the two stations to be on the opposite sides of winter inversions for a few days or weeks when such inversions are present. The correlation of 27 annual mean maximum temperatures is only 0.23 (r^2 = 0.05), and that of 28 annual mean minimum temperatures is just 0.59 (r^2 = 0.35).

It is clear from even these few examples that care should be taken when aggregating climate data for analysis of climate variability. Geographic domains—whether climate divisions, river basins, grid squares, or other delineations—may hide meaningful spatial patterns. Some geographic domains may be suitable in some months or seasons but less useful in other seasons. Conversely, spatial aggregations that often would make sense for climate studies may consist of noncontiguous bands stratified by elevation. The optimal way to determine how to lump data would be through use of seasonal and elevational structure of temporal correlations.

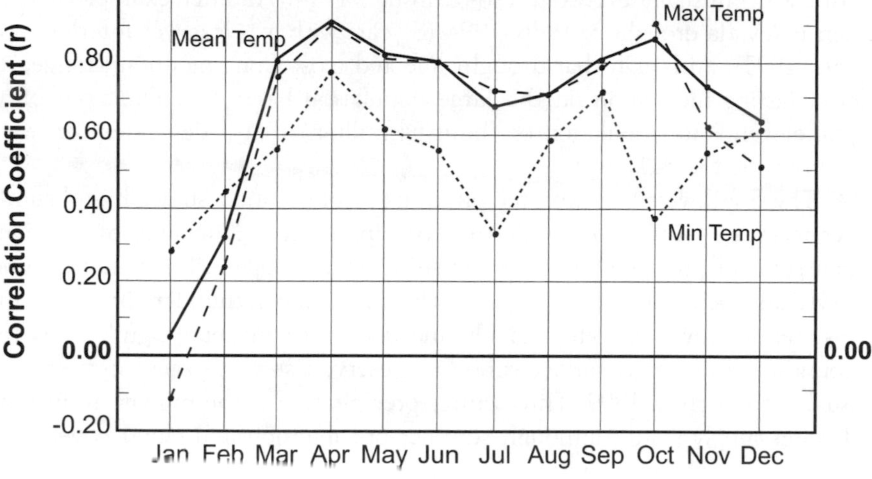

2.5 Correlation coefficients, by month, between Wolf Creek Pass and Alamosa, Colorado, for mean monthly maximum temperature, mean monthly minimum temperature, and mean monthly temperature, 1958–1999.

Unfortunately, we lack sufficient long time series of data of acceptable quality from dense networks spanning a wide range of elevations, in part because of the presumed redundancy of stations that are close together. We probably will never be able to obtain the quantity, quality, and density of data that would tell us clearly whether we need to be monitoring climate with better spatial resolution in mountainous regions. On the other hand, the absence of persuasive observational evidence that we actually do need such detailed information will reduce the motivation to retain dense networks of montane monitoring sites, many of which would operate under difficult circumstances. Nonetheless, we should continue to build toward the observational base that would resolve questions about variability of climate (see Chapter 4 for some specific suggestions).

DROUGHT

The West has not suffered from a major, large-scale drought in more than 40 years (Edwards and McKee 1997), although one may be developing as of

2002. The last such drought occurred in the 1950s; it affected primarily the Southwest. Droughts of short duration or small scale have occurred in various subregions since the 1950s, however. For example, the Columbia River basin was unusually dry from 1984 through 1994. Another example is the Sierra Nevada drought of 1986–1994 (8 years with a 1-year wet interlude in 1992–1993). Although that drought affected mostly one mountain range, it left a lasting impression on the large population base (30 million out of a total eleven-state population of about 54 million) that is dependent on water from the Sierra Nevada.

The Southwest has two precipitation seasons: winter snowfall and summer monsoon (Plates 5, 7). Winter precipitation is much more effective than summer precipitation in replenishing water supplies. Snowmelt occurs slowly and percolates into the soil, whereas summer thunderstorm rain in this area usually evaporates quickly and does not contribute significantly to groundwater recharge, surface runoff, or reservoir storage except in very unusual years such as 1999. Thus winter precipitation is the primary indicator of water-supply status, although summer precipitation and cloud cover can alter both supply and demand.

Drought in the Southwest in the 1950s extended from 1952 to 1957 (Figure 2.6). The winters of 1954–1955 and 1955–1956 were the two driest consecutive winters in the twentieth century (see Chapter 3). Figure 2.6 also shows that the Southwest has subsequently become increasingly wetter and that the wettest period in the past 105 years extended from the late 1970s to the early 1990s. This was also a period of great population growth. The winters of 1995–1996, 1998–1999, and 1999–2000 (all La Niña; discussed in the next section) were dry, and the year from July 1995 to June 1996 was the third-driest on record (Figure 2.7). The summer monsoon has shown little overall trend (Figure 2.8). Nevertheless, in the Southwest the regional water allocation and distribution system has not been tested by a major drought within the residence time of a large fraction of the current population.

EL NIÑO AND LA NIÑA

Oceanic conditions on the equator between Peru and the International Dateline have a significant effect on the winter climate of the western United States (Chapter 1). The ocean water there is sometimes warmer than average (El Niño) or cooler than average (La Niña). In the atmosphere, horizontal differences of barometric pressure—as shown, for example, by pressure at Tahiti Island minus pressure at Darwin, Australia—indicate the status of the Southern Oscillation. The oceanic and atmospheric components are often collectively referred to as ENSO (El Niño–Southern Oscillation). The equatorial Pacific spends about a quarter of the time in the El Niño phase, a quarter in the La Niña phase, and about half the time in neither phase.

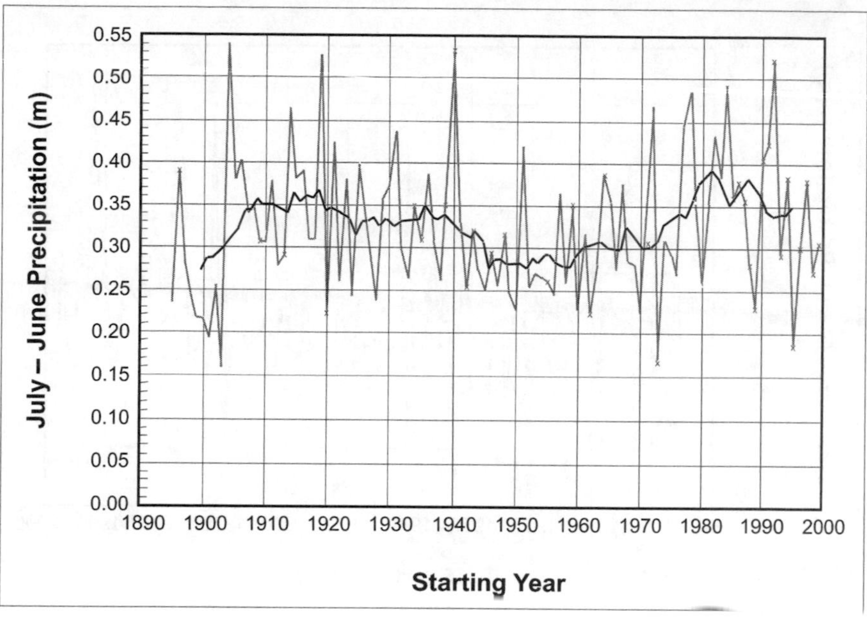

2.6 Statewide average winter-centered precipitation for Arizona, 1895–1999, for the twelve-month period July through June.

The Great Basin and much of the intermontane West experience a mixed and ambiguous response to El Niño and La Niña. Strong responses to both the warm and cold phases are centered, however, primarily on two areas: the Columbia River basin in the Pacific Northwest and the U.S./Mexican border in the desert Southwest (McGuirk 1982; Yarnal and Diaz 1986; Redmond and Koch 1991; Livezey et al. 1997). El Niño usually brings dry winters to the Pacific Northwest and wet winters to the Southwest. La Niña usually brings wet winters to the Pacific Northwest and dry winters to the Southwest. The latter relationship is probably the most consistent climate relationship in the United States; none of the fourteen strongest La Niña events since 1933 has produced a wet winter in Arizona.

Recent studies show that in the West the precipitation signal associated with ENSO is greatly accentuated in the streamflow records (Cayan et al. 1999; Kahya and Dracup 1993; Dracup and Kahya 1994; Piechota et al. 1997). In addition, two lags extend the lead time to as much as one year. The first is the lag between summer or autumn indicators of the Southern Oscillation and subsequent October–March precipitation. The second is the hydrologic lag between the time when precipitation falls and when it reaches streams.

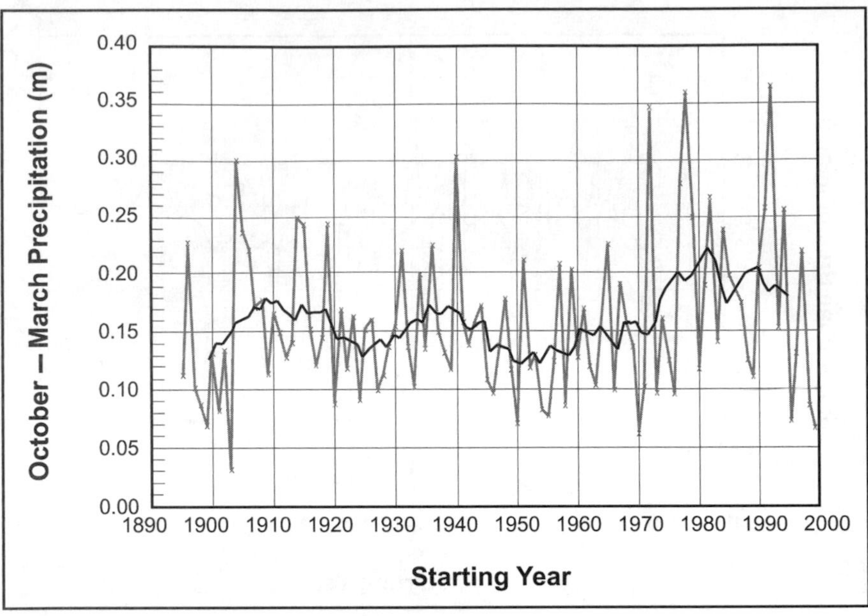

2.7 Statewide average winter-centered precipitation for Arizona, 1895–1999, for the cool half-year period October–March.

Thus during El Niño or La Niña conditions (about half the years), information is often available about potential runoff volumes, peaks, and timing far enough in advance to allow adjustments in water-management strategies. The credibility of such forecasts has risen to the point that water managers and others are beginning to use them cautiously in their decision-making framework (see Chapter 5). Although predictive capability is restricted to about thirteen months of lead time or less, it is nonetheless of great interest. When neither El Niño nor La Niña is present, our prospects for confident seasonal forecasts are considerably diminished.

McCabe and Dettinger (1999) have shown that the relationship between the Southern Oscillation Index and western winter precipitation appears to be modulated by some other factor, having varied from weaker (1920–1950) to stronger during the past 100 years (especially in the past several decades). The relationship was strong throughout the entire twentieth century in the desert Southwest, although the core area slowly migrated west from New Mexico to Arizona. The relationship varied more in the Pacific Northwest; it faded to near invisibility between 1920 and 1950. Mantua et al. (1997) have identified a 20- to 30-year fluctuation in the North Pacific

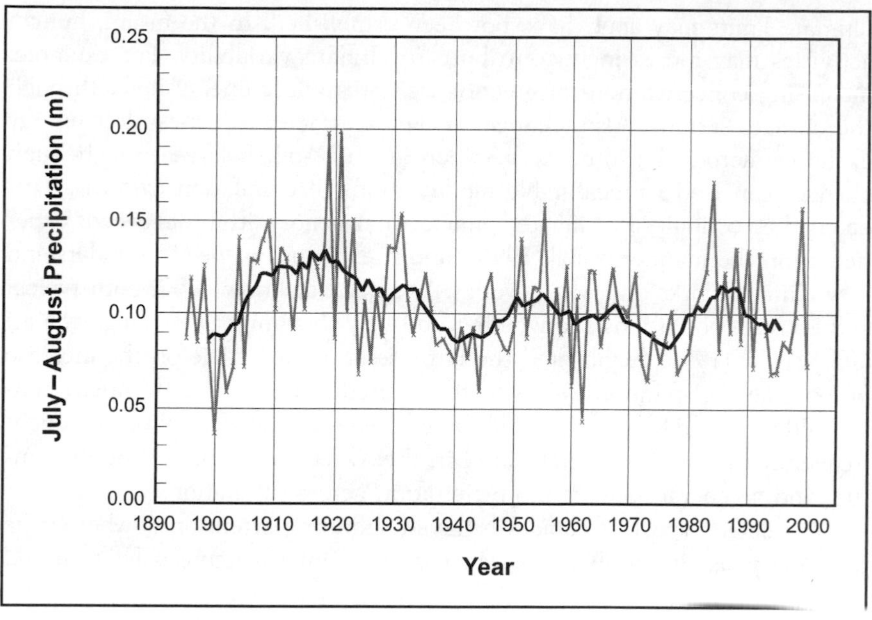

2.8 Statewide average precipitation for Arizona, 1895–1999, for the main portion of the monsoon season, July and August.

(Pacific Decadal Oscillation) that appears to have effects in the western United States, particularly the Pacific Northwest.

CLIMATE STATIONARITY

Climate fluctuates continuously on all timescales. The systematic observational records that serve as the primary basis for our understanding of climate system function extend over only about 100 years (in some places a little more). We have about 50 years of upper-air measurements, which yield the three-dimensional picture. Proxy data—indirect evidence in the form of physical and biological consequences of climatic processes (essentially filtered climate data), preserved in a variety of ways—reveal past climate (Chapter 3). They also provide a much-needed climatic context for our observational data.

In a recent essay, Bryson (1997) asserted that the history of climate can be described as a nonstationary time series. Contemporary climatologists generally agree with this view, according to which climate never repeats itself exactly. Thus we have not seen all possible climatic phenomena. The degree to which climate or components thereof can be predicted and what ultimate limits may apply have not been established. In the future, human activities may increasingly contribute to climate variability. For example, increasing concentrations of greenhouse gases affect energy flows through the climate system. Also, changes in land-surface properties alter energy exchange across the interface between air and soil. If large enough, such changes can lead to measurable modifications of circulation patterns. This opens the possibility for climate behavior that is not within our recent experience (in the manner of Bak 1996), and it explains our need to understand how climate works and why it varies. We need to know about both typical and extreme conditions and whether and why they might be changing. Karl and Knight (1998) highlighted the latter issue in an article on the increase of extreme precipitation events in the United States in the twentieth century. Edwards and McKee (1997) likewise noted the general decrease in drought frequency in the United States and in the West during the twentieth century and a general increase in precipitation across the nation.

A related aspect of climate behavior requiring attention is whether its temporal variability itself varies. Much societal infrastructure is built around variations in climate. The proper sizing of reservoirs, for example, to ensure adequate and reliable supplies takes into account expected climate anomalies.

Over a hundred-year interval at a typical climate station, annual precipitation usually varies from roughly half to double the long-term average; individual seasons vary more. For example, as the twentieth century progressed, the October–March total at Tahoe City, California, at the outlet of Lake Tahoe, showed increasing year-to-year fluctuations. In addition, moving four-year averages (the approximate residence time for the largest reservoirs) also showed increasing fluctuations (Figure 2.9). The last quarter century in particular has brought much more variability. Since 1976, the central Sierra Nevada has experienced its driest and wettest individual years on record, as well as its driest and wettest four-year periods on record, and an extended drought. On November 30, 1992, Lake Tahoe dropped to its lowest recorded level (6,220 feet, 1,896 m) and by June 25, 1996, it had rebounded 8.84 feet (2.69 m) to within 0.01 foot (3 mm) of its legally imposed maximum.

An important question is the extent to which our records of past climate are relevant to decisions that have consequences well into the future. As an example, expensive and long-lived structural solutions (dams, canals, treatment facilities) to water problems require design information that represents the best estimate of anticipated future conditions. Currently, predictions and projections about anthropogenically induced climate change are much too uncertain to provide unambiguous guidance. Regional detail is greatly needed, especially in a topographically diverse domain. For now, it seems safest to take a cue from present climate and assume that future climate will be similarly complex.

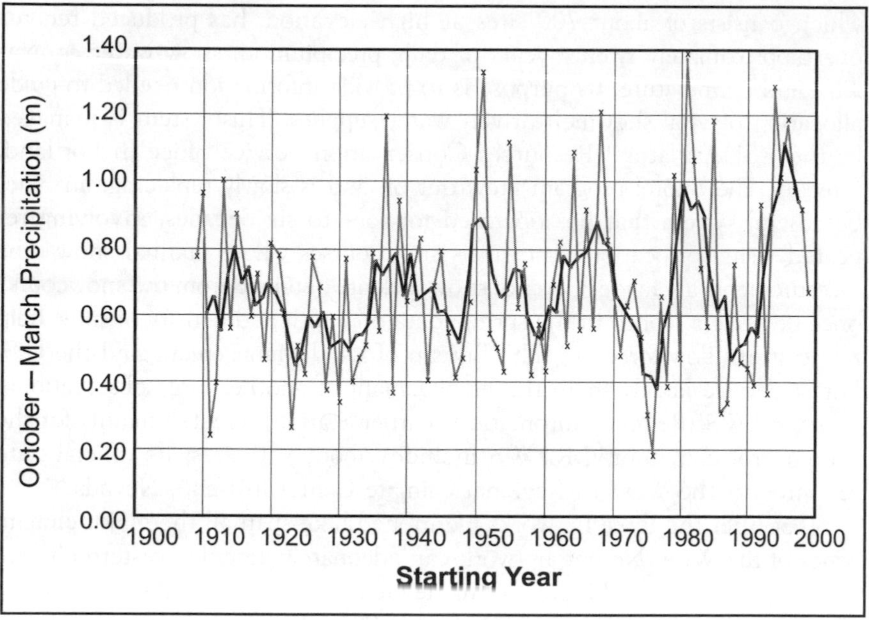

2.9 Annual and multiyear fluctuations in October–March precipitation (1909–1910 through 1999–2000) at Tahoe City, California. Much of this falls as snow (thin line, individual winters; thick line, four-year running mean).

OBSERVATIONAL ISSUES

Our sources of climate information include the Cooperative Network managed for the past sixty years by the National Weather Service and in prior decades by the U.S. Department of Agriculture (USDA). This largely volunteer network yields daily measurements at approximately 2,200 stations in the 1.174 million square miles (3.04×10^6 km^2) covered by the eleven western states, or 534 square miles (1,383 km^2) per station on average, equivalent to a square 23 miles (37 km) on a side. Because sites tend to be located near population centers, the network has a bias toward lower elevations. The United States has no formal national climate-observing system; the Cooperative Network is more akin to a confederacy of programs set up for other purposes. The network will, however, continue to serve as the backbone of our climate-observing efforts (Chapter 4).

About 50 percent of the West is under federal stewardship. In recent years, specialized networks covering only this region have been deployed for resource-management purposes. The Snotel (Snowpack Telemetry) network, which consists of about 700 sites at high elevation, has produced records over approximately twenty years of daily precipitation, snow-water equivalent, and temperature. Its purpose is to provide information needed to guide allocation of vital snowmelt-driven water supplies. This system is managed by the USDA Natural Resources Conservation Service office in Portland, Oregon. The Snotel program grew out of, and is slowly replacing, an older but useful system that has operated for four to six decades, involving repeated visits to locations known as snowcourses where manual snow-tube measurements are made. Cores of snowpack are obtained from the snowcourse once or twice a month in the spring; data on depth and density of snow help refine streamflow forecasts. The Bureau of Land Management and the U.S. Forest Service largely share the management of another large observational effort, RAWS (Remote Automatic Weather Station), created mainly for the fire control community. RAWS includes about 930 sites; its official data repository is the Western Regional Climate Center in Reno, Nevada.

Together the three networks mentioned here span all the major climate zones of the West. No one network can adequately describe western climate or its variations; combinations are needed. The average elevation of the Cooperative Network sites is 3,300 feet (1,005 m), that of the RAWS sites is 4,600 feet (1,402 m), and that of the Snotel sites is 7,100 feet (2,164 m). Sites extend from sea level to 11,600 feet (2,536 m). A number of smaller networks, typically including 50–100 sites or fewer, operate for special purposes in various states or regions (Redmond 1994).

Despite ample motivation to steer toward fuller integration, the climate data from different sources serving many purposes do not readily lend themselves to intermingling unless the idiosyncrasies of each source are addressed (Redmond 1994, 2000a, 2000b). Many studies, especially of temporal variation, cannot take place without complete and consistent records, and gaps are more likely when a human presence is required every day in a remote or demanding situation. For studies of temporal behavior, especially of trends, consistency in measurement procedures is vital. Unfortunately, automated precipitation sensors cannot yet match the quality or maintain the consistency of manual methods. Manual stations are much cheaper to operate than automatic ones but are often poorly maintained. The Cooperative Network, which is extremely useful for analyzing climate change and variability, has long been plagued by such inattention. Automated stations are more expensive per site and thus lack the desired density, and managers need frequent reminders about the value of the data. Despite repeated creative attempts, a major remaining problem is the lack of an accurate, robust, inexpensive, reliable automated gauge that can measure frozen precipitation to high resolution while unattended in harsh climates—sometimes for months—as well as the tried-and-true shielded 8-inch manual gauge (NRC 1998).

Remote sensing (radars, satellites, aircraft) frequently is suggested as a solution to the station density problem. Terrain blockage, the necessity of calibration, differences between snow and rain, bright bands from melting snowflakes, military chaff, and lack of ground truth, as well as cost and data volume, ensure that this approach is no panacea. Furthermore, the records are too short (less than a decade or two long) to be used in analyzing temporal variability, and consistency of observation is very difficult to achieve.

Measurements of variables other than temperature and precipitation—including soil moisture and solar irradiance—are greatly needed. Quantitative daily data on the status of vegetation are also much in demand. Detailed annual time series based on proxy records obtained from paleoclimatic sources are much needed to provide context for contemporary observations. These include data from tree rings, lake cores, flood deposits, corals, and other well-resolved evidence of past climates (Mann et al. 2000).

Current programs have left us with a large number of climate records at specific sites. Although some are very long, most are rather short (5–20 years). Ambitious projects are under way to use tools such as PRISM (Daly et al. 1999) to interpolate the station records from the Cooperative Network and from Snotel stations and create long-term, gridded data sets of high resolution. The lack of dense, long-term data sets, however, does place fundamental limits on the accuracy of spatial interpolation procedures and necessitates certain assumptions about the stability of relationships between climate and elevation that cannot easily be tested for validity.

For assessing climate variability and change, a particular need exists to retain as many long-term high-elevation climate-observing sites as possible, given that high elevations are important sources of water and sensitive indicators of climate change. A small network of mountaintop research stations does exist in the western United States, but more sites would be very welcome (Diaz et al. 1997).

SUMMARY

The past fifteen years have produced remarkable advances in description of climate in the West, a necessary prelude to analysis of climate function. As the burgeoning population of the West attempts to achieve a more sustainable relationship with its rugged landscape, decisions that are part of the process will inevitably be based on assumptions about climate. Thus the better we understand climate, the better our decisions can be.

WEB ACCESS TO WESTERN CLIMATE INFORMATION

Western regional and subregional maps of precipitation and percentage of annual average for each month, as well as maps for all contiguous states, are available from the Western Regional Climate Center and can be found at www.wrcc.dri.edu/precip.html. Statewide average precipitation and temperature time series are now available. The century-long record of data for all 344 U.S. climate divisions (with the caveats discussed earlier) can be found at www.wrcc.dri.edu/spi/divplot1map.html (same months each year) and www.wrcc.dri.edu/spi/divplot2map.html (consecutive months). Northern Hemisphere background circulation patterns, derived from very versatile Web-based software at the Climate Diagnostics Center, can be found at www.wrcc.dri.edu/monitor/climatology/circulation.html.

Correlation graphs (Figures 2.3–2.5) can be found at www.wrcc.dri.edu/monitor/misc/mfrsxtt.gif (Medford versus Sexton Summit), www.wrcc.dri.edu/monitor/misc/oregontv.gif (Oregon coast versus inland), and www.wrcc.dri.edu/monitor/misc/wcpalat.gif (Wolf Creek Pass versus Alamosa). A variety of other station-specific information, including visual and tabular displays of the annual cycle on a daily basis and time series information for about 2,800 western locations, can be found at www.wrcc.dri.edu/climsum.html (click on the state of interest).

Other western climate information can be found through the main page of the Western Regional Climate Center: www.wrcc.dri.edu. Two pages at the Climate Diagnostics Center contain links to other pages: www.cdc.noaa.gov/PublicData/web_tools.html (guide to Web tools) and www.cdc.noaa.gov/ClimateInfo (selected Web resources). El Niño and La Niña information can be found at www.wrcc.dri.edu/enso/enso.html (ENSO with a western emphasis) and www.cpc.ncep.noaa.gov (good links to much information). Global and national information from the National Climatic Data Center can be found at www.ncdc.noaa.gov (main page) and www.ncdc.noaa.gov/extremes.html (global, national trends). Drought information from the National Drought Mitigation Center can be found at www.drought.unl.edu (main page) and www.drought.unl.edu/dm/monitor.html (national drought monitor).

ACKNOWLEDGMENTS

This work was supported by the Western Regional Climate Center and the California Applications Project. The author benefited from two anonymous reviews and very helpful comments by Mark Serreze.

REFERENCES

Bak, P. 1996. *How Nature Works: The Science of Self-Organized Criticality.* Springer-Verlag, New York.

Bryson, R. A. 1997. The paradigm of climatology: An essay. *Bulletin of the American Meteorological Society* 78(3): 449–455.

Cayan, D. R., K. T. Redmond, and L. G. Riddle. 1999. ENSO and hydrologic extremes in the western United States. *Journal of Climatology* 12(9): 2881–2893.

Daly, C., T.G.F. Kittel, A. McNab, J. A. Royle, W. P. Gibson, T. Parzybok, N. Rosenbloom, G. H. Taylor, and H. Fisher. 1999. Development of a 102-Year High-Resolution Climate Data Set for the Coterminous United States. In *Proceedings, 10th Symposium on Global Change Studies.* 79th Annual Meeting of the American Meteorological Society, Dallas, Texas, January 10–15, pp. 480–483.

Daly, C., and R. P. Neilson. 1992. A Digital Topographic Approach to Modeling the Distribution of Precipitation in Mountainous Terrain. In *Interdisciplinary Approaches in Hydrology and Hydrogeology.* M. E. Jones and A. Laenen, eds. American Institute of Hydrology, St. Paul, pp. 437–454.

Daly, C., R. P. Neilson, and D. L. Phillips. 1994. A statistical-topographic model for mapping climatological precipitation over mountainous terrain. *Journal of Applied Meteorology* 33: 140–158.

Daly, C., G. H. Taylor, and W. P. Gibson. 1997. The PRISM Approach to Mapping Precipitation and Temperature. In *Proceedings, 10th AMS Conference on Applied Climatology.* American Meteorological Society, Reno, Nevada, October 20–23, pp. 10–12.

Diaz, H. F., M. Beniston, and R. S. Bradley, eds. 1997. *Climatic Change at High Elevation Sites.* Kluwer Academic Publishers, Dordrecht, Netherlands.

Dracup, J. A., and E. Kahya. 1994. The relationships between U.S. streamflow and La Niña events. *Water Resources Research* 30: 2133–2141.

Edwards, D. C., and T. B. McKee. 1997. *Characteristics of 20th Century Drought in the United States at Multiple Time Scales.* Colorado State University, Fort Collins, Department of Atmospheric Science Paper 634, Climatology Report No. 97-2.

Farnes, P. E. 1995. Estimating Monthly Distribution of Average Annual Precipitation in Mountainous Areas of Montana. In *Proceedings, Western Snow Conference.* Sparks, Nevada, April 17–19, pp. 78–87.

Houghton, J. G. 1969. Characteristics of Rainfall in the Great Basin. Ph.D. dissertation, Desert Research Institute, University of Nevada, Reno.

Kahya, E., and J. A. Dracup. 1993. U.S. streamflow patterns in relation to the El Niño/Southern Oscillation. *Water Resources Research* 29: 2491–2503.

Karl, T. R., and R. W. Knight. 1998. Secular trends of precipitation amount, frequency, and intensity in the United States. *Bulletin of the American Meteorological Society* 79: 231–241.

Livezey, R. E., M. Matsutani, A. Leetma, H. Rui, M. Ji, and A. Kumar. 1997. Teleconnective response of the Pacific–North American region atmosphere to large central equatorial Pacific SST anomalies. *Journal of Climatology* 10: 1787–1820.

Mann, M. E., E. Gille, R. S. Bradley, M. K. Hughes, J. T. Overpeck, F. T. Keimig, and W. Gross. 2000. *Global Temperature Patterns in Past Centuries: An Interactive Presentation.* Earth Interactions 4, Paper 4-004 <http://earthinteractions.org>.

Mantua, N. J., S. T. Hare, Y. Zhang, J. M. Wallace, and R. C. Francis. 1997. A Pacific interdecadal oscillation with impacts on salmon productions. *Bulletin of the American Meteorological Society* 78: 1069–1079.

McCabe, G. J., and M. D. Dettinger. 1999. Decadal variations in the strength of ENSO teleconnections with precipitation in the western United States. *International Journal of Climatology* 19: 1399–1410.

McGuirk, J. P. 1982. A century of precipitation variability along the Pacific Coast of North America and its impact. *Climatology Change* 4: 41–56.

National Research Council. 1998. *Toward a New National Weather Service: Future of the National Weather Service Cooperative Observer Network.* National Academy Press, Washington, DC.

Piechota, T. C., J. A. Dracup, and R. G. Fovell. 1997. Western U.S. streamflow and atmospheric circulation patterns during El Niño–Southern Oscillation. *Journal of Hydrology* 201: 249–271.

Redmond, K. T. 1994. An Integrated Climate Monitoring System for the West. In *Proceedings, Drought Management in a Changing West: New Directions for Water Policy*. D. A. Wilhite and D. A. Wood, eds. International Drought Information Center, Portland, Oregon, May 10–13, pp. 1–13.

———. 2000a. Integrated Climate Monitoring for Drought Detection. In *Drought: A Global Assessment,* Vol. 1. D. A. Wilhite, ed. Routledge Hazards and Disasters Series. Routledge, London, pp. 145–158.

———. 2000b. Climate monitoring: Taking the long view. *Water Resources Impact* 2(4) (July 2000): 7–10.

Redmond, K. T., and R. W. Koch. 1991. Surface climate and streamflow variability in the western United States and their relationship to large-scale circulation indices. *Water Resources Research* 27: 2381–2399.

Yarnal, B., and H. F. Diaz. 1986. Relationships between extremes of the Southern Oscillation and the winter climate of the Anglo-American Pacific Coast. *Journal of Climatology* 6: 197–219.

Dendrochronological Evidence for Long-Term Hydroclimatic Variability

THREE

CONNIE A. WOODHOUSE

THE ROLE OF PALEOCLIMATOLOGY IN REGIONAL ASSESSMENTS

It is important to place twentieth-century climate events, such as the Dust Bowl drought of the 1930s, into perspective so that their relative severity can be assessed. This can be done to an extent with the existing instrumental record. For example, although the southern Great Plains drought in 1998 had a severe effect on the economies of Texas and Oklahoma, costing $6 billion in Texas alone, the physical characteristics of that drought (spatial extent and duration) pale in comparison to the 1950s drought (Plate 8). The Dust Bowl drought of the 1930s was even more severe in terms of coverage and duration; it encompassed almost 70 percent of the country in its worst year and lasted seven to eight years (Riebsame et al. 1991). How unusual were the droughts of the 1930s and 1950s? Records for instruments that measure indicators of drought (primarily rain gauges) exist for less than 100 years in much of the central and western United States. This is a period of insufficient length for answering questions about the frequency of unusual droughts.

Along with placing extreme twentieth-century events into a broader temporal context, it is also important to determine whether the instrumental record of climate variability, which includes these extreme events, is representative of long-term natural climatic variability. Because twentieth-century records are used as the basis for water-resource planning and management, this is an important question and can be answered only with information about climate for an extended period of time prior to the use of weather instruments.

Paleoclimatic data from a number of kinds of environmental proxies or indicators of climate can provide information about past climates. Variations in climatic conditions can be reflected in tree-ring widths and densities, the chemical characteristics of layers of ice in glaciers and ice caps, fossil material in lake and ocean sediments, or other natural physical records. Once the relationship between climate and an environmental indicator is determined, it is possible to use the record as a proxy for climate variability in the past. The proxy record may be calibrated with climate data to generate a model that can be used in reconstructing a record of past climate.

DENDROCHRONOLOGY: THE STUDY OF TREE RINGS

Tree rings are the best source of terrestrial temperate-region paleoclimatic data on timescales of hundreds to several thousands of years, although a limited number of even longer (up to 12,000 years) tree-ring records exist. The National Oceanographic and Atmospheric Administration World Data Center for Paleoclimatology (http://www.ngdc.noaa.gov/paleo) houses the International Tree-Ring Data Bank, which contains contributions of tree-ring data from dendrochronologists around the world (Grissino-Mayer and Fritts 1997). In the United States alone, more than 400 tree-ring collections exist for the time period 1700–1979 (Figure 3.1). The spatial coverage of tree-ring data is quite good for the western United States, although there are several areas of deficiency. In some regions, such as western Colorado, most tree-ring collections were last made in the 1970s and are now out of date. In other areas, such as the western Great Plains, the scarcity of trees has inhibited the collection of tree-ring samples, although work now in progress will improve coverage.

The basic unit of a tree-ring study is a tree-ring chronology. A tree-ring chronology is generated by averaging tree-ring measurements (usually widths or densities) of dated rings from a number of tree-ring samples, usually in the form of cores about 5 mm in diameter, from one location. The sampling site is selected according to the climate variable of interest. In studies that focus on drought, a site is selected where trees are typically stressed by drought. If temperature variations are of interest, trees at high elevation or high latitude are sampled because growth of trees in these locations is sensitive to

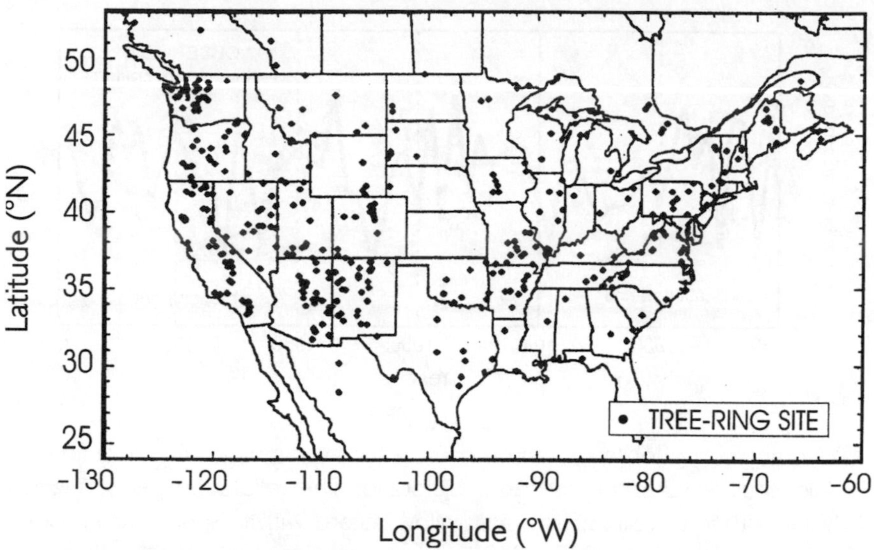

3.1 Distribution of tree-ring chronology sites that fall within the period 1700–1979 (modified from Cook et al. 1999).

temperature, which dictates the length of the growing season. Usually twenty or more trees are sampled so that the variations unique to individual trees do not have excessive influence (Fritts 1976; Cook and Kairiukstis 1990).

In reconstructing climate from tree rings, variations in year-to-year growth as represented by one or more tree-ring chronologies are calibrated with an instrumental climate record by use of regression analysis. The analysis results in a transfer function (a function that uses climate as a dependent variable and tree-ring data as an independent variable), which is then used to reconstruct the climate variable for the length of the tree-ring record. The skill of the reconstruction is evaluated by comparison of the instrumental climate values and values produced by the regression (Figure 3.2). Climate conditions most limiting to tree growth (such as dry conditions) tend to be duplicated more accurately than extremes of the opposite sign (such as wet conditions). In general, the effects of extreme conditions are muted by regression analysis. Consequently, reconstructions are usually a conservative estimate of climate variability.

DENDROCHRONOLOGICAL EVIDENCE OF HYDROCLIMATIC VARIABILITY

The network of chronologies shown in Figure 3.1 was used to reconstruct drought, as measured by the Palmer Drought Severity Index (PDSI, a widely

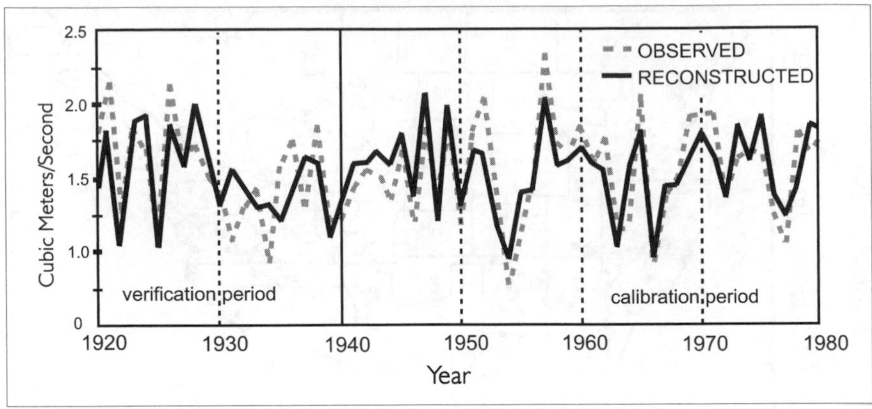

3.2 Comparison of observed and reconstructed average annual stream flow for Middle Boulder Creek, Nederland, Colorado. The calibration period (1940–1980) was used to calibrate the streamflow record with tree-ring chronologies. The verification period (1920–1939) was used to verify the calibration model.

used measure of drought that incorporates precipitation, air temperature, and soil moisture) from 1700 to 1979 for a set of grid points that covered the conterminous United States (Cook et al. 1999). Maps were generated of the spatial patterns of drought for each year of this time period (http://www.ngdc.noaa.gov/paleo/pdsi.html). In Plate 9, the observed instrumental and reconstructed patterns of drought for the worst year of droughts in the 1930s (1934) and the 1950s (1956) can be compared. The tree rings accurately duplicate the general spatial patterns and duration of drought but do not fully capture the severity of drought. Thus, as mentioned previously, they are conservative estimators of drought.

With the PDSI maps it is possible to assess the probability of the 1930s and 1950s droughts. Over the past three centuries, four droughts—each lasting two to six years—appear to have been comparable to the 1950s drought. The worst years of these four droughts are shown in Plate 10. Although the spatial extent of drought in 1934 is not equaled in any of the reconstructions, the effect of the 1930s drought on specific regions in the intermountain West was probably matched by several droughts in the past 300 years.

Prior to 1700, spatial coverage of tree-ring records is more limited, but a number of longer chronologies do exist that make possible longer reconstructions. One long reconstruction is of annual rainfall for north-central

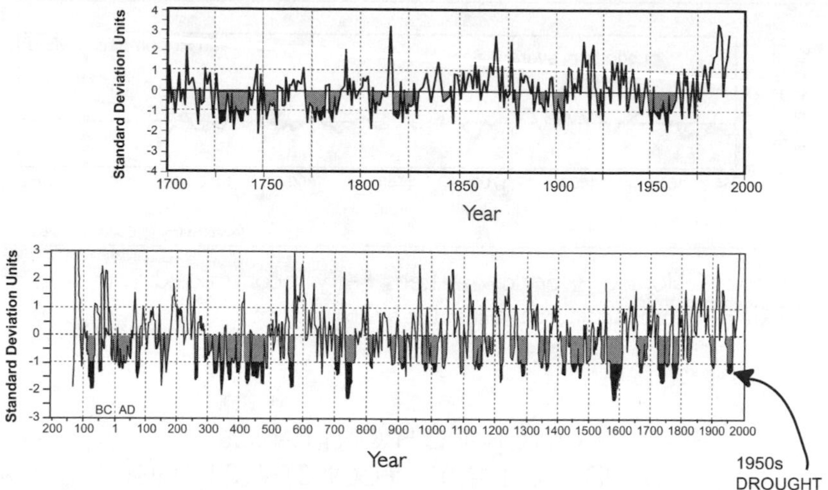

3.3 Annual rainfall for western New Mexico, as reconstructed from tree rings. The top graph shows these reconstructions for A.D. 1700–1992, whereas the bottom shows the entire period (136 B.C. to A.D. 1992). Data are from Grissino-Mayer 1996.

New Mexico (Grissino-Mayer 1996). This reconstruction, which extends to 136 B.C., allows comparison of the 1950s drought—which was quite severe in New Mexico—with droughts of the past two millennia. Although several droughts of a magnitude similar to the 1950s drought have occurred in the past three centuries, none has been conspicuously more severe (Figure 3.3). When the 1950s drought is assessed in the context of the last 2,000 years, however, it becomes evident that much more severe droughts have occurred in the past (Figure 3.3). A drought in the late sixteenth century is especially notable for its severity and relatively recent occurrence. It is found in a large number of proxy records besides the one for New Mexico. Multiple records help to chronicle its length, severity, and spatial extent. For example, this drought can be seen in the upper Colorado River flow reconstruction of Stockton and Jacoby (1976; Figure 3.4); it corresponds to one of the lowest periods of flow in 500 years. This drought illustrates the problem of basing resource management solely on the instrumental record. The 1922 Colorado River Compact was based on a flow record that began in the early twentieth century, the first two decades of which were the most anomalously wet in the entire 500-year time span.

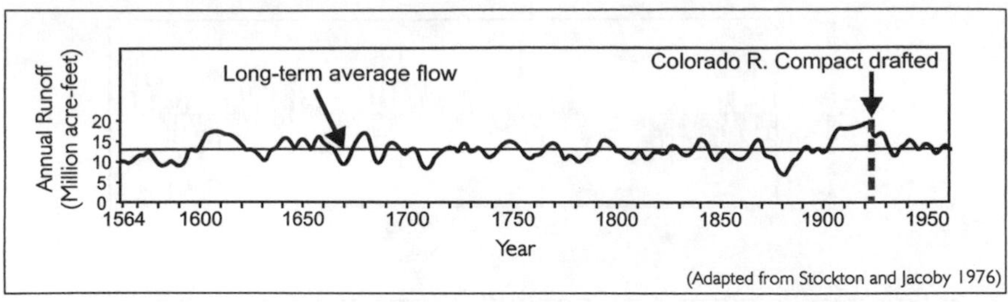

3.4 Upper Colorado River flow at Lees Ferry reconstructed from tree rings, 1564–1961 (from Stockton and Jacoby 1976).

COLORADO FRONT RANGE DENDROCLIMATIC RECONSTRUCTIONS

Two reconstructions of hydroclimatic variability have been generated for the Colorado Front Range. Reconstructions of average annual stream flow for Middle Boulder Creek at Nederland, Colorado, and regional spring (March–May) precipitation for ten stations in the South Platte basin (Woodhouse 1999) demonstrate the ability of tree rings to provide information on fine temporal and spatial scales. Because Middle Boulder Creek is mostly unregulated, its flow reflects climatic influences on flow for the entire water year. Annual discharge is mainly related to the amount of winter and spring snowpack, but the annual hydrographs reflect summer precipitation as well. The gauging station and 93 km^2 basin lie above 2,500 m elevation. In contrast, the South Platte spring precipitation record reflects precipitation falling during the time of year of maximum precipitation for the South Platte River basin. Most of the weather stations from which data were averaged to create the record for spring precipitation are located at elevations less than 1,800 m. Thus the record represents conditions at a lower elevation and over a larger region than the streamflow record for Middle Boulder Creek.

The Middle Boulder Creek stream flow and South Platte precipitation show some similarities as well as marked differences. For example, the 1960s drought—the longest and most extreme period of drought in the precipitation record—is less notable in the flow record. In the flow record the 1950s show the most extreme low flow, and the 1930s show the most persistent period of low flow (Figure 3.5).

The reconstructions of South Platte spring precipitation and Middle Boulder Creek flow are of good quality (Table 3.1; see also Woodhouse 2001), and they reproduce the instrumental records with some accuracy (Figure 3.6). Overall, dry periods in the precipitation record coincide with periods of below-average flow, indicating some coherence between the two variables, as

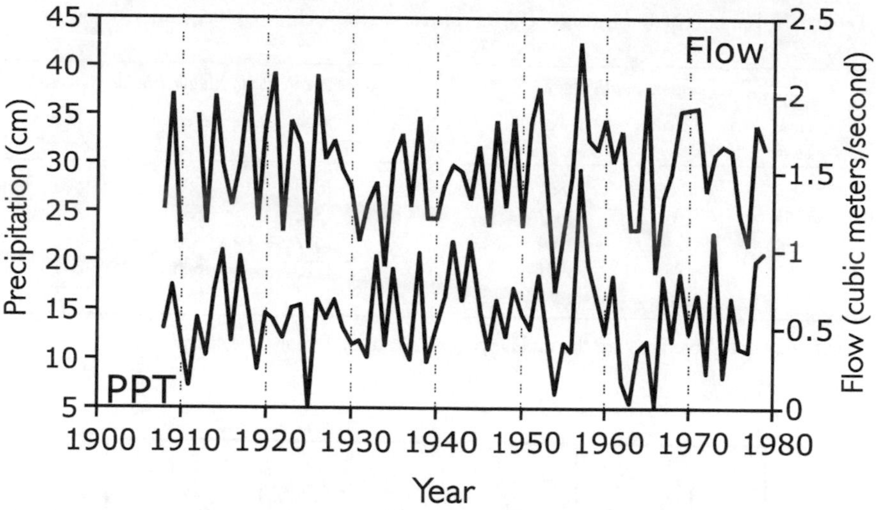

3.5 The top line is average annual flow for Middle Boulder Creek, as gauged at Nederland, Colorado, from 1908 to 1979. The bottom line is total spring precipitation, averaged for ten stations in the South Platte River basin, 1908–1979

Table 3.1—Statistics for precipitation and streamflow reconstructions. Correlation (r) and explained variance (R^2) values are used to evaluate the relationships between observed and reconstructed values.

South Platte Regional Spring Precipitation	*Middle Boulder Creek Average Annual Stream Flow*
calibration period: r = 0.657	calibration period: r = 0.812
verification period: r = 0.664	verification period: r = 0.671
variance explained: $R^2_{adj} = 0.404$	variance explained: $R^2_{adj} = 0.620$

would be expected. There are, however, instances when a drought in the precipitation record is not reflected in the flow record and vice versa. Table 3.2 shows the five single years of lowest precipitation or lowest flow, three-year averages, and five-year averages (1725–1979). The ranking shows that the two records share two of the extremes (1954 and 1843), but years of extreme conditions are different for the two variables over three- or five-year averaging intervals. Years of lowest spring precipitation in the South Platte basin are primarily clustered in the early 1860s and early to mid-1960s, whereas periods of lowest flow in Middle Boulder Creek occur in the mid-1840s and 1880s. These results suggest that droughts primarily affecting

Table 3.2—Ranking of driest years or clusters of years as shown by low precipitation or low stream flow.

Spring Precipitation, South Platte			*Stream Flow, Middle Boulder Creek*		
Single Year	*3-Year Average**	*5-Year Average**	*Single Year*	*3-Year Average**	*5-Year Average**
1963	1862	1861	1880	1886	1846
1863	1963	1862	1842	1887	1887
1954	1730	1964	1887	1880	1886
1801	1955	1731	1770	1846	1844
1842	1964	1965	1954	1879	1888

*Years shown for averages are the middle year of 3-year or 5-year time spans.

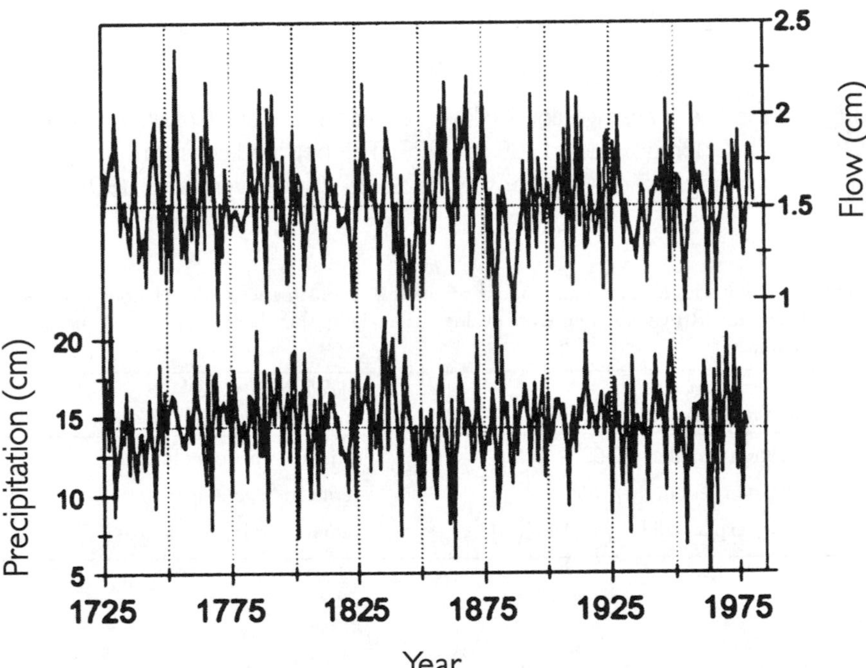

3.6 Reconstructions of Middle Boulder Creek average annual flow (TOP) and South Platte regional spring precipitation (BOTTOM) for 1725–1979. Reconstructions have also been smoothed with a 5-weight binomial filter.

winter snowpack (the main source of annual flow) in the Middle Boulder Creek watershed may not always affect regional spring precipitation, whereas a deficit in spring rainfall is not always linked to low average annual flow at the Nederland gauge, even though spring is typically the time of greatest snowfall accumulation.

Inconsistencies between sites are likely related to topographic differences. Heavy snow at high elevations in spring may not coincide with high precipitation at lower elevations. Similarly, spring upslope storms may provide moisture to lower elevations but not to higher elevations. The flow record also appears to indicate more periods of persistent dry conditions than the precipitation record, suggesting that factors controlling drought at high elevations primarily in winter and spring are more persistent from year to year than those influencing spring precipitation at low elevations.

The two reconstructions appear to resolve geographic differences and can be used to explore the record of past drought. The longer records provide a broader temporal context for evaluating twentieth-century droughts. These records suggest that the twentieth-century South Platte precipitation record is a fairly good representation of the past 200 years. The dry years in the 1960s are well matched by a period in the 1860s, although the value for 1963 is the lowest in the 254-year record. The twentieth-century flow record is not quite as representative of the reconstructed long term record. The most severe and persistent low-flow episodes of the twentieth century appear to have been exceeded several times in the nineteenth century (1840s, late 1880s). The extended record is still rather limited, however. Evidence from other proxy data in the central and western United States suggests that a drought in the late sixteenth century lasted up to two decades (Woodhouse and Overpeck 1998). It is not yet clear whether this drought also affected Front Range stream flow and precipitation in the South Platte basin.

Tree-ring data are useful for extending records of climatic and hydrologic variability into the past. These extended records provide information about a more complete range of natural variability than provided by instrumental records, which are mostly limited to the twentieth and twenty-first centuries. Although past climate is not necessarily an analog to the future, records of past variability can be used as a guide to the range of variability that may be expected in the future. Tree-ring reconstructions of past climate could be used by water-resource managers to establish an expected range of natural variability.

REFERENCES

Cook, E. R., and L. A. Kairiukstis. 1990. *Methods of Dendrochronology: Applications in the Environmental Sciences.* Kluwer Academic Publishers, Dordrecht.

Cook, E. R., D. M. Meko, D. W. Stahle, and M. K. Cleaveland. 1999. Drought reconstructions for the continental United States. *Journal of Climate* 12: 1145–1162.

Fritts, H. C. 1976. *Tree Rings and Climate.* Academic, London.

Grissino-Mayer, H. D. 1996. A 2129-Year Reconstruction of Precipitation for Northwestern New Mexico, U.S.A. In *Tree Rings, Environment, and Humanity.* J. S.

Dean, D. M. Meko, and T. W. Swetnam, eds. Radiocarbon, Tucson, AZ, pp. 191–204.

Grissino-Mayer, H. D., and H. C. Fritts. 1997. The International Tree-Ring Data Bank: An enhanced global database serving the global scientific community. *Holocene* 7: 235–238.

Riebsame, W. E., S. A. Changnon, and T. R. Karl. 1991. *Drought and Natural Resources Management in the United States: Impacts and Implications of the 1987–89 Drought.* Westview, Boulder.

Stockton, C. W., and G. C. Jacoby. 1976. Long-Term Surface Water Supply and Streamflow Levels in the Upper Colorado River Basin. Lake Powell Research Project Bulletin No. 18. Institute of Geophysics and Planetary Physics, University of California, Los Angeles.

Woodhouse, C. A. 1999. Artificial neural networks and dendroclimatic reconstructions: An example from the Front Range, Colorado, USA. *Holocene* 9: 521–529.

———. 2001. A tree-ring reconstruction of streamflow for the Colorado Front Range. *Journal of the American Water Resources Association* 37: 561–569.

Woodhouse, C. A., and J. T. Overpeck. 1998. 2000 years of drought variability in the central United States. *Bulletin of the American Meteorological Society* 79: 2693–2714.

Acquisition, Management, and Dissemination of Climate Data

FOUR

Rene Reitsma
Phil Pasteris

Archives of data on climate, if reflective of rigorous standards and accompanied by appropriate supplementary documentation, can be used for multiple purposes ranging from testing of hypotheses about climate functions to improving regional resource management. Although numerous data archives exist, experience has demonstrated difficulties with their use and the need for various improvements in collection and archiving of data, as well as with data retrieval and manipulation. These issues were considered as part of a workshop in Boulder, Colorado, on June 21–22, 1999, that included contributions from individuals familiar with collection, archiving, analysis, and application of climate data. This chapter is a summary of the workshop's main conclusions.

HIGHER AND MORE COMPATIBLE RESOLUTIONS ARE REQUIRED

Climate data are often temporally or spatially incompatible with their intended use. For instance, spatial and temporal resolution of climate variables often do not match resolutions for land use or topography. This common mismatch in scale or resolution suggests that future efforts to match data sets more effectively would be rewarding. Standards and cross-referenced data sets are needed.

Needs for high-resolution climate data persist. Present resolution does not reflect increased understanding of the interaction between climate and other environmental variables. Often, variability of weather and climate correlates closely with small-scale spatial variability in relief (see Chapter 2) or land use, yet most climate data sets provide resolution inadequate for analysis of these influences. As a consequence, statistical techniques often are used to infer probabilistically the values for dependent (climatic) variables from a series of observed data on independent variables (topography, land use) at coarser levels of resolution. An example of this approach is the spatial climate modeling conducted by Oregon State University through the use of PRISM (Precipitation Regression on Independent Slopes Method; Daly et al. 1994). This technique has been used to model precipitation and temperature at a 4-km grid spacing throughout the United States and portions of Europe and Asia. The National Climatic Data Center has chosen PRISM as the spatial engine to produce the Climate Atlas for the United States for the period 1971–2000 (see http://nndc.noaa.gov/?http://ols.nndc.noaa.gov/plolstore/plsql/olstore.prodspecific?prodnum=C00519-CDR-A0001).

Although a statistical approach is perhaps essential given the incompatibilities of data sets and lack of resolution for climate data, it suffers from problems inherent in statistical models. For example, at the workshop one scientist mentioned inaccuracy of statistically inferred precipitation for the Sangre de Cristo Mountains in Colorado and New Mexico. Those responsible for generating the data by use of modeling replied that the Sangre de Cristo Mountains are special in their relief and topography, hence are not well represented by the regression model. This example illustrates a common problem with statistical modeling—the inference of an individual estimate from a population of such estimates requires that the casual relationships be the same for the individual estimate as for the underlying population. This requirement is known as "distributive reliability" (Harré 1978).

THE NEED FOR BETTER METADATA

Metadata on climate consist of information on climate data sets. Examples include station identity (site name, aliases, and all identification numbers), location (coordinates, elevation, geopolitical placement, topography), equipment or instrumentation and its exposure, observation and dissemination practices, data inventory, and temporal changes in any of this information (Lazar et al. 1999). Although extensive and rigorous standards for metadata are in force, some disagreement exists as to the degree of compliance with these standards. Some workshop participants viewed the provision of good metadata to be an integral part of the task of providing data and believed that metadata should be regulated by standards that apply to other data. Others argued that market mechanisms could encourage proper development of metadata. Under the latter approach, the provision of better metadata would lead to competitive advantage and hence would allow a given source to capture a greater share of the data market. Because most climate data are provided by governmental agencies, however, implementation of a market-based strategy might be difficult.

RESOLUTION AND METADATA

Climate data sets are not always optimally suited to the purpose for which they are used. For instance, in the case of emergency management of floods resulting from excessive rainfall, data sets must be quickly filtered and searched for relevant information. Such a search might be conditional (search X for Y, where Y is larger than Z) or could involve aggregation or disaggregation (P is the maximum of Q for area M and time period N). Such operations are likely to involve use of metadata for a combination of data sets with varying spatial and temporal resolution. Metadata also allow evaluation of the quality of the data.

It is through metadata that large amounts of useful information can be added to data sets. A recent and very promising development in this area is the release of Extensible Markup Language (XML; Bosak and Bray 1999). XML, a restricted form of Standard Generalized Markup Language (SGML; International Organization for Standardization HTML Syntax Number 8879), can be used to equip Web-based documents with additional metadata for use by special-purpose Web browsers. Through the use of XML it would be possible to develop Web-based facilities for interactive access of metadata and cross-referencing of metadata and core data.

Another approach to better use of combined data and metadata is the Web-based metadata engine prototyped for the Unified Climate Access Network (UCAN; Pasteris et al. 1997; Southern Regional Climate Center 1999). UCAN allows users to locate climate data through a wide variety of data keys based on geographic (latitude, longitude, elevation), political (e.g., state, county), or data elements (e.g., temperature, precipitation, evaporation). UCAN links a relational database containing metadata to data files stored in a NetCDF (Network Common Data Form) format. This segmented database concept supports a rapid search and update of database keys via the relational database, which determines the appropriate time series to extract. The NetCDF files are optimized for rapid extraction of climate data ranging from the entire period of record to intervals as short as a month (see http://www.unidata.ucar.edu/packages/netcdf/ and http://www.unidata.ucar.edu/packages/netcdf/docs.html).

HYPERDATABASES

Climate data are still too disparate and too difficult to combine. For example, a strong need exists for rapid access of nationwide reservoir and streamflow information for the United States from a single location or portal on the Internet. At present, users must visit a large number of web sites to obtain such data. In general, a growing need exists for tools, which could be called hyperdatabases, that allow users to obtain data from various locations and backgrounds and to combine them without having to know their specific representations and formats. Hyperdatabases could be facilities on the Internet through which one could retrieve, search, and combine information from databases into data sets for specific purposes. The hyperdatabases themselves would be empty (i.e., they would not contain any climate data). Instead, they would serve as brokers of climate data. An example of a facility that contains some of these characteristics is the multiorganizational, multiagency UCAN (Pasteris et al. 1997; Southern Regional Climate Center 1999).

HYPERDATABASES AS REGIONAL DATA ARCHIVES FOR USE IN APPLICATIONS

Applications of data to management often require that regional data be extracted from larger data sets. Regional information obtained in this way supports national and state organizations responsible for natural-resource management. Such a hierarchical system compounds some of the difficulties already mentioned. Yet demonstrable utility of data analysis is what many organizations are currently striving to achieve. For example, initiatives such as the Regional Earth Science Applications Centers (National Space and Aeronautics Administration 2002) or the Front Range Infrastructure Resources Project (United States Geological Survey 1999) are aimed at the beneficial application of the large amounts of data collected and maintained at and above the regional scale.

Several approaches can be taken to the challenge of achieving utility. One of the most frequently used is to make data available to the public through user-friendly, Web-based user interfaces. Average daily number of visits to the web site then is used to indicate the success of the endeavor and hence the usefulness of the data. Frequency of use of a web site, however, may not be an ideal measure of data usefulness or data use.

EFFICIENCY AND EFFECTIVENESS

Thomas Landauer (1995) offers convincing evidence that the large investments in information technology during the 1980s and early 1990s did not lead to significant increases in productivity until at least 1995. Certainly, such investments have had a strong role in improving companies' competitiveness (many people like to be able to go to the Web and check where their package is en route from A to B, even though the package does not get to B any faster), but investments in information technology—massive as they have been—have not, according to Landauer, led to significant increases in productivity except in the telecommunications industry.

Noting Landauer's impressive body of evidence, one is likely to wonder whether similar results would apply to environmental management, both private and governmental. Obviously, defining the concept of productivity in terms of observable variables is a serious challenge (Cook and Campbell 1979; DeLeone and McLean 1992), but should we not ask ourselves whether our ever-expanding data networks and databases actually make a difference in effectiveness (doing things better) or in efficiency (doing things with less effort)? Such an evaluation is very much needed as a guide to further investments in information technology.

SPEED UP THE COOPERATIVE NETWORK

Increasing the data generation speed of the so-called Cooperative Network, which consists of many thousands of volunteers collecting climate data for public use (Chapter 2), would be beneficial. A pilot study of this issue encompassing about 250 stations (Karl 1999) has led to formation of the U.S. Climate Reference Network (CRN), a network of climate stations being developed as part of a National Oceanic and Atmospheric Administration initiative. The primary goal of its implementation is to provide future long-term homogeneous observations of temperature and precipitation that can be coupled to past long-term observations for the detection and attribution of present and future climate change. Data from the CRN will be used in operational climate monitoring activities and for placing current climate anomalies into a historical perspective. The CRN will also provide the United States with a reference network that meets the requirements of the Global Climate Observing System. If fully implemented, the network will consist of about 250 stations nationwide. Implementation of the CRN is contingent on the availability of funding (see http://www.ncdc.noaa.gov/oa/climate/research/crn/crnnetman.html).

Fast access to bad data may not be in the best interest of users; data made available to users must be appropriately annotated. Also, there is more value in consistency than in accuracy of data (i.e., systematic errors in data can be accounted for relatively easily, whereas random errors are much more difficult to deal with). Undocumented systematic changes can, however, introduce significant bias. One example is the measurement of snowfall, which is greatly influenced by exposure and shielding.

SUMMARY

The workshop on climate data brought agreement on a number of issues among participants with expertise in data management, acquisition, and use. Two problems of broad concern are weakness of spatial and temporal resolution for climate data and incompatibility of data from different sources. The resolution problem can be only partially resolved by statistical approaches; more intensive climate monitoring would be required to provide spatial resolution compatible with current analytical needs. Several kinds of solutions are possible for the incompatibility problem, but no consensus has as yet been reached on which of these is most feasible. Metadata, which consist of site-specific information supplementary to the climate data themselves, are of much greater importance than has generally been appreciated and must be an important consideration in the collection and archiving of data. An obvious need exists for the development of hyperdatabases, which would consist of publicly accessible climate information brokering software with links to many different sources of climate data. Hyperdatabases would facilitate the creation of regional databases and their use in applications. Some efforts along these lines are in progress, but they need to be accelerated and broadened. In general, investments in information technology have not been well evaluated with respect to their effect on efficiency and utility of climate data; such an evaluation should be undertaken. Finally, the Cooperative Network is a key source of information on climate data in the United States, but its usefulness is hampered by slow archiving rates, which should be increased. The need for speed in archiving data should be balanced with a continuing need to maintain consistency of data collection, which is crucially important for time-series analysis.

A suitable long-term goal for users of climate data in the United States is a complete history of climate data arranged as a four-dimensional matrix based on latitude, longitude, elevation, and time. The internal consistency and comprehensive nature of such a database would greatly stimulate analysis of climate data.

REFERENCES

Bosak, J., and T. Bray. 1999. XML and the second-generation Web. *Scientific American* 281: 79–83.

Cook, T. D., and D. T. Campbell. 1979. *Quasi-Experimentation: Design and Analysis Issues for Field Settings.* Houghton Mifflin, Boston.

Daly, C., R. P. Neilson, and D. L. Phillips. 1994. A statistical-topographic model for mapping climatological precipitation over mountainous terrain. *Journal of Applied Meteorology* 33: 140–158.

DeLeone, W. H., and E. R. McLean. 1992. Information system success: The quest for the dependent variable. *Information Systems Research* 3: 60–95.

Harré, R. 1978. Accounts, Actions and Meanings—The Practice of Participatory Psychology. In *The Social Context of Method.* M. Brenner and P. Marsh, eds. Croom Helm, London, pp. 46–65.

Karl, T. R. 1999. *Adequacy of Climate Observing Systems.* National Academy Press, Washington, DC.

Landauer, T. K. 1995. *The Trouble With Computers: Usefulness, Usability, and Productivity.* MIT Press, Cambridge, MA.

Lazar, A., C. Barthel, and D. Kuiper. 1999. Metadata Strategies for the Efficient Management and Exchange of Data Throughout a Unified Climate Access Network. 11th Conference on Applied Climatology, American Meteorological Society. Dallas, Texas, January 10–15.

National Space and Aeronautics Administration. 2002. *RESAC Program Overview* <http://www.esad.ssc.nasa.gov/resac/>.

Pasteris, P., R. Reinhardt, K. Robbins, and C. Perot. 1997. UCAN-Climate Information for the Next Century. First Symposium on Integrated Observing Systems, American Meteorological Society. Long Beach, California, February.

Southern Regional Climate Center. 1999. UCAN Unified Climate Access Network <http://www.srcc.lsu.edu/ucan.net/UCAN.html>.

United States Geological Survey. 1999. Front Range Infrastructure Resources Project <http://greenwood.cr.usgs.gov/pub/fact-sheets/fs-0109-98>

Linkages Between Prediction of Climate and Hydrology

PART TWO

Prediction of climate even at very high levels of accuracy is of little direct use to water managers without a reliable linkage between climate predictions and hydrologic predictions. Connections between climate and hydrology have been available for some time in relatively crude forms, such as the correlation between snowpack on specific dates and amount of runoff in the following months. Presently, a number of intensive efforts are under way to improve the sophistication and scope of the linkage between climate and hydrology. Clark and colleagues (Chapter 5) give an excellent overview of the present methods for forecasting water supply from climate information and provide specific information on future prospects for improving hydrologic forecasting. Chapter 6, by Strzepek and Yates, is an authoritative set of predictions on potential for change in water availability for various major drainage basins in the western United States. Although the predictions are largely hypothetical, they are very informative in a number of respects because the basins of the West respond differently—even in a qualitative sense—to climate change. Strzepek and Yates describe some of the most important mechanisms and show conclusively that we cannot expect monolithic or uniform effects of climatic change on availability of water. In Chapter 7, Lettenmaier and Hamlet show how long-range forecasts such as those associated with synoptic climate variation can be harnessed in the use of hydrologic prediction and give specific examples drawn from the Northwest. Chapter 8, by Labadie and Garcia, is an evaluation, based on a workshop, of the potential for future advances to make predictions of use to water managers.

Use of Weather and Climate Information in Forecasting Water Supply in the Western United States

FIVE

MARTYN P. CLARK
LAUREN E. HAY
GREGORY J. MCCABE
GEORGE H. LEAVESLEY
MARK C. SERREZE
ROBERT L. WILBY

Water in the West is allocated among diverse uses and is subject to mounting demand caused by a growing population and changes in institutional practices (Pulwarty 1995; Diaz and Anderson 1995). Demand coupled with climate variability and potential climate change presents formidable challenges for water managers. For example, the U.S. Bureau of Reclamation has indicated that if the West were to experience a drought similar to that of 1931–1940, the water needs of the lower Colorado River basin would not be met (el-Ashry and Gibbons 1988). These concerns have stimulated attempts to develop better water-management tools and improved information on snowpack. A key goal is to improve hydrologic forecasts.

Currently, hydrologic forecasts are made with the extended streamflow prediction (ESP) procedure (Day 1985). ESP is based on a hydrologic model that is calibrated with observed precipitation and temperature data up to the beginning of the forecast interval and then run with an ensemble of temperature and precipitation data from every past year in the historical record. The model thus provides an ensemble of possible outcomes given the antecedent conditions (e.g., soil moisture, water equivalent of the accumulated snowpack) at the start of the forecast. Because the methodology of ESP weights equally the history of each past year, it yields a wide range of possible outcomes.

Although advances in climate research in recent decades have led to potentially useful weather predictions and climate outlooks (e.g., Kalnay et al. 1998; Barnston et al. 2000), this information is not used in ESP modeling. The use of historical data in ESP modeling means that accuracy of the forecasts is entirely dependent on the effect of basin initial conditions on future runoff. Replacing the ensemble of historical data used in ESP modeling with an ensemble of weather forecasts and climate outlooks (i.e., using weather forecasts and climate outlooks as input to hydrologic models) may reduce the range of possible outcomes and increase the accuracy of hydrologic forecasts.

This chapter examines ways in which information on weather and climate can be useful in managing water resources in the western United States. We focus on two timescales of variability. On intraseasonal timescales (1–2 weeks) we examine the possible use of output from global-scale atmospheric forecast models in generating short-term streamflow forecasts. On seasonal timescales we assess the role of El Niño and La Niña events in shaping the seasonal snowpack, which may then be used in forecasting seasonal hydrology.

INTRASEASONAL HYDROLOGIC FORECASTS

This section addresses the hypothesis that coupling short-range atmospheric forecast models with hydrologic models can improve upon the traditional ESP methods (e.g., Day 1985). As a demonstration of this approach, over 2,500 individual eight-day forecasts from the National Center for Environmental Prediction (NCEP) reanalysis project (described later) are used as input to the U.S. Geological Survey's (USGS) Precipitation Runoff Modeling System (PRMS) to forecast stream flow in the Animas River basin in southwestern Colorado (Figure 5.1). The Animas River basin has a drainage area of 1,820 km^2 at elevations that range from 2,000 to 4,000 m. The surface hydrology of the Animas River basin is dominated by snowmelt, which is typical of most small basins in the mountainous areas of the continental western United States. This basin therefore provides a good test of the feasibility of using atmospheric forecasts for predicting runoff in that region.

NCEP Reanalysis

The NCEP reanalysis project (Kalnay et al. 1996) produced a fifty-one-year (1948–1998) record of global atmospheric fields derived from a Numerical Weather Prediction model that was kept unchanged over the analysis period. Use of fixed conditions in the reanalysis eliminates pseudoclimatic jumps in the climate time series caused by upgrades in the modeling system

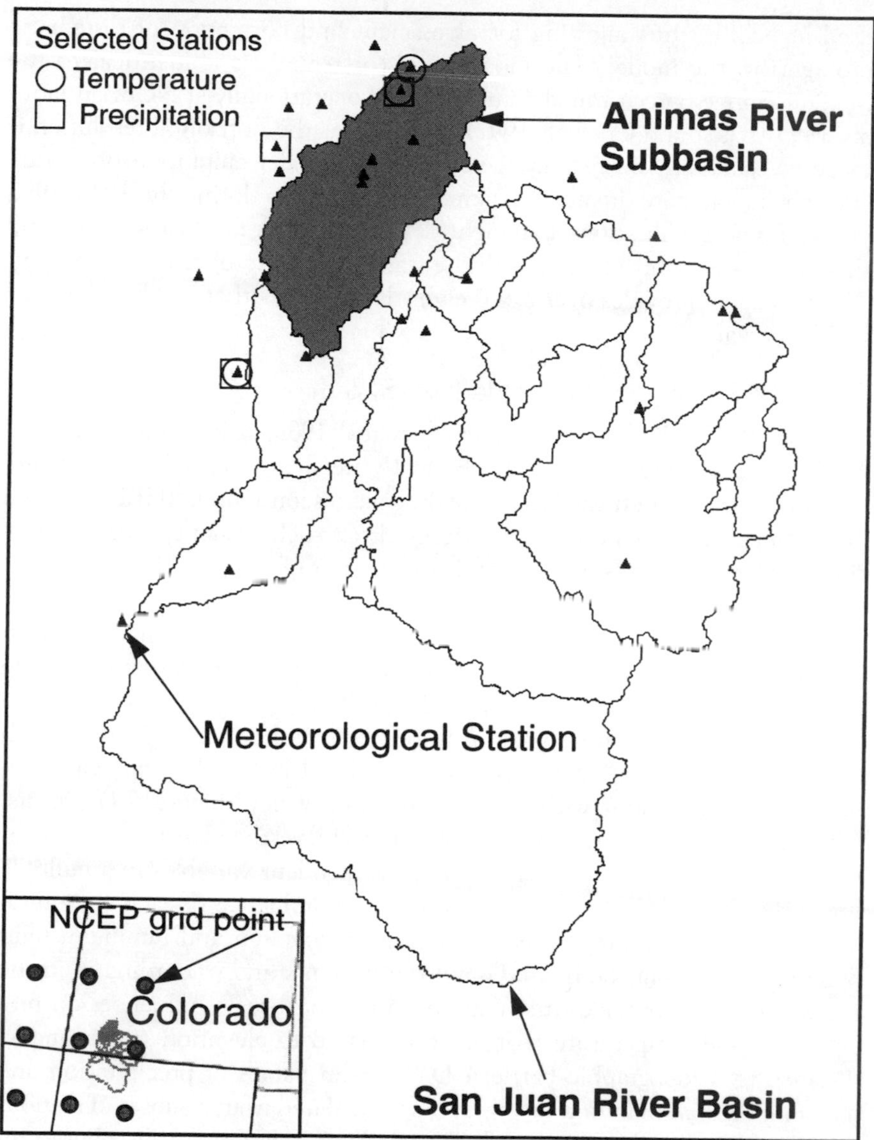

5.1 Map showing the meteorological station network (small triangles) of the San Juan River basin and the Animas River subbasin, Colorado. Locations of stations providing temperature data (circles) and precipitation data (squares) for modeling are shown.

used at NCEP, thus allowing for assessment and correction of systematic problems in the model. The model used for reanalysis is identical to the Medium Range Forecast model implemented operationally at NCEP in January 1995 (Basist and Chelliah 1997), except that the horizontal resolution is twice as coarse in the reanalysis version. The model employs a horizontal grid spacing of approximately 210 km. Every five days during the 1948–1998 period, a single realization of an eight-day atmospheric forecast was run, and the output was archived along with the standard (day zero) reanalysis output. This procedure provides over 2,500 eight-day forecasts that can be compared with observations.

The USGS Precipitation Runoff Modeling System

The PRMS hydrologic model (Leavesley at al. 1983; Leavesley and Stannard 1995) is a distributed-parameter, physical watershed model. Parameters are spatially distributed on the basis of hydrologic response units (HRUs), which are distinguished on the basis of characteristics such as slope, aspect, elevation, vegetation type, soil type, and distribution of precipitation. A typical area for an HRU is 5 km^2. Each HRU is assumed to be homogeneous in its hydrologic response. The PRMS model generally is run with daily data on precipitation and maximum and minimum temperature, which are available for most climate stations across the United States.

Because the station network in the Animas Valley is not dense enough to allow direct measurements on each individual HRU, the daily values at meteorological stations within and around the valley (Figure 5.1) are distributed to the HRUs within the basin. To achieve this, latitude (x), longitude (y), and elevation (z) were used as independent variables in a multiple linear regression (MLR) model to establish the influence of each station on mean spatial variations in precipitation and maximum and minimum temperature throughout the basin. Use of the station x and y coordinates in the regression model provides information on the local-scale influences on precipitation and temperature that are not related to elevation (for example, distance to a topographic barrier). Daily mean values of precipitation and maximum and minimum temperature, as calculated from a subset of stations in the region, are then used with the xyz MLR relations to distribute precipitation and temperature over the basin according to the mean values of x, y, and z for each HRU. A separate regression model was developed for each month to account for seasonal variations in the relationships of topography with precipitation and temperature. The distributed data are used to compute an energy and water balance for each HRU. The sum of the water balances of each HRU, weighted by unit area, produces the daily watershed response. In basins the size of the Animas (1,820 km^2), the travel of water through the channel network typically occurs on timescales of less than one day. For this reason, we chose not to include flow routing routines in our hydrologic simulations.

Snow is the major form of precipitation and source of stream flow in the Animas River basin. PRMS simulates the accumulation and depletion of a snowpack on each HRU. Snowpack is maintained and modified as both a water reservoir and a dynamic heat reservoir. Snowmelt does not occur until the snow surface temperature warms to 0°C, and outflow at the base of the snowpack does not occur until the entire snowpack warms to 0°C. A water balance is computed daily, and an energy balance is computed twice each day. Lack of data on wind speed, humidity, and radiation means that the energy balance computations must be greatly simplified. These computations include estimates of incoming shortwave and longwave radiation and the heat content of precipitation. The turbulent heat transfers of latent and sensible heat are approximated. Solar radiation is distributed to each HRU on the basis of slope and aspect. Neither shortwave nor longwave radiation is measured at the climate stations in this region, so these variables are estimated from their empirical relationships with temperature and precipitation.

Precipitation and snowmelt ultimately reach the channel network through surface runoff, subsurface flow, or groundwater. The first opportunity for precipitation and snowmelt to generate stream flow is via surface runoff, which takes place if the net water input exceeds the infiltration capacity of the soil over the entire HRU or if some areas of an HRU become completely saturated (this typically arises in lower regions of an HRU where topography causes an accumulation of water). PRMS simulates both of these processes. In the Animas River basin the combination of highly pervious soils (maximum infiltration rate is assumed to be 50 mm/day) and moderately low daily precipitation and snowmelt means that only a small fraction of the basin area is completely saturated. Thus contributions from surface runoff are small, and almost all runoff in the Animas River basin is derived from subsurface and groundwater flow.

The dominance of subsurface flow results in a time lag between surface-water inputs and the hydrologic response of the basin. In the PRMS model, water percolates to deep soil zones and groundwater reservoirs after the water-holding capacity of the upper soil layer is exceeded. Water recharges the groundwater reservoir at an assumed rate of 3 mm/day. Water in excess of this rate becomes inflow to the subsurface, which drives subsurface flow. Subsurface flow is the rapid movement of water through the matrix and preferential flow paths composing the soil and unsaturated-zone profile to the stream channel. Subsurface flow increases nonlinearly with the amount of water stored in the subsurface reservoir. The response time of the catchment is therefore faster if subsurface storage is high. For hydrologic forecasting, accurate initialization of subsurface conditions is important. Base flow from the groundwater reservoir is computed as a linear function of groundwater storage and occurs on timescales of days to weeks, thus introducing additional lags into the hydrologic system.

Hydrologic Forecast Procedures

Assuming a perfect hydrologic model, the accuracy of hydrologic forecasts depends on the skill of atmospheric forecasts and the accuracy with which initial conditions can be specified over the basin. The most important initial conditions for hydrologic forecasts are the water equivalent of snow and the soil moisture on each HRU and the amount of water stored in the subsurface and groundwater reservoirs. We establish these initial conditions by forcing PRMS with distributed station observations of precipitation and maximum and minimum temperature for the time period of the NCEP reanalysis, starting three years prior to the day before the first forecast date. The state variables (e.g., water equivalent of snow, soil moisture, subsurface storage) are saved for every five days and then used as initial conditions for the eight-day forecasts.

The hydrologic simulations obtained with distributed station data are used to assess the accuracy of the hydrologic forecasts. Taking modeled values as truth assumes a perfect hydrologic model and allows us to focus attention on hydrologic effects of errors in the atmospheric forecasts. If observed runoff is used as truth, situations may arise for which errors in the atmospheric forecasts are of opposing sign to errors in the hydrologic model simulations, thus tending to cancel them. This would provide a misleading perception of reliability for the hydrologic forecasts.

Following the simulations of runoff based on station data, two hydrologic forecast experiments were performed. In the first experiment, eight-day hydrologic forecasts were run using constant precipitation and maximum and minimum temperatures computed from historical station data for each forecast period. This is termed the climatology experiment. It provides output analogous in many respects to the mean response in ESP simulations and quantifies the skill that is possible when forecasts are based on historical station data. The second experiment forces the PRMS hydrologic modeling system with the eight-day forecasts of precipitation and maximum and minimum temperature from an average of the nine NCEP model grid points surrounding the Animas River basin. This is termed the forecast experiment.

At the NCEP model grid points overlying the Animas basin, the raw NCEP forecasts show systematic biases in predicting temperature and precipitation. Winter and spring temperatures are too low, summer temperatures are too high, precipitation in late winter is too high, and precipitation in summer and early autumn is too low. These biases are markedly different at various forecast lead times (i.e., biases for one-day forecasts differ from the biases for eight-day forecasts). As a pre-processing step, we removed all systematic biases before the NCEP forecasts were used in the PRMS hydrologic modeling system. Biases in maximum and minimum temperatures were removed by computing a monthly climatology of the NCEP temperatures for each forecast lead time, subtracting the forecast value from that climatology (to produce a daily anomaly value), and adding the daily anomaly to the corresponding monthly station climatology of maximum and minimum temperature. Because precipitation data are not normally distributed, correcting for constant bias over the entire data range is not valid. To circumvent this problem, we classified the precipitation data by deciles and performed the bias correction outlined earlier independently for each decile. The bias corrections only apply to systematic biases in the NCEP fields; they do not account for biases associated with specific weather regimes (e.g., precipitation may be underestimated during the passage of a cold front and overestimated when high pressure is dominant). The bias corrections are possible only because the NCEP model was held constant throughout the reanalysis period. In an operational setting the model is frequently upgraded, and such corrections (as well as more sophisticated statistical procedures such as statistical downscaling) are not possible.

Surface-Water Hydrology of the Animas River Basin

Monthly variations in the mean and standard deviation of the three variables used as input to PRMS (maximum and minimum temperatures and precipitation) and of the three modeled variables (snowmelt, actual evapotranspiration [ET], and runoff) are presented in Figure 5.2. Figure 5.2, which is based on the simulations using station data, describes the influence of precipitation and temperature on the hydrology of the Animas River basin, as well as seasonal variations in the hydrologic response of the basin. Temperatures reflect the midlatitude, continental location of the region. Mean maximum temperatures are close to 0°C from November to March, and minimum temperatures are below freezing most of the year. Temperatures rise significantly in summer months, when mean basin-average maximum temperatures are close to 20°C. Both minimum and maximum temperatures are most variable in winter months. Precipitation is relatively constant across seasons but tends to be low in early summer and high during autumn. Most winter precipitation falls as snow and is stored in the snowpack until spring. Modeled snowmelt is highest and most variable in April, May, and June. Although temperatures are generally too low for melt to occur in midwinter, a secondary melt occurs in October and November when snow cover is transient. Actual ET is highest throughout the summer, especially in May and June, when the amount of soil moisture is high. The seasonal cycle of runoff is similar to that of snowmelt, illustrating the importance of snow for the

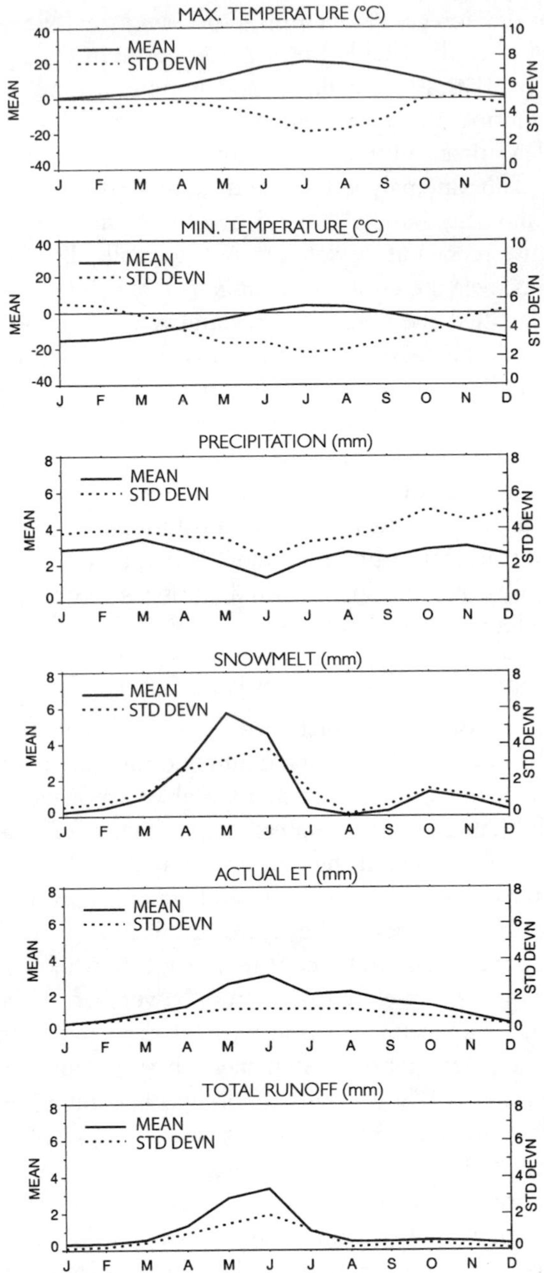

5.2 Seasonal cycles of the long-term monthly mean and daily standard deviation of maximum and minimum temperature and precipitation (inputs to PRMS-distributed hydrological model; see text) as well as snowmelt, actual evapotranspiration (ET), and total runoff (PRMS model outputs) for the Animas basin.

hydrology of the Animas River basin. Runoff is highest and most variable in May and June.

Hydrologic Forecasts Using Climatology

In the climatology experiment, forecast errors in maximum and minimum temperatures and precipitation are similar for different forecast lead times because these variables are set to constant mean values on the basis of historical data for each forecast period. The small changes in error with forecast time occur because, with the forecast runs spaced five days apart, different data are used to verify forecasts at different lead times. Forecasts of snowmelt, actual ET, and runoff, as computed by the model, were lowest at the start of the forecast cycle because of the influence of initial conditions on forecasts. To facilitate direct comparison with the NCEP forecasts, the climatological forecasts were run every five days. Results for the climatology experiment are presented in the left column of Plate 11.

Hydrologic Forecasts With NCEP Output

The accuracy of the forecasts based on output from the NCEP model is presented in the middle column of Plate 11. The right column is the difference between the forecast and the climatology experiment (middle column subtracted from the left column) and represents the improvement provided by using forecasts instead of climatology.

Only in some cases is the accuracy of the NCEP forecasts superior to that of forecasts based on climatology (middle and right columns of Plate 11). Generally, forecast accuracy is limited by the coarse horizontal resolution of the NCEP model (for example, precipitation varies on the subgrid scale) and deficiencies in modeling of physical phenomena. For example, summer precipitation may be poorly represented because of inadequacies in parameterization of convective processes. The poor forecasts of precipitation limit the use of output for river basins where the surface hydrology is dominated by rain. In snowmelt-dominated river systems typical of the western United States, however, short-term variation in runoff is influenced more by variation in temperature than by variation in precipitation. Thus accurate hydrologic predictions are more likely. In the Animas River basin, because the highest accuracy in forecasts of maximum temperature coincides with the spring melt (Plate 11), accurate predictions of runoff are feasible.

Improvements in the modeled variables through use of NCEP forecasts are readily apparent from the last three graphs of the right column in Plate 11. During the spring melt period, errors in forecasts of snowmelt are low for the first four days of the forecast cycle because of the high accuracy in forecasts of maximum temperature for March, April, and May. Likewise, reductions in forecast errors of actual ET are evident in late spring. In midsummer, however, NCEP forecasts of ET are less accurate than forecasts of ET based on climatology. This is consistent with the poor forecast accuracy for maximum temperature at this time of year. In terms of runoff, reductions in forecast errors during the spring melt period are remarkable. Forecast errors at the end of the forecast period (seven to eight days) are almost halved when the hydrologic model is forced with the NCEP output. Forecast accuracy for runoff at such long lead times exceeds the forecast accuracy for maximum temperature and snowmelt because of the natural lags and integrating effects of hydrologic systems.

Summary and Discussion

The NCEP atmospheric model, when coupled with the PRMS-distributed hydrologic model, provides forecasts of runoff with errors much lower than those of hydrologic forecasts based on climatology. In addition, accuracy of runoff forecasts is evident at longer lead times (about four days) than for the forecasts of maximum temperature and snowmelt because of the natural lags and integrating effects of hydrologic systems, particularly with regard to subsurface flow. Greater accuracy at longer lead times underscores the importance of specifying initial conditions accurately over the basin at the start of the forecast period. Although we do not currently assimilate satellite data into the hydrologic modeling, we anticipate that in the future satellite data will provide more accurate estimates of initial conditions for hydrologic forecasts.

The accuracy of runoff forecasts in this study is possible because the hydrology of the Animas River basin is dominated by snowmelt, which is influenced predominantly by temperature. In other river basins where the hydrology is more heavily influenced by rainfall, the accuracy of precipitation forecasts will be more important. Because the precipitation forecasts from global-scale models are rather poor (particularly on the small spatial scales used in hydrologic applications), use of raw, global-scale forecasts is unlikely to provide reliable forecasts of runoff for rainfall-dominated hydrologic systems. In the Animas River basin, skillful predictions of maximum temperature happen to coincide with the timing of snowmelt. Predictions of maximum and minimum temperatures do not improve upon climatology during summer, and predictions of maximum temperature are little better than climatology in winter.

Because some of the largest forecast errors can be attributed to the coarse horizontal resolution of the NCEP model, it may be prudent to explore methods that resolve subgrid-scale information in the forecast fields by statistical downscaling through the use of Model Output Statistics (MOS; Wilks 1995; Wilby et al. 1999). MOS is based on empirical relations between features reliably simulated as global-scale forecast models at grid-box scales (e.g.,

500 hectoPascals [hPa] geopotential height) and surface characteristics at subgrid scales (e.g., occurrence and amount of precipitation). An alternative approach is through dynamical downscaling, whereby a dynamically based regional-scale climate model is nested within the global-scale forecast model. Although the computational requirements of such an approach are demanding, rapid increase in computer power over the past decade has allowed regional climate models to become a major tool in short-term numerical weather prediction. Comparisons of ten years of dynamically downscaled NCEP output from the RegCM2 regional climate model run at a horizontal resolution of 50 km, and statistically downscaled NCEP output using MOS for simulating the surface hydrology of the Animas River basin shows that the statistical approach performs slightly better than the dynamical approach (Wilby et al. 2000) or the raw NCEP output. Because precipitation and temperature variations often occur on spatial scales much smaller than 50 km, however, it may be premature to rule out the utility of regional climate models. It is likely that nesting a series of regional climate models to scales of less than a square kilometer is necessary to resolve adequately the subgrid-scale variations important for hydrologic modeling.

SEASONAL HYDROLOGIC FORECASTS

At seasonal timescales, predictive skill can be derived from knowledge of ways in which the slowly varying components of the climate system (e.g., tropical sea-surface temperatures, continental soil moisture, and albedo) alter the probability of extremes in precipitation and temperature. In the snow-fed river systems of the western United States, many water-management decisions are based solely on the amount of water stored in the seasonal snowpack at various times during the accumulation season, whereas relatively little attention is paid to mechanisms by which low-frequency climate variations may influence patterns of snow accumulation. One of the most widely studied low-frequency variations in the climate system is the El Niño–Southern Oscillation (ENSO). Here we focus on the Columbia and Colorado River basins and show how El Niño and La Niña events modulate the seasonal evolution of the montane snowpack in these basins.

Effects of ENSO on Western Water Resources

The El Niño–Southern Oscillation describes quasi-periodic variations in sea-surface temperature (SST) of the tropical Pacific Ocean and associated pressure oscillations between Tahiti and Darwin, Australia (Rasmusson and Wallace 1983; Chapter 1). El Niño (warm) events are characterized by above-normal SSTs in the eastern tropical Pacific Ocean and increased convection and precipitation near and east of the International Dateline. During La Niña (cold) events, SST anomalies generally oppose those of El Niño events, but the negative anomalies in tropical convection and precipitation are of small magnitude and do not extend as far east as the positive anomalies of El Niño (Hoerling et al. 1997). The tropical anomalies perturb midlatitude atmospheric circulation patterns. During warm events, enhanced tropical convection results in intensification of the Hadley circulation and a strengthening and eastward extension of the Pacific subtropical jet. This is associated with a deepening of the Aleutian Low in the North Pacific Ocean and amplification of the northern branch of the tropospheric wave train over North America, resulting in a characteristic "split flow" (e.g., Bjerknes 1969; Horel and Wallace 1981; Rasmusson and Wallace 1983). The midlatitude response to cold events generally involves weakening of the subtropical jet and damping of the wave train over North America. Because of the lack of symmetry in tropical convection patterns between warm and cold events, however, sea-level pressure and anomalies at 500 hPa height during warm events are shifted, on average, 35° east of those in cold events (Hoerling et al. 1997).

In an analysis of hydrologic effects of ENSO over the western United States, Cayan and Webb (1992) computed correlations between the Southern Oscillation Index (SOI)—a simple measure of the phase and strength of ENSO—and April 1 snow-water equivalent (SWE) and between the SOI and annual runoff. They showed that SWE and annual runoff are positively correlated with the SOI in the northwestern United States but are negatively correlated with the SOI in the southwestern United States. This suggests that El Niño events lead to decreased SWE and annual runoff in the Northwest and to increased SWE and annual runoff in the Southwest and that La Niña events have the opposite effects. Other studies provide similar findings (Cayan and Peterson 1990; Koch et al. 1991; Redmond and Koch 1991; Cayan 1996). In El Niño years the amplified northern branch of the wave train over North America is associated with higher temperatures over the northwestern United States and Canada and a northward shift of the storm track toward Alaska, which decreases mean precipitation over the Pacific Northwest. To the south, the strengthened subtropical jet entrains more moisture from the Pacific Ocean, which, when combined with the increased cyclonic shear on the jet's poleward flank, increases the likelihood of precipitation over the southwestern United States. In La Niña years the stronger zonal flow allows more storm systems to penetrate into the northwestern United States, resulting in increased precipitation over this area. Over the southwestern United States, the weakened subtropical jet decreases the likelihood of precipitation.

Data

We base our analysis on SWE data collected manually on a monthly basis during winter and spring at several hundred permanent snowcourses maintained by the United States Department of Agriculture's cooperative snow survey program. The program is coordinated by the Natural Resources Conservation Service, which has partners conducting measurements in Arizona, Colorado, Idaho, Montana, Nevada, New Mexico, Oregon, Utah, Washington, and Wyoming. In addition, the California Department of Water Resources has an independent program. The number of snowcourse locations grew from fewer than 10 in the 1910s to almost 2,000 in the 1970s. Snowcourse measurements decreased in the 1980s and 1990s as a result of the development of the automated Snowpack Telemetry (Snotel) system (Serreze et al. 1999). SWE is generally measured on or about the beginning of each month from January through June. Additional measurements may be taken in the middle of the month if knowledge of snowpack conditions is deemed critical. Snowcourse measurements are most frequently taken at the beginning of April, which is the peak for SWE in much of the West. The frequency and timing of measurements vary considerably with the locality, the nature of the snowpack, difficulty of access, and cost (NRCS 1988).

Snow depth and SWE are measured by pushing a tube through the snowpack to the ground surface and extracting a core. SWE is determined by weighing the tube with its snow core and subtracting the weight of the empty tube. Between five and ten cores are taken at regular intervals along each snowcourse site. An average of all samples for a site is the SWE value for that site. Generally, courses are about 300 m long and are situated in small meadows protected from the wind (NRCS 1988). Possible problems with snowcourse measurements include changes in vegetation, which may change patterns of snow accumulation along the snowcourse, and errors in data entry. For present purposes, quality control of the data involves only removal of obviously impossible cases in which snowpack depth was reported as being less than its water equivalent. The effects of remaining errors are reduced through our use of basinwide SWE averages (described later).

Definition of El Niño and La Niña Events

The first step in our analysis is to identify El Niño and La Niña events. Any such classification is somewhat arbitrary. Our definition is based on mean winter (November through April) values of the Niño 3.4 index, which is the average SST anomaly in the central equatorial Pacific Ocean (5°N–5°S; 120°W–170°W), where tropical convection is most sensitive to SST variations (Hoerling et al. 1997; Trenberth 1997). Each year over the period 1951–1996 was ranked in terms of the mean wintertime anomaly value (Table 5.1). The ten warmest and ten coldest years were extracted for analysis. Attention is restricted to years after 1951, as the SST data for earlier years are somewhat unreliable. Sensitivity tests were conducted by substituting the

Table 5.1—Years ranked in terms of the magnitude of the Niño 3.4 index (area-averaged SST over the region 120°W–170°W, 5°S–5°N) and the Southern Oscillation Index (sea-level pressure difference between Tahiti and Darwin). The SOI was multiplied by –1, which makes it ordinally consistent with the Niño 3.4 index. The years are defined as the date at the end of winter (i.e., the winter of 1982–1983 is taken as 1983).

Rank	*Index NINO3.4*	*Index SOI*	*Rank*	*Index NINO3.4*	*Index SOI*	*Rank*	*Index NINO3.4*	*Index SOI*
1	1989	1974	17	1963	1957	32	1978	1959
2	1974	1971	18	1962	1984	33	1980	1981
3	1971	1989	19	1981	1961	34	1952	1994
4	1976	1976	20	1961	1968	35	1964	1969
5	1956	1956	21	1957	1986	36	1977	1995
6	1985	1951	22	1960	1954	37	1988	1973
7	1955	1967	23	1990	1979	38	1970	1952
8	1996	1962	24	1982	1977	39	1995	1990
9	1951	1963	25	1979	1965	40	1969	1958
10	1984	1960	26	1994	1964	41	1966	1966
11	1975	1955	27	1954	1988	42	1987	1993
12	1968	1975	28	1953	1991	43	1973	1978
13	1965	1982	29	1991	1970	44	1958	1987
14	1986	1985	30	1993	1980	45	1992	1992
15	1967	1972	31	1959	1953	46	1983	1983
16	1972	1996						

Southern Oscillation Index for the Niño 3.4 index (Table 5.1) and by varying the number of years designated as El Niño or La Niña events.

Computing Basinwide Estimates of SWE

Basinwide estimates of SWE were computed for each of the major subbasins in the Columbia and Colorado watersheds. This was done using snowcourse measurements taken on or near February 1, March 1, April 1, and May 1 for all years between 1951 and 1996. Snowcourses were assigned to drainage basins by use of hydrologic unit codes (Seaber et al. 1987). For the Columbia River, the relevant subbasins within the United States are the upper Columbia and Yakima, Pend Oreille and Kootenai, Snake River, and lower Columbia (Figure 5.3). The relevant subbasins in the Colorado River basin include the upper Green; the Colorado headwaters, the White River, and the Yampa; the lower Green and Lake Powell; the San Juan, Gunnison, and Dolores Rivers; and the lower Colorado (Figure 5.3). Descriptions of these basins are provided in Table 5.2.

ENSO-SWE Associations for the Columbia River Basin

Composite anomalies and inter-ENSO variability in SWE for the subbasins in the Columbia River are illustrated in Figure 5.4. The bars represent the

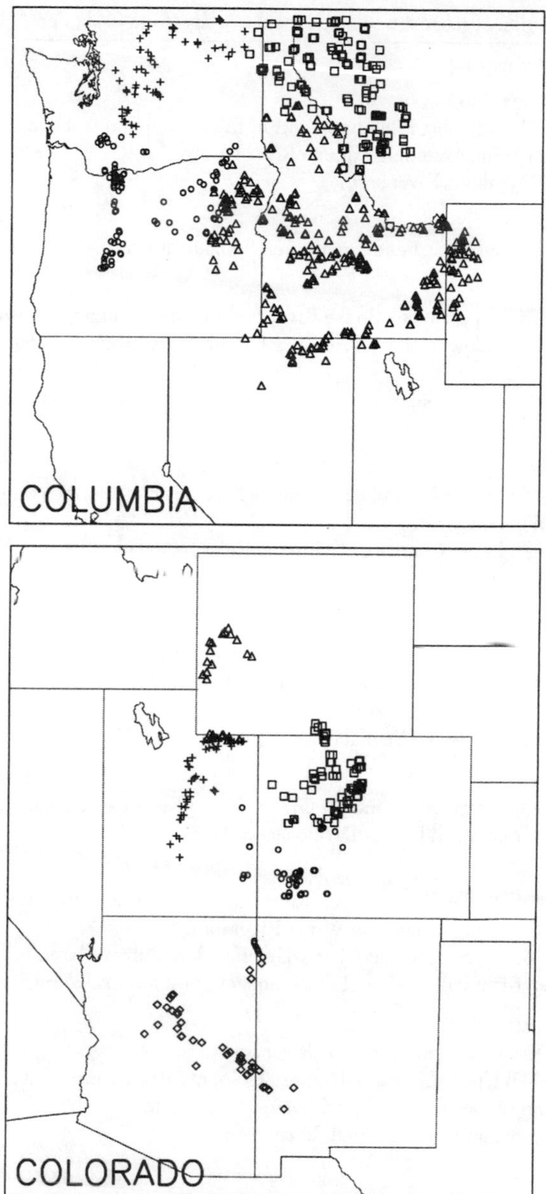

5.3 Snowcourse sites in the Columbia and Colorado River basins. Subbasins in the Columbia River basin are the upper Columbia and Yakima (+), the Pend-Oreille and Kootenai (squares), the Snake (triangles), and the lower Columbia (o). Subbasins in the Colorado are the upper Green (triangles), the Colorado headwaters (squares), the lower Green (+), the San Juan (o), and the lower Colorado (diamonds).

Table 5.2—Description of subbasins selected for this study (from Seaber et al. 1987).

COLUMBIA RIVER BASIN

UPPER COLUMBIA AND YAKIMA

Subregion 1702: Columbia River basin within the United States above the confluence with Snake River basin, excluding Yakima River basin

Subregion 1703: Yakima River basin

PEND OREILLE AND KOOTENAI

Subregion 1701: Kootenai, Pend Oreille, and Spokane River basins within the United States

SNAKE

Subregion 1704: Upper Snake: Snake River basin to and including Clover Creek basin

Subregion 1705: Middle Snake: Snake River basin below Clover Creek basin to Hells Canyon Dam

Subregion 1706: Lower Snake: Snake River basin below Hells Canyon Dam to its confluence with the Columbia

LOWER COLUMBIA

Subregion 1707: Middle Columbia: Columbia River basin below the confluence with Snake River basin to Bonneville Dam

Subregion 1708: Lower Columbia: Columbia River basin below Bonneville Dam, excluding Willamette basin

COLORADO RIVER BASIN

COLORADO HEADWATERS AND THE WHITE AND YAMPA

Subregion 1401: Colorado Headwaters: Colorado River basin to but excluding Bitter Creek basin and excluding Gunnison River basin

Subregion 1405: White and Yampa River basins

UPPER GREEN

Subregion 1404: Great Divide–Upper Green: Green River basin above the confluence with Yampa River basin, and Great Divide closed basin

LOWER GREEN AND LAKE POWELL

Subregion 1406: Lower Green: Green River basin below the confluence with Yampa River basin but excluding Yampa and White River basins

Subregion 1407: Upper Colorado–Dirty Devil: Colorado River basin below the confluence with Green River basin to Lees Ferry compact point but excluding San Juan River basin

GUNNISON/DOLORES/SAN JUAN

Subregion 1402: Gunnison: Gunnison River basin

Subregion 1403: Upper Colorado–Dolores: Colorado River basin from and including Bitter Creek basin to the confluence with Green River basin

Subregion 1408: San Juan: San Juan River basin

LOWER COLORADO

Subregion 1502: Little Colorado River basin

Subregion 1504: Upper Gila: Gila River basin above Coolidge Dam, including Animas Valley closed basin

Subregion 1506: Salt River basin

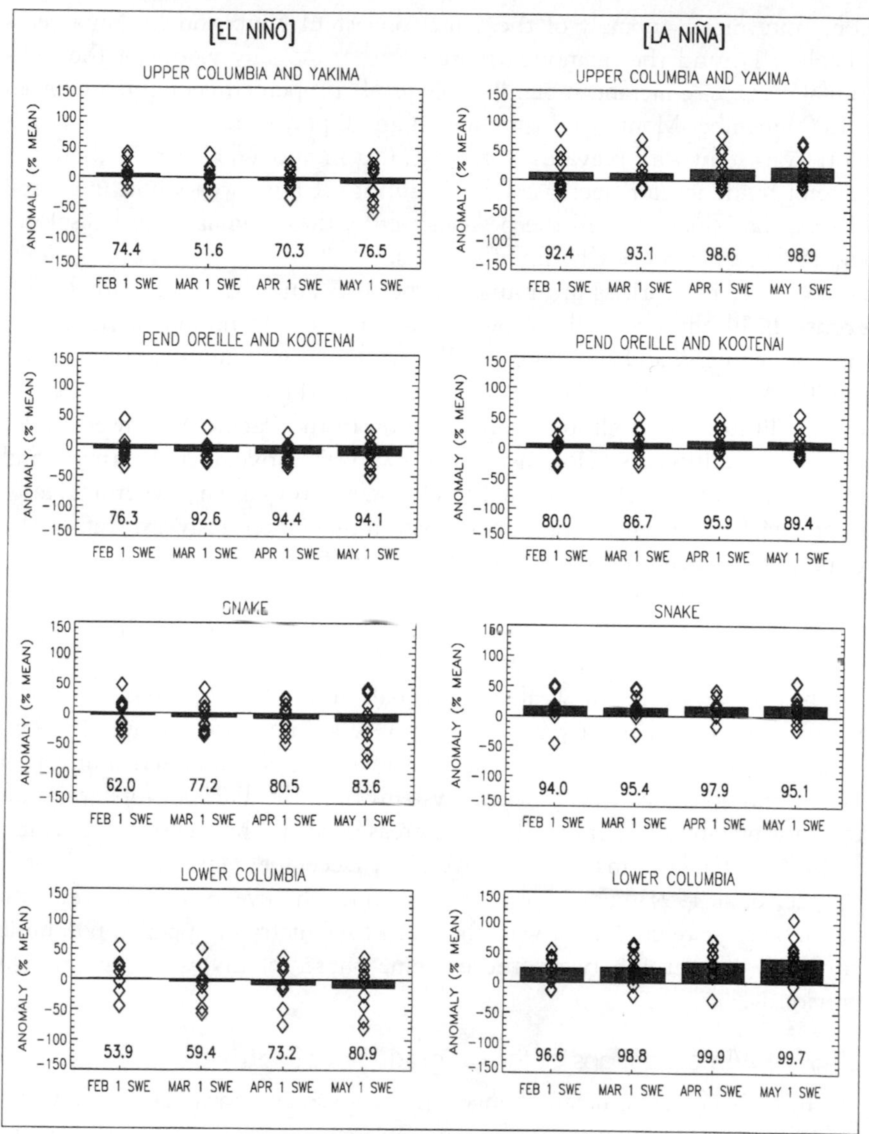

5.4 Anomalous snow-water equivalent (SWE) (expressed as a percentage of the mean) in El Niño and La Niña years for major subbasins in the Columbia River basin. The bars represent the mean of the ten strongest El Niño or La Niña years (based on the Niño 3.4 index; Table 5.1), and the diamonds represent the ten years used to compute the means. The numbers at the bottom of each plot indicate the statistical significance of deviations as determined by a 1,000-member Monte Carlo simulation.

composite mean anomaly of the ten strongest El Niño and La Niña years (Table 5.1), and the diamonds represent the anomaly values of the individual composite members. Significance levels (in percentage) obtained from a 1,000-member Monte Carlo simulation are displayed at the bottom of each plot. Consistent with previous work (e.g., Cayan and Webb 1992), there is a general tendency for decreased SWE during El Niño years and increased SWE in La Niña years. In the broadest sense, these signals reflect displacement of the storm track by ENSO events.

Some nonlinearities are found in the association of SWE with ENSO events. In El Niño years the snowpack is very close to normal at the beginning of February, and decreases in SWE at the accumulation season are not significant at the 95 percent confidence level (Figure 5.4). Livezey et al. (1997) illustrate that although precipitation in the Pacific Northwest under El Niño conditions is below normal in October, November, February, and March, it is actually above normal in December and January, when increases in precipitation occur in conjunction with a midwinter eastward shift of the tropospheric wave train (Hoerling and Kumar 2000). Since most snow in the Pacific Northwest mountains falls in midwinter (Serreze et al. 1999), increase of precipitation in December and January largely cancels the effects of decreased precipitation in other months.

In comparison to the weak SWE signals during El Niño years, increases in SWE during La Niña years are relatively strong (often above the 95% confidence level). Stronger zonal flow over western North America under La Niña conditions steers more storm systems over the Pacific Northwest, increasing precipitation there. These increases occur throughout the winter and are particularly marked in November, December, January, and March (Livezey et al. 1997). The stronger signals in La Niña years as compared with El Niño years are consistent with the lack of symmetry in upper tropospheric circulation patterns between the extreme phases of ENSO, as mentioned previously.

ENSO-SWE Associations for the Colorado River Basin

In El Niño years, mean changes in the seasonal snowpack of the Colorado River basin (Figure 5.5) show a transition between drier-than-average conditions in the north (best expressed in the upper Green) and wetter-than-average conditions in the southwest (best expressed in the lower Colorado). Broadly opposing patterns of La Niña years are consistent with results from previous work (e.g., Cayan and Webb 1992). Signals in the upper Green (Figure 5.5) are consistent with those of the Columbia River basin (Figure 5.4), where decreases in SWE during El Niño years reflect the amplified tropospheric wave train and associated decreases in precipitation; opposite patterns appear with La Niña. In the lower Colorado basin the significant

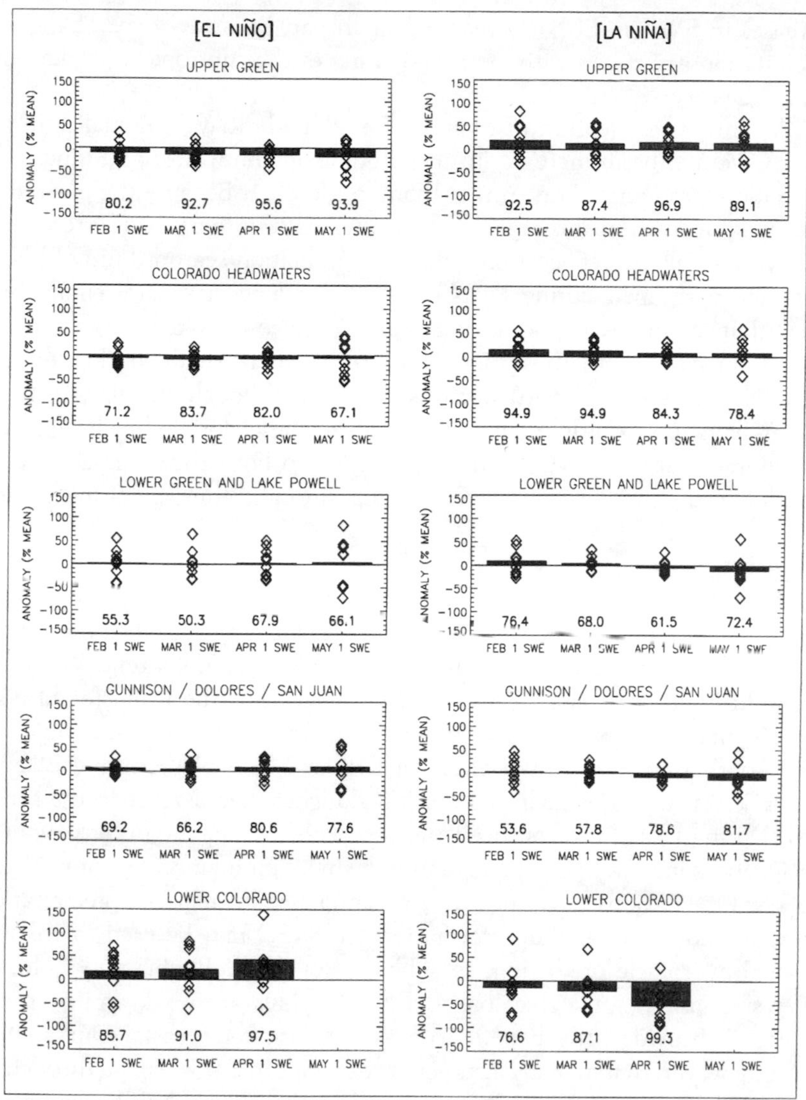

5.5 Anomalous SWE (expressed as a percentage of the mean) in El Niño and La Niña years for major subbasins in the Colorado River basin. The bars represent the mean of the ten strongest El Niño or La Niña years (based on the Niño 3.4 index; Table 5.1), and the diamonds represent the ten years used to compute the means. The April 1 SWE anomaly in the lower Colorado during the 1972–1973 El Niño event is off the scale at 216 percent of the mean (not plotted). The numbers at the bottom of each plot indicate the statistical significance of deviations as determined by a 1,000-member Monte Carlo simulation.

increases in SWE in El Niño years occur in part because of the changes in precipitation associated with a strengthening of the subtropical jet; La Niña causes the reverse.

An interesting feature of Figure 5.5 is that the SWE anomalies in the lower Colorado basin increase in magnitude over the winter and tend to be much more pronounced on April 1 than on March 1. Because the mean date of maximum SWE in this region is typically near February 20 (Serreze et al. 1999), these signals reflect a prolonged accumulation season during El Niño years and early melt during La Niña years. Evidence indicates similar seasonal changes in the Colorado headwaters, the lower Green, and the San Juan River basins, but these signals are much more subdued (Figure 5.5). Sensitivity tests (not shown) demonstrate that although the signals in the lower Colorado are fairly robust, the signals in the Colorado headwaters, the lower Green, and the San Juan are sensitive to both the number of years included in the El Niño and La Niña composites and the type of index used.

DISCUSSION

Effects of ENSO on snowpack, and thus on water resources, over the Columbia River basin differ from the effects of ENSO farther south in the Colorado River basin. Over the Columbia River basin SWE generally tends to be low during El Niño years and high in La Niña years, but the trend for El Niño years is much less pronounced. Over the Colorado River basin there is a north-south transition in the effect of ENSO. Mean SWE during El Niño years is lower than average in the north and higher than average in the southwest. During La Niña years the signal is reversed. Over the lower Colorado, precipitation and SWE anomalies tend to be more pronounced in spring, when the ENSO variations have the strongest influence on regional precipitation.

Our study suggests that information on ENSO may be used to enhance seasonal hydrologic predictions in subbasins of the Columbia and Colorado River systems. In some river basins ENSO signals are weak and by themselves will provide little benefit to seasonal runoff forecasts. Our analysis shows, however, that in many instances the change in ENSO signal throughout the accumulation season is often greater than the mean ENSO signal at the end of that season. For example, SWE is high in the Colorado headwaters basin on February 1 in La Niña years but is much less so (and not significant even at the 90% level) on April 1. Because on average over two-thirds of the snowpack has accumulated by the beginning of February (Serreze et al. 1999), monitoring the water equivalent of the snowpack throughout the accumulation season and combining this information with knowledge of seasonal changes in the ENSO signal may improve predictions of water supply.

Variations in Tropical Pacific SSTs are not the only slowly varying component of the atmosphere-land-ocean system that may improve prediction of seasonal climate. Other low-frequency variations may be important, such as variations in midlatitude sea-surface temperature and continental-scale variations in land-surface snow mass, land-surface albedo, and soil moisture. For example, Gutzler and Preston (1997) show that large-scale variations in snow mass over the western United States have a significant relationship to spring and summer precipitation over New Mexico. The hypothesized mechanism, which is similar to that invoked for snow-monsoon relationships over Eurasia, is that the energy required to melt above-average snow mass and evaporate the associated higher amounts of soil moisture causes a delay in seasonal warming of the landmass. This modulates the land-ocean temperature contrast and the summer monsoonal circulation over the southwestern United States. Expansion of seasonal climate forecasts to include these additional influences may result in improved hydrologic predictions on seasonal timescales.

SUMMARY AND CONCLUSIONS

This chapter explains the utility of weather and climate information for improving intraseasonal and seasonal hydrologic predictions. We demonstrate for the Animas River in Colorado that atmospheric forecasts provide predictions of runoff with lower forecast errors than those obtained by current practice, which uses historical station data to characterize expected future weather. Because the surface hydrology of the Animas basin is dominated by snowmelt, variations in temperature are much more important than variations in precipitation for short-term runoff forecasts. Temperature forecasts are more reliable than precipitation forecasts, thereby lending enhanced accuracy to hydrologic forecasts that use output from numerical weather prediction models. In rainfall-dominated river basins, however, raw, global-scale atmospheric forecasts may be of little benefit; improvements in forecast skill may be realized only through statistical or dynamical downscaling methodologies.

For seasonal hydrologic forecasts, El Niño and La Niña events significantly modulate the seasonal snowpack evolution in many major subbasins of the Columbia and Colorado River systems. In the Columbia River basin the general tendency is for decreased SWE during El Niño years and increased SWE in La Niña years, but changes during La Niña years are much more pronounced. Over the Colorado River basin, drier-than-average conditions occur in the north and wetter-than-average conditions occur in the southwest during El Niño, and opposite trends occur during La Niña. Over the lower Colorado basin, SWE anomalies tend to be larger in spring, when the El Niño and La Niña events have the strongest influence on regional precipitation. The analysis suggests that ENSO information may be useful for improving seasonal hydrologic predictions in many subbasins of the Columbia and Colorado River systems. In some basins the ENSO signals are weak and by themselves provide little benefit for seasonal runoff forecasts. In many instances, however, the change in ENSO signal during the accumulation season is greater than the mean ENSO signal at the end of that season. Therefore, monitoring the water equivalent of the snowpack during the accumulation season and combining this information with knowledge of seasonal changes in the ENSO signal may provide improved predictions of water supply.

ACKNOWLEDGMENTS

This research was supported by the NOAA Regional Assessment Program (NOAA Cooperative Agreement NA67RJ0153), the NASA RESAC Program (NAG13-99005), and a research grant from the National Science Foundation (EAR-9634329). We gratefully acknowledge comments from Martin P. Hoerling, Allan H. Frei, and Roger G. Barry.

REFERENCES

Barnston, A. G., Y. He, and D. A. Unger. 2000. A forecast product that maximizes utility for state-of-the-art seasonal climate prediction. *Bulletin of the American Meteorological Society* 81: 1271–1280.

Basist, A. N., and M. Chelliah. 1997. Comparisons of tropospheric temperatures derived from the NCEP/NCAR reanalysis, NCEP operational analyses, and the microwave sounding unit. *Bulletin of the American Meteorological Society* 78: 1431–1447.

Bjerknes, J. 1969. Atmospheric teleconnections from the equatorial Pacific. *Monthly Weather Review* 97: 163–172.

Cayan, D. R. 1996. Interannual climate variability and snowpack in the western United States. *Journal of Climate* 9: 928–948.

Cayan, D. R., and D. H. Peterson. 1990. The Influence of North Pacific Circulation on Streamflow in the West. In *Aspects of Climate Variability in the Pacific and Western Americas,* Geophysics Monograph Series, Vol. 55. D. H. Peterson, ed. American Geophysical Union, Washington, DC, pp. 375–398.

Cayan, D. R., and R. H. Webb. 1992. El Niño/Southern Oscillation and Streamflow in the Western United States. In *El Niño: Historical and Paleoclimatic Aspects of the Southern Oscillation.* H. Diaz and V. Markgraf, eds. Cambridge University Press, Cambridge, pp. 29–68.

Day, G. N. 1985. Extended streamflow forecasting using NWSRFS. *Journal of Water Resources Planning and Management* 111: 157–170.

Diaz, H. F., and C. A. Anderson. 1995. Precipitation trends and water consumption related to population in the southwestern United States. *Water Resources Research* 31: 713–730.

el-Ashry, M., and D. Gibbons. 1988. *Water and Arid Lands of the Western United States.* Cambridge University Press, New York.

Gutzler, D. S., and J. W. Preston. 1997. Evidence for a relationship between spring snow cover in North America and summer rainfall in New Mexico. *Geophysical Research Letters* 24: 2207–2210.

Hoerling, M. P., and A. Kumar. 2000. Understanding and Predicting Extratropical Teleconnections Related to ENSO. In *El Niño and the Southern Oscillation: Multi-Scale Variability and Global and Regional Impacts.* H. Diaz and V. Markgraf, eds. Cambridge University Press, Cambridge, pp. 57–88.

Hoerling, M. P., A. Kumar, and M. Zhong. 1997. El Niño, La Niña, and the nonlinearity of their teleconnections. *Journal of Climate* 10: 1769–1786.

Horel, J. D., and J. M. Wallace. 1981. Planetary scale atmospheric phenomena associated with the Southern Oscillation. *Monthly Weather Review* 109: 813–829.

Kalnay, E., M. Kanimitsu, R. Kistler, W. Collins, D. Deaven, L. Gandin, M. Iredell, S. Saha, G. White, J. Wollen, Y. Zhu, M. Chelliah, W. Ebisuzaki, W. Higgins, J. Janowiak, K-C. Mo, C. Ropeleweski, J. Wang, A. Leetmaa, R. Reynolds, R. Jenne, and D. Joseph. 1996. The NCEP/NCAR 40-year reanalysis project. *Bulletin of the American Meteorological Society* 77: 437–471.

Kalnay, E., S. J. Lord, and R. D. McPherson. 1998. Maturity of operational numerical weather prediction: Medium range. *Bulletin of the American Meteorological Society* 79: 2753–2769.

Koch, R. W., C. F. Buzzard, and D. M. Johnson. 1991. Variation of Snow Water Equivalent and Streamflow in Relation to the El Niño/Southern Oscillation. *Proceedings of the 19th Annual Meeting of the Western Snow Conference.* Juneau, Alaska, April 12–15.

Leavesley, G. H., R. W. Lichty, B. M. Troutman, and L. G. Saindon. 1983. *Precipitation-Runoff Modeling System: User's Manual.* U.S. Geological Survey Water Investigation Report 83-4238.

Leavesley, G. H., and L. G. Stannard. 1995. The Precipitation-Runoff Modeling System—PRMS. In *Computer Models of Watershed Hydrology.* V. P. Singh, ed. Water Resources Publications, Highlands Ranch, CO, pp. 281–310.

Livezey, R. E., M. Masutani, A. Leetmaa, H. Rui, M. Ji, and A. Kumar. 1997. Teleconnective response of the Pacific–North American region atmosphere to large Central Equatorial Pacific SST anomalies. *Journal of Climate* 10: 1787–1820.

Natural Resources Conservation Service (NRCS). 1988. *Snow Surveys and Water Supply Forecasting.* United States Department of Agriculture Natural Resources Conservation Service, Agricultural Information Bulletin 536.

Pulwarty, R. S. 1995. Adaptive Management of the Colorado River: The Role of Science and Scientists. In *U.S. Department of Interior Glen Canyon Environmental Studies: Aquatic, Geomorphic and Climatic Integration Group Report, Flagstaff, AZ.* T. Melis and D. Wegner, eds. U.S. Government Printing Office, Washington, DC, pp. 36–49.

Rasmusson, E. M., and J. M. Wallace. 1983. Meteorological aspects of the El Niño/ Southern Oscillation. *Science* 222: 1195–1202.

Redmond, K. T., and R. W. Koch. 1991. Surface climate and streamflow variability in the western United States and their relationship to large-scale circulation indices. *Water Resources Research* 27: 2381–2399.

Seaber, P. R., F. P. Kapinos, and G. L. Knapp. 1987. *Hydrologic Unit Maps.* U.S. Geological Survey Water-Supply Paper 2294.

Serreze, M. C., M. P. Clark, R. L. Armstrong, D. A. McGinnis, and R. S. Pulwarty. 1999. Characteristics of the western U.S. snowpack from SNOTEL data. *Water Resources Research* 35: 2145–2160.

Trenberth, K. E. 1997. The definition of El Niño. *Bulletin of the American Meteorological Society* 78: 2771–2777.

Wilby, R. L., L. E. Hay, W. J. Gutowski Jr., R. W. Arritt, E. S. Takle, Z. Pan, G. H. Leavesley, and M. P. Clark. 2000. Hydrologic responses to dynamically and statistically downscaled General Circulation Model Output. *Geophysical Research Letters* 27: 1198–1202.

Wilby, R. L., L. E. Hay, and G. H. Leavesley. 1999. A comparison of downscaled and raw GCM output: Implications for climate change scenarios in the San Juan River basin, Colorado. *Journal of Hydrology* 225: 67–91.

Wilks, D. S. 1995. *Statistical Methods in the Atmospheric Sciences: An Introduction.* Academic Press, San Diego.

Assessing the Effects of Climate Change on the Water Resources of the Western United States

SIX

KENNETH M. STRZEPEK
DAVID N. YATES

This chapter describes the effect of climate change on river runoff over the western United States, with the recognition that water resources play a key role in socioeconomic development in this region. The present analysis addresses the sensitivity of hydrologic resources, aquatic ecosystems, and water-management institutions to climate change. A variety of modeling approaches were used in the analysis, the goal of which is to provide decision makers with an understanding of the types of hydrologic effects that climate change may impose on hydrologic systems, management of water, and environmental resources in the western United States.

Few regional-scale investigations of the effects of climate change on runoff in the western states have been undertaken (Wolock and McCabe 1999). Significant research has been done, however, on basin-scale climate change (Kaczmarek 1996; Arnell et al. 1996; Miller 1997). It is impossible to synthesize basin-scale studies into a regional analysis because of the inconsistencies of scenarios. A comprehensive, detailed regional analysis would require significant time, resources, and data. An alternative approach that other researchers have used to assess effects of regional climate change on

Box 6.1—The water-balance model

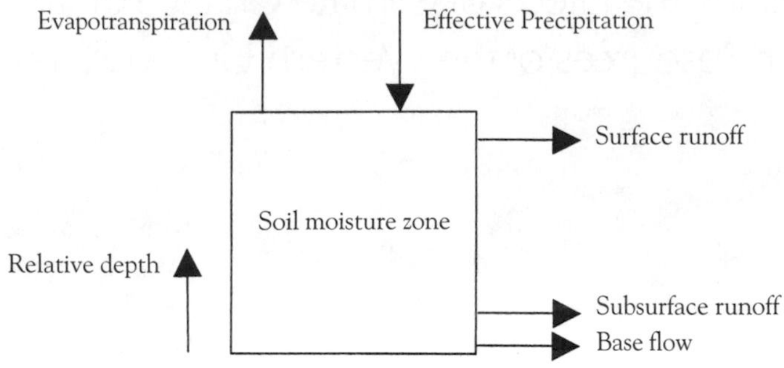

Model Equations

$$S_{max}\frac{dz}{dt} = P_{eff}(t) - R_u(z,t) - R_s(z,t) - Ev(PET, z, t) - R_b \qquad (1)$$

$$R_{sat} = \alpha z^{\gamma} \qquad (2)$$

$$R_{uns} = \begin{cases} z^{\varepsilon}(P_{eff} - R_b), & P_{eff} > R_b \\ 0, & P_{eff} \leq R_b \end{cases} \qquad (3)$$

where
S_{max} = maximum storage capacity
z = relative storage ($0 \leq z \leq 1$)
P_{eff} = effective precipitation
R_u = runoff in the unsaturated zone
R_s = runoff in the saturated zone
Ev = evapotranspiration
R_b = base flow

Model Inputs

1. Precipitation and snow. Monthly precipitation comes from the grided climate databases as mm of water. An important component of the hydrologic cycle for some basins is snowmelt. The water balance includes a temperature-index snowmelt model that creates an "adjusted" effective precipitation (i.e., precipitation that generates runoff). The three parameters for each grid are S_{max}, α, and ε.

2. Potential evapotranspiration. Potential evapotranspiration is modeled by the Priestly-Taylor method (Priestly and Taylor 1972), which was chosen because of its simplicity and the evidence supporting such an empirical relationship on a regional basis, as needed for river-basin modeling.

runoff is the use of less detailed and less comprehensive modeling techniques. For example, Schaake (1990) performed a regional analysis of the southeastern United States as a means of providing insight into the sensitivity of the region's runoff to changes in climate. He used a nonlinear monthly water balance with fixed parameters for the entire region. Yates (1996) extended Schaake's monthly nonlinear regional concept to account for spatial variability in parameters of the nonlinear water-balance model; he was able to relate the three parameters of a water-balance model (Box 6.1) to land use and land cover on a grid of 0.5° latitude by 0.5° longitude.

The grid-based water-balance model allows for a rapid examination of the potential consequences of climate change for runoff over a range of a few thousand km^2 to continental scale on the basis of data obtained directly from gridded climate databases and gridded climate-model results. The hydrologic simulations shown here are for natural conditions; no water-management component is included in the simulations. Given the uncertainty and large spatial scales of climate models (2° by 3° grids), 0.5° by 0.5° gridded regional hydrologic models provide adequate precision for modeling the effects of future regional climate scenarios only at the scale of large river basins (Strzepek and Yates 1997).

METHODS

A consistent regional analysis of the effects of climate change on the hydrologic resources of the western United States was conducted with a water-balance model based on a geographic information system. The hydrologic analysis was performed on a 0.5° by 0.5° spatial scale covering the continental United States west of the 90th meridian, which includes twenty-two western states and ten major river basins. The gridded water-balance model was run with a set of twelve monthly averaged climate variables, including various aspects of precipitation, temperature, water vapor pressure, and cloudiness. The data set was developed by the National Center for Atmospheric Research's Vegetation Mapping and Analysis Project (VEMAP; Kittel et al. 1995). Before effects of climate change were simulated, the water-balance model was validated with nineteen medium-sized river basins in the region that were minimally affected by management. Because spatially distributed VEMAP climate data were given only as monthly means, simulations required that initial and final soil-moisture storage be nearly equal. This condition, which was accomplished through iteration, implies that the system is in hydrologic equilibrium.

Average monthly runoff from the U.S. Geological Survey streamflow database was compared to modeled runoff for the nineteen basins under base conditions (no change in climate). The model overpredicted the total historical discharge of the nineteen basins by about 10 percent but generally reflected the observed temporal distribution of flow (Figure 6.1). The general correlation of modeled and observed runoff indicates that the water-balance model adequately captures the major hydrologic processes at both the regional and basin scales.

The water-balance model was run a second time with the same parameters but with input derived from a general circulation model (GCM) for two scenarios: 1xCO_2 and 2xCO_2. GCM simulations are not predictions; they give possible future climates that are internally consistent spatially. They have become a standard for developing regional climate-change scenarios for regional impact assessments.

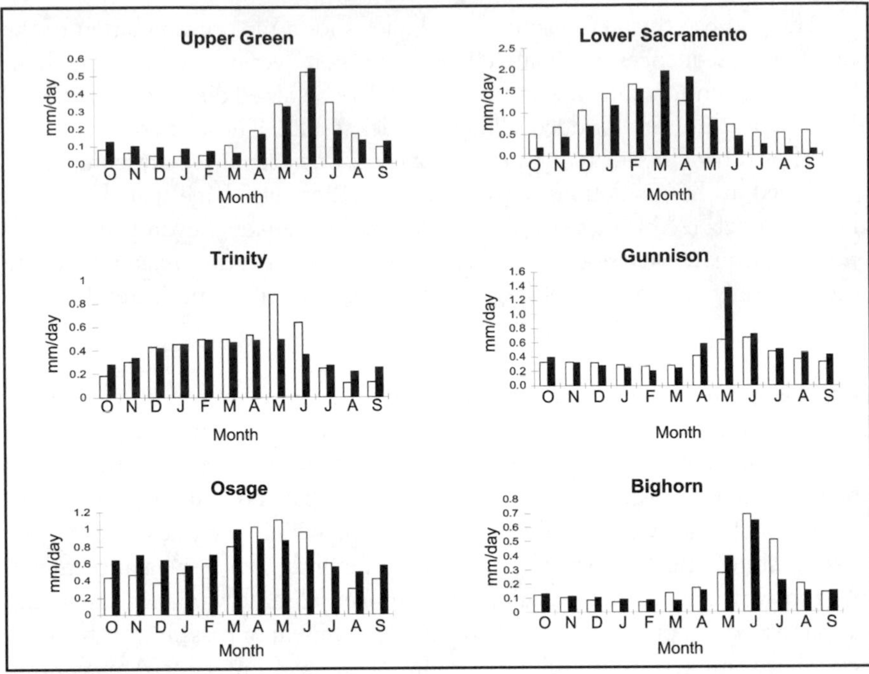

6.1 Comparison of observed and modeled discharge for six basins in the western United States.

Two GCM models were used to develop the two climate scenarios: the Goddard Institute for Space Science (GISS) model and the Geophysical Fluid Dynamics Laboratory (GFDL-R30) model at 2.22° by 3.75° resolution (Hansen et al. 1984; Manabe and Wetherald 1990; Wetherald and Manabe 1990). For each GCM, the traditional approach of extracting equilibrium temperature and precipitation from the $1xCO_2$ and the $2xCO_2$ was used; temperature change ($2xCO_2 - 1xCO_2$) and precipitation ratios ($2xCO_2/1xCO_2$) were applied to the historical climate data to derive the climate data for scenarios of interest.

The GISS scenario suggests an increase of 2.5°C in mean annual temperature globally and a 4.7°C increase over the western United States, as well as a 10 percent increase in global precipitation and an 8 percent average increase in precipitation over the western United States. The GFDL scenario suggests an increase of mean annual temperature of 3.0°C globally and 4.2°C over the western United States, as well as a 15 percent increase in global precipitation and a 30 percent increase in precipitation over the western United States. Figure 6.2 shows the relative change in average annual precipitation over the western United States for the two scenarios; Figure 6.3

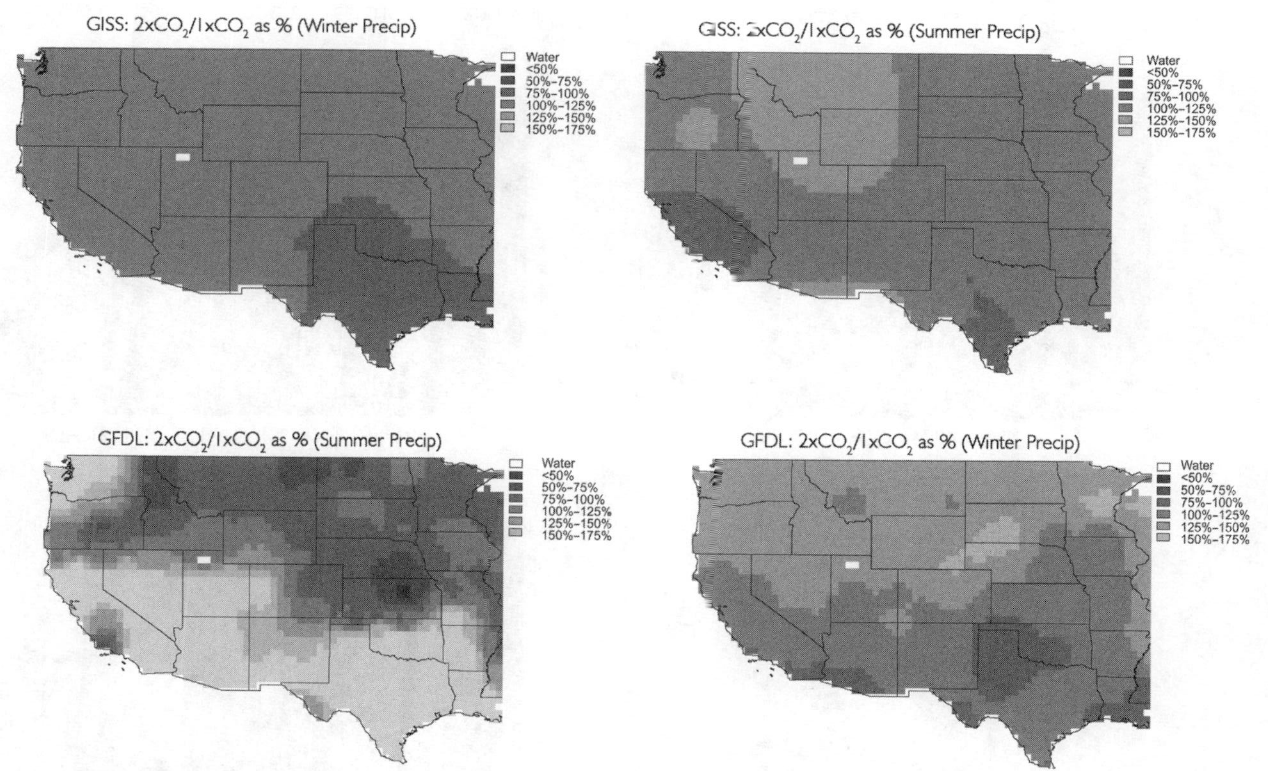

6.2 Average change in precipitation for winter (DJF) and summer (JJA), given as percent ratios of ($2xCO_2/1xCO_2$) for Goddard Institute for Space Science (GISS) and Geophysical Fluid Dynamics Laboratory (GFDL) simulations.

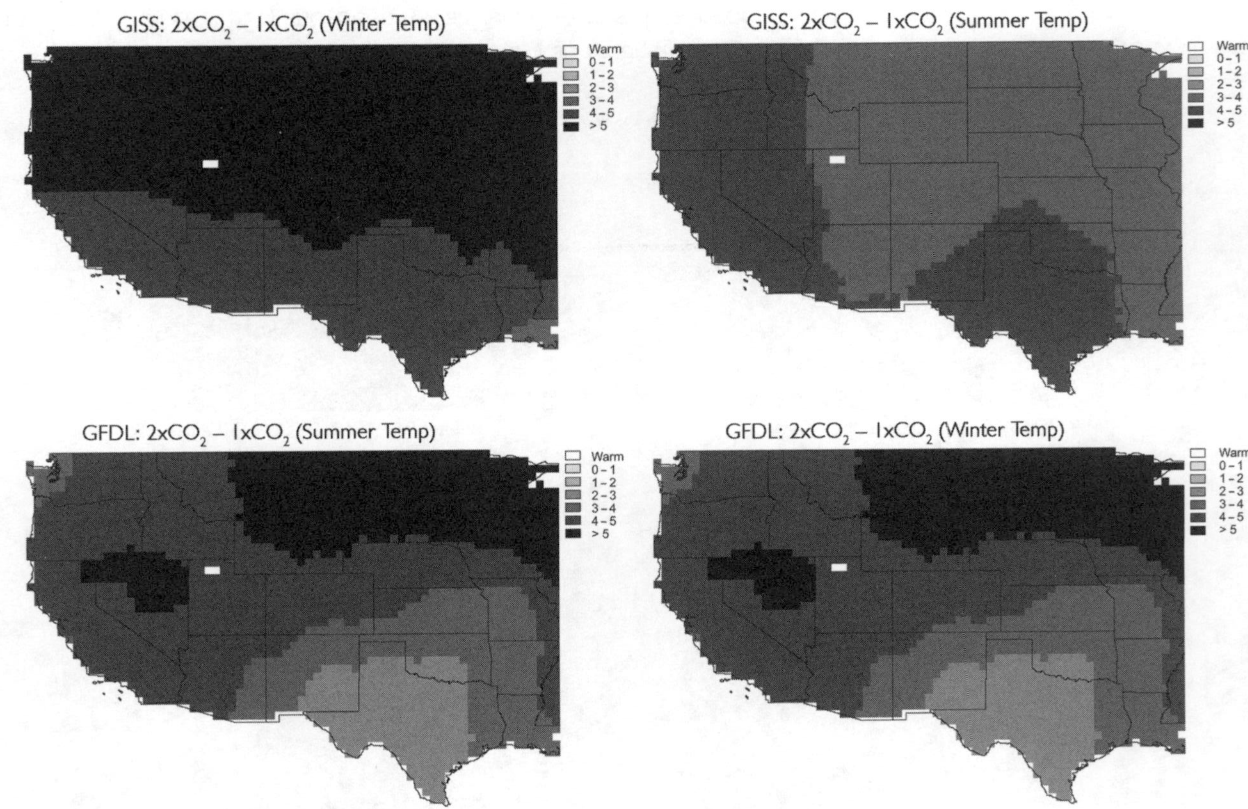

6.3 Average change in temperature (summer and winter) given as $2xCO_2 - 1xCO_2$ for GISS and GFDL simulations.

shows the annual average increase in temperature over the same region for the GISS and GFDL simulations. The GFDL scenario generally exhibited greater longitudinal variability. Estimated runoff from the hydrologic simulation for the two climate-change scenarios was compared to simulated current-day runoff.

ANALYSIS

Results: Annual Means

Figures 6.4 and 6.5 show hydrologic simulations for the major river basins. Table 6.1 is a summary of the results presented in Figures 6.4 and 6.5 for the major river basins, and Table 6.2 is a summary of the results for specific states. The values in Tables 6.1 and 6.2 are the sums of runoff from each grid cell in a basin or state for each scenario, expressed as a ratio to historical runoff for each basin or state.

For the GISS scenario (4.7°C increase in annual temperature and 8% increase in precipitation), four river basins show decreases in runoff of up to 15 percent, and six basins have increases of up to 23 percent (Table 6.1). Twelve states show decreases in runoff as high as 16 percent, and ten states show increases of up to 27 percent (Table 6.2). Because grids were used to represent area, aggregated results for states and river basins do not match exactly.

Integrated results show a 12 percent increase in runoff over the western United States. The northern Pacific Coast shows the greatest increase in runoff, and the Southwest and the central plains states show conditions drier than present ones. If the two basins of the northern Pacific Coast are excluded in the aggregation of total runoff, a net decrease in runoff of 2 percent is predicted over the remainder of the West.

The GFDL scenario (4.2°C increase in annual temperature and 30% increase in precipitation) is cooler and wetter than the GISS scenario. No river basins show decreases in runoff, and predicted increases in runoff are as high as 166 percent (Table 6.1). Only Nebraska shows a decrease (8%), and twenty-one states show increases of up to 157 percent (Table 6.2). The Great Plains show a slight drying, but the rest of the region—especially the rivers of Texas and the lower Colorado River—shows major increases in runoff. The GFDL results, averaged over the western United States, suggest a 45 percent increase in regional runoff.

These results have three important implications for the water resources of the western United States: (1) high spring flows with potential flooding and lower flows in late summer may be more serious consequences of climate change than drought for most regions in the western United States (cf. Chapter 5); (2) the effects will vary widely across the West; (3) the differences between the results for the two GCM scenarios are based more on differences in precipitation than in temperature.

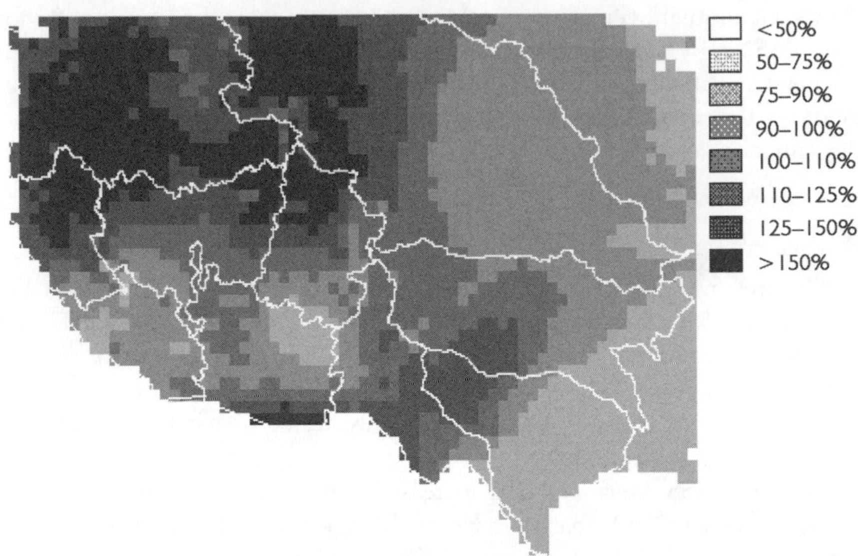

6.4 Ratio of $2xCO_2$ GISS-modeled runoff to modeled runoff based on historical climate for the major river basins.

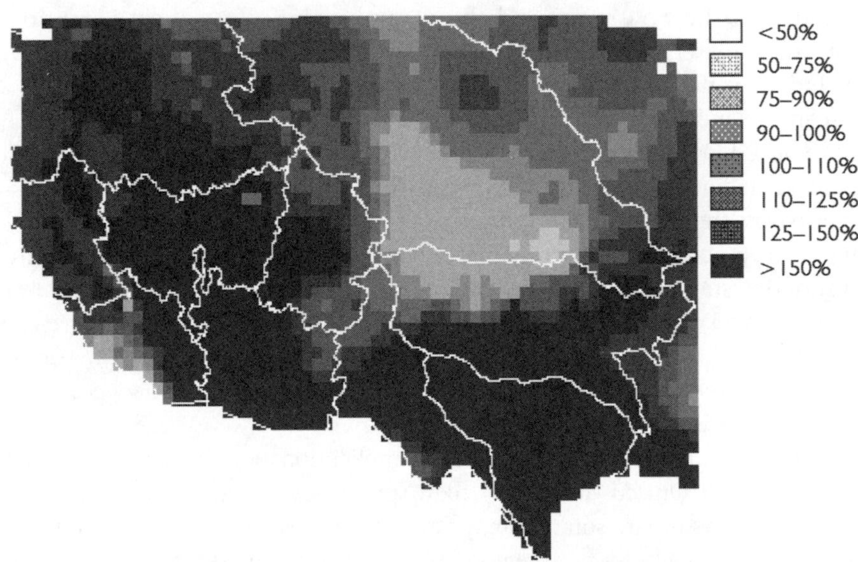

6.5 Ratio of $2xCO_2$ GFDL-modeled runoff to modeled runoff based on historical climate for the major river basins.

Table 6.1—Ratio of modeled annual runoff based on $2xCO_2$ GCM climate to annual runoff as modeled for current climate for the major western river basins.

Basin/Region	*Simulation Type* GISS	GFDL
Arkansas-Red	0.91	1.59
California-North	1.22	1.37
California-South	0.87	1.15
Columbia	1.23	1.39
Grand Basin	1.10	1.66
Lower Colorado	0.98	2.46
Missouri	1.05	1.17
Rio Grande	1.01	1.35
Texas	0.85	2.65
Upper Colorado	1.08	1.30
All basins	1.12	1.45

Table 6.2—Ratio of modeled annual runoff based on $2xCO_2$ GCM climate to annual runoff as modeled for current climate for the western states.

State	*Simulation Type* GISS	GFDL
Arizona	0.97	2.54
Arkansas	0.87	1.39
California	1.18	1.34
Colorado	1.04	1.17
Idaho	1.16	1.37
Iowa	0.92	1.14
Kansas	0.97	1.09
Louisiana	0.84	1.83
Minnesota	0.91	1.16
Mississippi	0.91	1.44
Montana	1.19	1.29
Nebraska	0.95	0.92
Nevada	1.13	2.06
New Mexico	1.03	1.41
North Dakota	0.98	1.14
Oklahoma	0.95	1.73
Oregon	1.27	1.40
South Dakota	0.94	1.07
Texas	0.86	2.57
Utah	1.20	1.64
Washington	1.24	1.42
Wyoming	1.19	1.20
All states	1.09	1.44

Precipitation is simulated less accurately than any other variable by GCMs. Consequently, the simulations cannot be viewed as predictions, but they are regionally consistent. The likelihood of any specific type of change is greatest in those regions where the two GCMs yield similar results. Thus the Great Plains states, which support the greatest agricultural production, are most likely to experience a decline in runoff.

Results: Temporal Trends

Temporal (seasonal) changes in runoff are as important as spatial changes for water-resource management in the western United States, especially because of the prior appropriation doctrine of water allocation in the West, according to which water rights are specified in terms of time and location (Chapter 14). Also, current water-diversion patterns are based largely on the demand for irrigation water. For irrigated crops, a shift in the timing of runoff relative to the growing season could have significant effects on mechanisms for storing and distributing water. In addition, flood control and flood protection are designed for peak flows, which are seasonal.

Figure 6.6 is a plot of the average monthly distribution of runoff for the entire western United States. It shows that both types of GCM simulations

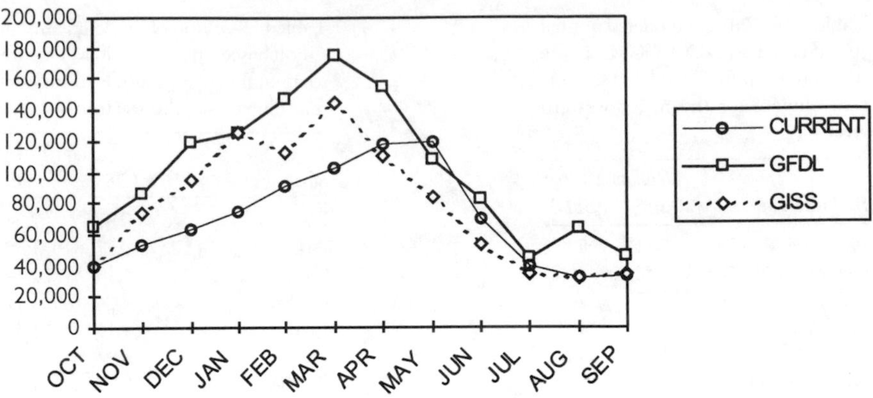

6.6 Monthly distribution of total runoff for the western United States for current and 2xCO_2 scenarios (units are 10^6 m^3).

show an increase in peak runoff and a shift toward earlier peaks as compared with current climatic conditions. The shift in peak flow (about two months) is explained by earlier snowmelt caused by higher temperatures. Because 70 percent of western runoff is from snowmelt, temperature significantly affects the timing of the peak runoff.

Current western water institutions are vulnerable to changes in the timing of runoff (Chapter 16). Water rights are issued for amounts of flow over a specified time. If a majority of river discharge from snowmelt occurs earlier in the year and irrigation demands do not change, there could be less water for holders of junior water rights. Thus some water users could be harmed, even if annual runoff does not change.

Figures 6.7 and 6.8 present temporal summaries of the effect of climate change on the month of peak flow for ten major basins of the West, ordered west to east. The western basins, which have generally larger snowmelt contributions, would experience peak flows one to two months earlier. In contrast, the central plains basins would have later peak flows (zero to two months later for the GISS scenario and two to five months later for the GFDL scenario). The change in timing of the peak flow would be reason for concern, as water allocations are based on both quantity and timing of flows. It is not clear whether western water institutions are flexible enough to deal with these potential changes (see Chapter 9).

Results: Extreme Events

The effect of climate change on water management also depends on changes in the magnitude of peak and low flows. Figures 6.9 and 6.10 are

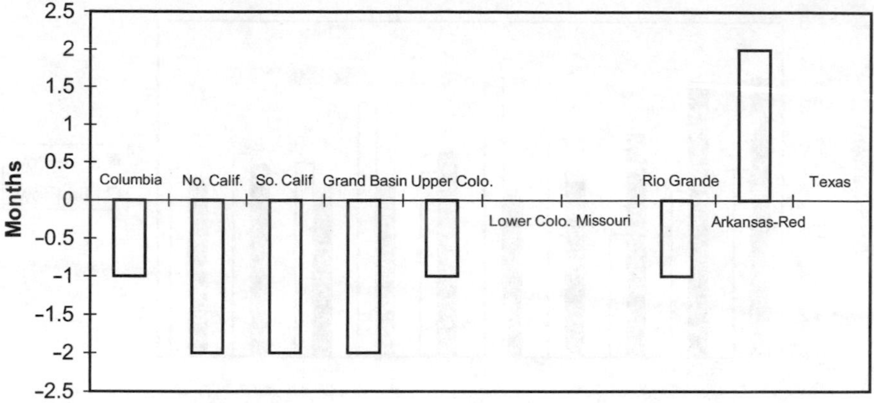

6.7 Effects of climate change on peak monthly runoff for GISS simulation. Change is shown as the number of months that peak discharge has shifted relative to the current peak month.

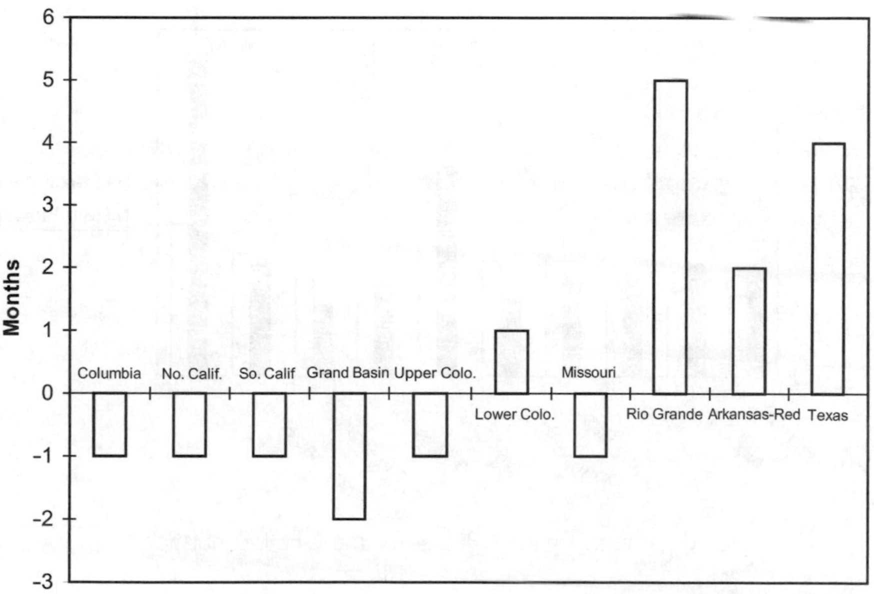

6.8 Same as Figure 6.7, but for the GFDL simulation.

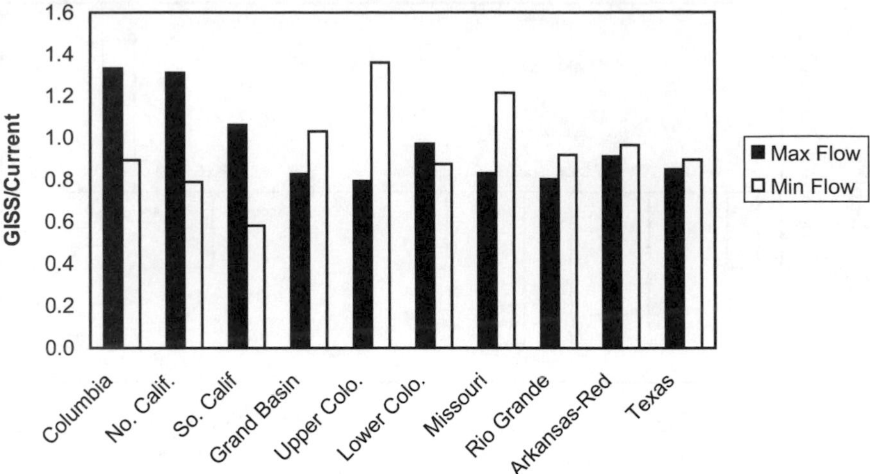

6.9 Effects of climate change on annual maximum and minimum flows for the GISS simulation; shown as the ratio of modeled to historical flows.

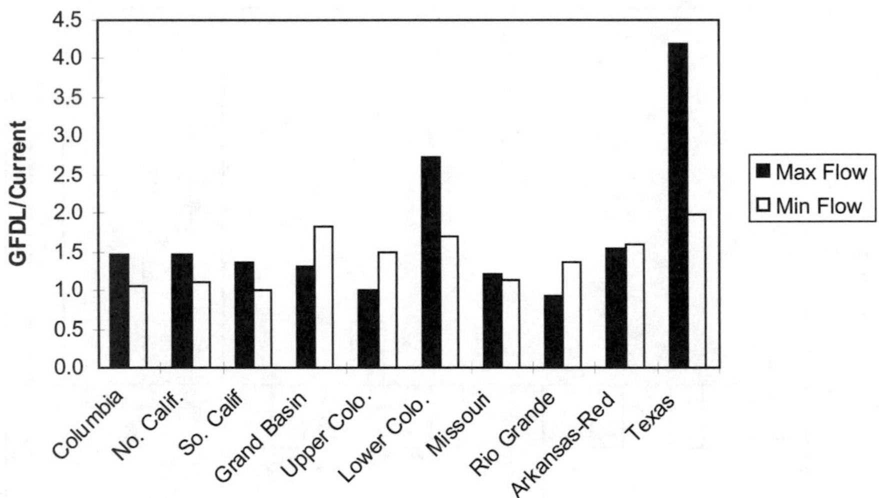

6.10 Same as Figure 6.9, but for the GFDL simulation.

summaries of the effect of climate change on the peak and low flows for the ten major basins. For the GFDL scenario (Figure 6.10), eight of the ten basins show an increase in peak flow, and several of these are extreme (lower Colorado and Texas). Consistent with the regional spatial analysis, however, the GISS scenario shows only three of ten basins with increased peak flows, and all of these are located along the Pacific Coast; the basins of the Rocky Mountains and the central plains all show a decrease in peak flow. These results suggest that flooding in the Pacific Northwest might be a consequence of climate change.

Equally important to water management are low flows, which especially affect aquatic ecosystems and water quality. The GFDL scenario shows nine basins with relative increases in low flows and one basin that remains the same. The GISS results, however, show the three Pacific Coast basins exhibiting an amplified seasonal cycle (increased peak flows and decreased low flows), whereas the remaining basins show change of –13 percent to +35 percent in low flows. These GCM scenarios suggest that more extreme conditions, including higher peaks and lower minimum flows, could occur. This sort of outcome (greater extremes of both high and low flows) has been a subject of previous speculation (Miller 1997).

DISCUSSION AND CONCLUSIONS

Western water management is dominated by large reservoirs that moderate damaging floods and store water for use. Changes in the timing of flows might require additional storage or changes in current reservoir operation to meet demands that do not match past or current streamflow patterns. In many cases the stored water is transferred out of the river basin to meet demands far away (e.g., Chapter 10). An important aspect of reservoir design is the variability of stream flow. Should variability increase as a result of changes in extreme events, the current water-resources infrastructure might not meet design criteria, in which case both infrastructural and water-management adaptations might be required.

Water allocation in the West is extremely sensitive to the temporal distribution of stream flow. Enormous stress will be placed on existing water institutions at the local, state, and federal levels should climatic change affect the timing of water supply or water demand. The ability of institutions to change along with the changing patterns of runoff needs to be explored.

SUMMARY

A regional assessment of the potential effects of climate change on the water resources of the western United States was performed by use of future scenarios for climate generated by global climate models incorporating CO_2 concentrations at double current levels. The effect of these potential future climates was assessed with a spatially distributed water-balance model. Quantitative results were viewed not as predictions but as indications of the sensitivity of hydrologic resources and water institutions to potential change in climate.

For two global climate models, effects of climate change on runoff vary across the western United States. Individual river basins change in annual runoff from +157 percent to –14 percent. Individual states show changes that range from +146 percent to –15 percent. Both models suggest that most river basins will experience an increase in peak flow and that snowmelt-dominated watersheds will shift to earlier peak flows. An amplification of the seasonal hydrologic cycle through increased peak flows and decreased low flows suggests new challenges for water managers and increased stress on aquatic ecosystems.

REFERENCES

Arnell, N. W., B. C. Bates, H. Lang, J. J. Magnuson, and P. Mulholland. 1996. Hydrology and Freshwater Ecology. In *IPCC Climate Change 1995. The Second Scientific Assessment.* R. T. Watson, M. C. Zinyowera, and R. H. Moss, eds. Cambridge University Press, Cambridge, pp. 325–364.

Hansen, J., A. Lacis, D. Rind, G. Russell, P. Stone, I. Fung, R. Ruedy, and J. Lerner. 1984. Climate Sensitivity: Analysis of Feedback Mechanisms. In *Climate Processes and Climate Sensitivity, AGU Geophysical Monograph* 29. J. E. Hansen and T. Takahashi, eds. American Geophysical Union, Washington, DC, pp. 130–163.

Kaczmarek, Z. 1996. Water Resources Management. In *IPCC Climate Change 1995. The Second Scientific Assessment.* R. T. Watson, M. C. Zinyowera, and R. H. Moss, eds. Cambridge University Press, Cambridge, pp. 469–486.

Kittel, T., N. Rosenbloom, T. Painter, D. Schimel, and VEMAP modeling participants. 1995. The VEMAP integrated database for modeling U.S. ecosystem/vegetation sensitivity to climate change. *Journal of Biogeography* 22: 857–862.

Manabe, S., and R. T. Wetherald. 1990. Reported in J.F.B. Mitchell, S. Manabe, V. Meleshko, and T. Tokioka. Equilibrium Climate Change and Its Implication for the Future. In *Climate Change: IPCC Scientific Assessment.* J. T. Houghton, G. J. Jenkins, J. J. Ephramus (eds). Cambridge University Press, Cambridge, UK, pp. 131–172.

Miller, K. A. 1997. *Climate Variability, Climate Change and Western Water.* Report to the Western Water Policy Review Commission. National Technical Information Service, Springfield, VA.

Priestly, C.H.B., and R. J. Taylor. 1972. On the assessment of surface heat flux and evaporation using large-scale parameters. *Monthly Weather Review* 100: 81–92.

Schaake, J. C. 1990. From Climate to Flow. In *Climate Change and U.S. Water Resources.* Paul E. Waggoner, ed. Wiley-Interscience, New York, pp. 177–206.

Strzepek, K. M., and D. N. Yates. 1997. Climate change impacts on the hydrologic resources of Europe: A simplified continental scale analysis. *Climatic Change* 36: 79–92.

Wetherald, R. T., and S. Manabe. 1990. Hydrologic sensitivity to CO_2-induced global warming. *Civil Engineering Practice Journal* 5: 33–36.

Wolock, D., and G. McCabe. 1999. Estimates of runoff using water-balance and atmospheric general circulation models. *Journal of the American Water Resources Association* 35: 1341–1350.

Yates, D. N. 1996. WATBAL: An integrated water balance model for climate impact assessment. *Water Resources Development* 12: 121–139.

SEVEN

Improving Water-Resource System Performance Through Long-Range Climate Forecasts: The Pacific Northwest Experience

Dennis P. Lettenmaier
Alan F. Hamlet

In the western United States, water is a scarce and often limiting resource. Growth of human population and changing operational priorities both within the region and elsewhere have created pressure for more efficient management of water. Many management efforts focus—with considerable justification—on managing water demand and developing better tools, such as water marketing, that have the potential to reduce conflicts over water. Management of water supply is more problematic, as the total amount of available reservoir storage for beneficial use is constrained; opportunities for creating additional storage facilities that might help align supply and demand in terms of point and time of use are minimal. Considerable opportunity exists, however, for more efficient management of both water supply and demand if more accurate forecasts of future runoff could be provided (e.g., Yeh et al. 1982).

Much of the annual runoff of most western rivers (by some accounts, over 70% for the West as a whole) originates as snowmelt. In many cases fairly accurate forecasts of runoff from snowmelt-dominated rivers can be made at or near the time of maximum snow accumulation in early spring

(Lettenmaier and Garen 1979). Prior to the beginning of winter snow accumulation, however, the accuracy of methods currently in use is no better than predictions based on climatology. Considerable potential appears to exist for development of streamflow forecasting methods that make use of improved climate information and forecasts with lead times of several months to a year. Such forecasts would exploit information about future climate derived from knowledge of teleconnections having to do largely with thermal inertia of the oceans (e.g., Barnston et al. 1994). The El Niño–Southern Oscillation (ENSO) is the best studied and understood of such teleconnections, but others such as the Pacific Decadal Oscillation (PDO) have the potential to be incorporated in so-called end-to-end forecast schemes that could have implications for water management. In this chapter we provide an overview of current work by the Climate Impacts Group (CIG) at the University of Washington related to the use of climate forecasts for water management.

BACKGROUND

Present research at CIG aimed at developing improved long-lead streamflow forecasts for water management in the Columbia River basin in the Pacific Northwest (PNW) is founded on recent advances in the understanding of recurring global and regional climate patterns and their effect on the observed climate of the PNW (Battisti and Sarachik 1995; Mantua et al. 1997; Dettinger et al. 1998; Cayan et al. 1999; Hamlet and Lettenmaier 1999). Advances in understanding global teleconnections as they affect weather and climate, particularly winter precipitation and temperature in the PNW, have in turn led to development of streamflow forecasting methods that make use of long-range climate forecasts. These methods, when carried to their full extent, link global- and regional-scale climate models, macroscale hydrologic simulation tools, and, potentially, reservoir models. A few examples of the testing and implementation of these methods within the Columbia River basin are described here.

We have developed two basic methods for producing forecasts based on long-lead climate forecasts. The first and most general is a full dynamic linking of ensemble forecasts generated by a climate model to force a hydrologic model, which in turn produces streamflow forecasts. The simulated stream flows may also be routed through a water-resources management model, which helps with interpretation of the streamflow forecasts in the context of water-management decisions. Figure 7.1 shows the system of linked models used in CIG's research on the Columbia River basin and its water-resources management system. This approach can provide a number of ensemble members that is limited only by computer resources. Hence, in principle the precision (but not necessarily the accuracy) of the results can be controlled. Furthermore, this method is not limited by the variability or spatial coverage of

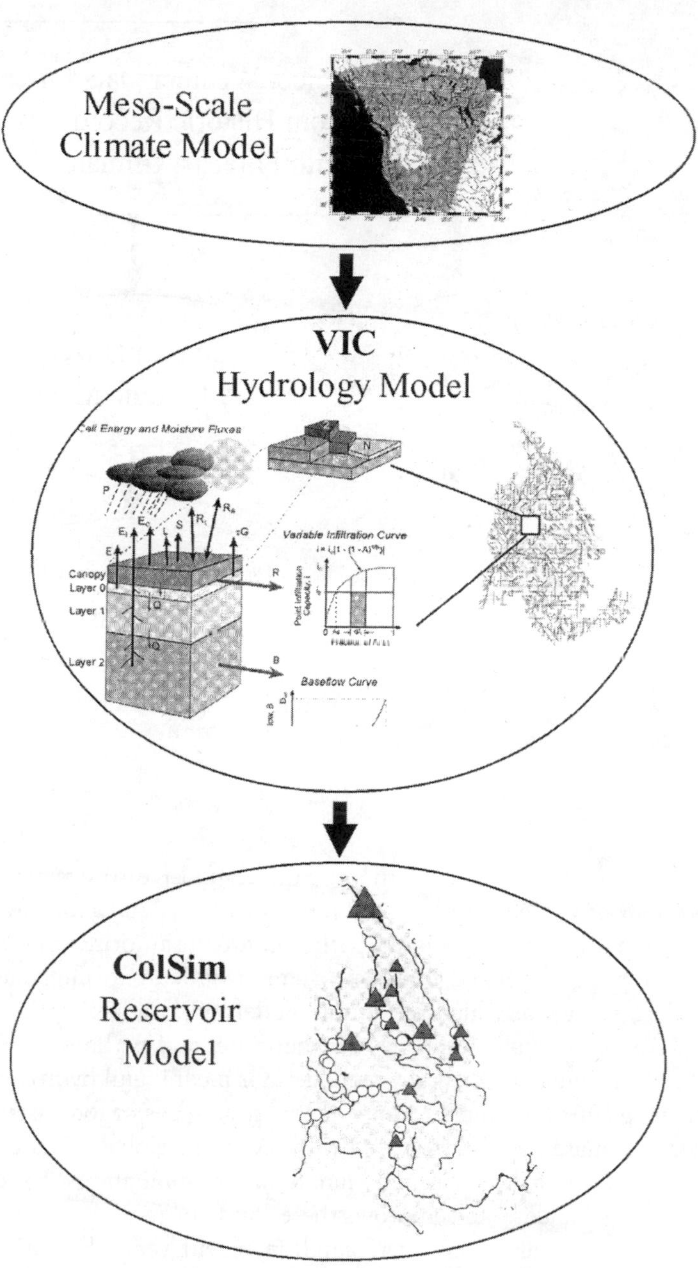

7.1 Schematic diagram for full dynamic linking of climate, hydrology, and water-resources models used for generating and interpreting streamflow forecasts. SOURCE: Hamlet and Lettenmaier (2000).

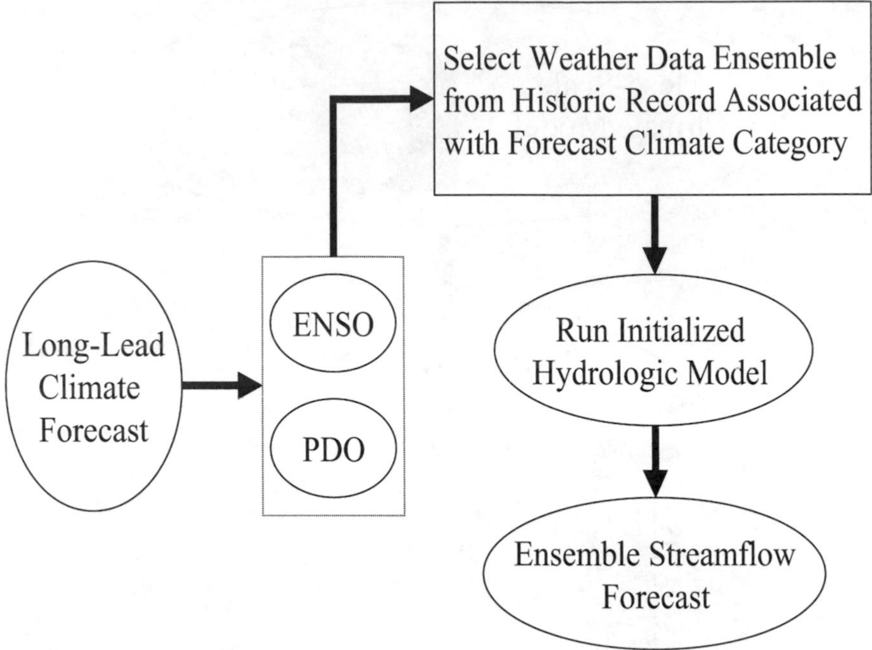

7.2 Schematic diagram of the resampling method for streamflow forecasting. Figure adapted from Hamlet and Lettenmaier (1999).

surface observations, and it is in theory fully responsive to observable driving conditions (e.g., sea-surface temperatures, sea ice, soil moisture) in the climate system that are associated with the forecast horizon. In practice, however, such a fully linked forecast system is difficult to implement. Climate models, particularly high-resolution models appropriate to the regional scale, are computationally intensive. Furthermore, and perhaps more important, bias and insufficient accuracy of models for meaningful hydrologic simulations have plagued attempts to implement these types of forecasts. Finally, the regional climate signals associated with recurrent global climate patterns (e.g., ENSO) are not present in all climate-model simulations. Experiments that attempt to validate and improve these kinds of tools for particular regions have been a major focus of research in recent years. Regional experiments for the PNW conducted by Leung et al. (1999), for example, will be discussed in this chapter.

The second forecasting method that has been explored for the PNW is outlined in Figure 7.2. It uses the same hydrologic tools shown in Figure 7.1, but the climatic driving data are derived without the use of models. Specific periods of time in the historical record are associated with recurrent climate patterns, and this information is then used to select (resample) data from the historical record based on a real-time climate forecast. One use of this method that will be discussed in more detail below is reported by Hamlet and Lettenmaier (1999). The advantages of this method are largely practical. Because observed driving data used for resampling contain realistic representations of observed climate variability for a particular region, the simulations tend to be realistic; cumulative bias is greatly reduced in the simulations because only the relatively small bias inherent in the predictions of the hydrologic model is present. Computer resources required for this forecasting method are modest compared with the more sophisticated method involving dynamically linked models. Problems with this method are that the ensemble forecasts are constrained in their range and statistical properties by the historical record. Thus conditions that occur infrequently or are unlike those in the historical record may not be well represented in the forecasts.

METHODS FOR CHARACTERIZING CLIMATE VARIABILITY AND RETROSPECTIVE DEFINITIONS OF CLIMATE STATES

In the PNW, winter climate variability in the twentieth century can be well characterized with numerical indexes associated with two global climate phenomena, the PDO (Mantua et al. 1997) and ENSO. One of the primary differences between these two phenomena is the timescale of variation. The PDO typically persists in a predominantly warm or cool state for several decades at a time, whereas ENSO varies on interannual timescales. In its effect on PNW climate, the PDO appears to determine the status of mean climate for several decades at a time, whereas ENSO determines the variability on shorter timescales about this mean. The warm phases of PDO and ENSO (El Niño) are typically warmer and dryer than normal in the PNW, whereas the cool phases (La Niña) are typically cooler and wetter. These tendencies are most pronounced when PDO and ENSO are in phase, that is, warm/warm or cool/cool (Hamlet and Lettenmaier 1999; Mote et al. 1999).

For retrospective tests of the forecasting method, objective definitions for the phase of each climate phenomenon must be specified. For this study interannual variation in the PDO (PDO index; see Mantua et al. 1997 for details) is not considered because it cannot be forecast with appreciable skill. Instead the observed phase of the PDO is defined as a persistent, bimodal, epochal function that is in cold phase from 1900 to 1924 and 1947 to 1976 and in warm phase from 1925 to 1946 and 1977 to 1996 (Mantua et al. 1997). Considerable debate exists at present as to whether the climate regime associated with the PDO shifted from warm phase to cold phase in 1996 or 1997. We believe, based on the observed regional climate and the streamflow record of the Columbia River, that such a transition probably occurred before 1997, and for forecast purposes we assume cold phase for 1998 and beyond.

ENSO phase is defined by Trenberth's (1997) NINO3.4 index averaged from December to February. A water year (October to September) associated with a winter index value (December–February mean) more than 0.5 standard deviations above the mean value for the period 1900–1996 is defined as a warm phase (El Niño). A year associated with a winter index value (December–February mean) more than 0.5 standard deviations below the long-term mean value is defined as cold phase (La Niña). All other years are defined as ENSO neutral.

DEFINITION OF CLIMATE CATEGORIES FOR FORECASTING AND ASSOCIATED PNW STREAMFLOW SIGNALS

The combinations of the two PDO states (warm, cool) and three ENSO states (warm, neutral, cool) define the six climate categories (CCs) used here for climate and streamflow forecasting. Figure 7.3 shows naturalized (i.e., with water-management effects removed) April–September average stream flow in the Columbia River at The Dalles, Oregon (an important river checkpoint used for water management), associated with each of these categories compared to the remainder of the data. Years shown as solid triangles are within the climate category, whereas years outside the climate category are shown as open circles. The darker line shows the average of the data within the climate category, whereas the lighter line shows the average of the data outside the climate category. A wide separation of the two horizontal lines demonstrates a strong signal, on average, associated with the climate indicator, whereas a small separation demonstrates a weak signal. The strongest signals correspond to the climate categories for which the PDO and ENSO are in phase (CC 1 and CC 6); other categories show relatively little average signal.

IMPROVEMENTS IN CLIMATE FORECASTING

Dramatic improvements in the ability to monitor, forecast, and interpret ENSO events have resulted from a number of recent developments within the climate-prediction community. Among the recent advances are the implementation of the National Oceanographic and Atmospheric Administration's TOGA (Tropical Ocean Global Atmosphere)-TAO (Tropical Atmosphere Ocean) program (e.g., McPhaden et al. 1998; Trenberth et al. 1998), general improvements in ocean-atmosphere simulation models that have improved simulations of the evolution of ENSO, and better interpretation of the resulting climate forecasts at the regional scale (e.g., Barnston et al. 1994; Mason et al. 1996). One result of the TOGA-TAO program has been the installation of an array of observational buoys in the tropical Pacific. This

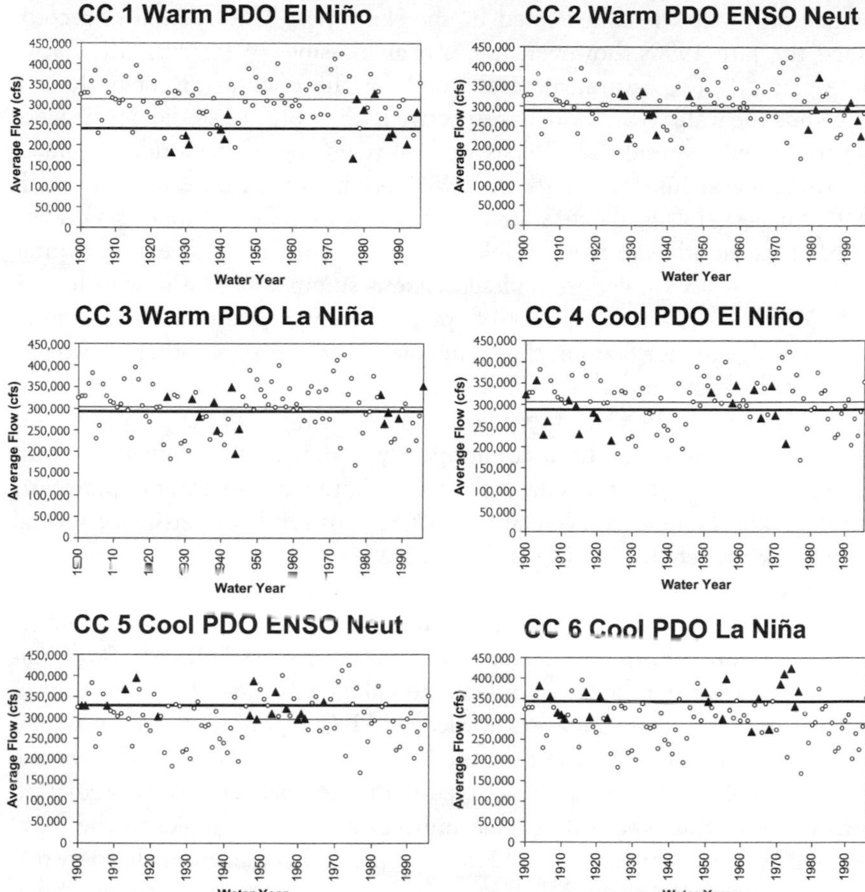

7.3 Scatterplots of April–September average virgin stream flow of the Columbia River at The Dalles, Oregon, for six climate categories (CC 1–CC 6). Dark triangles are years within the climate category; open circles are years outside the category. The darker line is the average for years within the climate category. The lighter line is the average of years outside the category. SOURCE: Hamlet and Lettenmaier (1999).

has been an important addition to the suite of tools used to forecast ENSO because it allows numerical ocean-atmosphere models to be initialized relatively accurately in real time, thus enhancing the skill of sea-surface temperature (SST) forecasts.

Validation of the forecast improvements resulting from implementation of the combined TAO observing system and state-of-the-art ocean-atmosphere simulation models is complicated by the short post-TAO period of record. Since the late 1990s, however, it has been possible to forecast the ENSO state correctly (e.g., warm, neutral, cool) for the coming winter by June 1 preceding the water year. More retrospective ENSO forecast studies are needed, but the excellent results in the past several years and earlier studies evaluating the general improvements in ENSO prediction skill (e.g., Latif et al. 1994) suggest that lead times for categorical winter ENSO forecasts on the order of six months are now feasible. Because of the linkages between winter climate, snow accumulation, and subsequent summer stream flow, a June 1 ENSO forecast yields a lead time of approximately twelve months from forecast to peak summer flows in the Columbia River basin and other snowmelt-dominated catchments of the PNW.

Although the PDO is important for forecasting PNW stream flow (see Figure 7.3), its physical dynamics are poorly understood in comparison with ENSO, and PDO cannot currently be predicted on an interannual basis (Mantua and Hare 2000). Because a PDO status tends to persist for several decades, however, the PDO can be used effectively in climate forecasting. The primary challenge in this context is the identification of transitions. Methods for identifying PDO regime shifts based on thresholds of summer stream flow are reported by Hamlet and Lettenmaier (1999). Although empirical prediction methods are difficult to validate because they occur infrequently (only three generally acknowledged PDO regime shifts are indicated in the historical record), these methods show some apparent skill over the observational record of about 100 years. Presumably, empirical prediction techniques of this type will become unnecessary as the physics of the PDO become better understood. In addition, proposals have been made to instrument the North Pacific as was done with the TAO array in the tropics. Such an observing system could perhaps be used to identify, in real time, the interannual signature of the PDO, which frequently persists on timescales of several months. The effects on PNW stream flow associated with the interannual PDO index are very similar to those shown in Figure 7.3 for the PDO epochs, which suggests that such a forecasting scheme could work well in practice. Perhaps more important, instrumenting the North Pacific would improve the availability of data that could eventually lead to a better understanding of North Pacific atmosphere-ocean interactions and perhaps even to numerical forecasts of the PDO using ocean-atmosphere models.

MACROSCALE HYDROLOGIC MODEL

The Variable Infiltration Capacity (VIC) hydrology model developed at the University of Washington and Princeton University (Liang et al. 1994) has been implemented at its highest resolution (⅛° latitude and longitude) over the Columbia River basin. Most of the experimental results discussed here, however, were produced with an earlier 1° latitude and longitude grid resolution (Nijssen et al. 1997). The VIC model has been widely used for simulation of regional-scale watersheds in North America (e.g., Abdulla et al. 1996) and Europe (e.g., Lohmann et al. 1998). Figure 7.4 shows a schematic of the physical processes simulated in the VIC model, and Figure 7.5 shows the channel network connectivity within the Columbia River basin as represented at ⅛° resolution, for which the VIC model is easily run on a desktop PC.

Because the details of the model and its previous implementations for the Columbia River at 1° and ¼° resolution (Nijssen et al. 1997; Matheussen et al. 2000) have been described well elsewhere, they will be omitted here. The ⅛° model is a higher-resolution version of the ¼° implementation described by Matheussen et al. (2000). The data for the hydrology model are derived from interpolated station data for the period 1948–1997, rescaled to long-term monthly means simulated by a statistical-topographic precipitation and temperature model (Daly et al. 1984).

SELECTED RESULTS FROM FORECASTING EXPERIMENTS THAT USE FULL DYNAMIC LINKING

Leung et al. (1999) used a mesoscale regional climate model (RCM) based on mesoscale model version 5 (MM 5; Anthes et al. 1987; Grell et al. 1993) driven by a medium-range forecasting (MRF) global climate model developed at NCEP (Ji et al. 1994; Kumar et al. 1994). The RCM produced ensemble climate simulations (24 members) for the month of January for (1) normal SST and (2) warm SST as would be typical of El Niño conditions. These January ensembles, after insertion into a repeating 1960 water-year temperature and precipitation time series, were used to drive the 1° VIC hydrologic model described earlier. The study showed that the climate simulations were not sufficiently accurate in their raw state to produce reasonable streamflow simulations. After a simple bias-correction scheme was applied, however, the simulations reproduced the effects of January climate on stream flow in the Columbia reasonably well (Figure 7.6). Furthermore, the simulated streamflow signals at The Dalles associated with typical warm ENSO (El Niño) January events were found to be realistically reproduced after bias correction of the climate simulations and use of the RCM to downscale the warm SST MRF simulations (Figure 7.7). These results are encouraging to the extent that they demonstrate that climate models can successfully simulate the warm ENSO climate signal in the PNW from first principles and that streamflow simulations based on bias-corrected climate simulations can be both reasonable in absolute value and accurate in representation of the typical effects of warm ENSO conditions on stream flows.

An extension of the work reported by Leung et al. (1999) has evaluated some preliminary six-month simulations (October–March) for normal SST

7.4 Variable Infiltration Capacity (VIC) model schematic. SOURCE: Hamlet and Lettenmaier (2000).

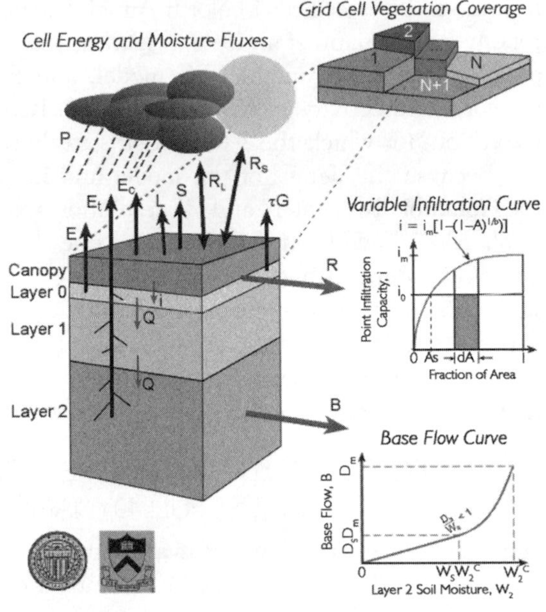

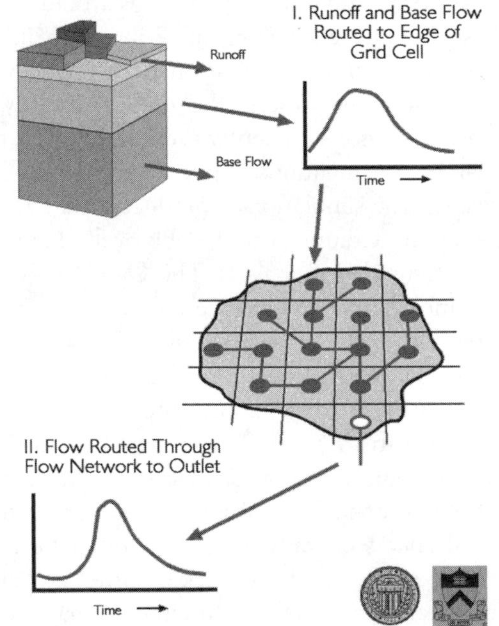

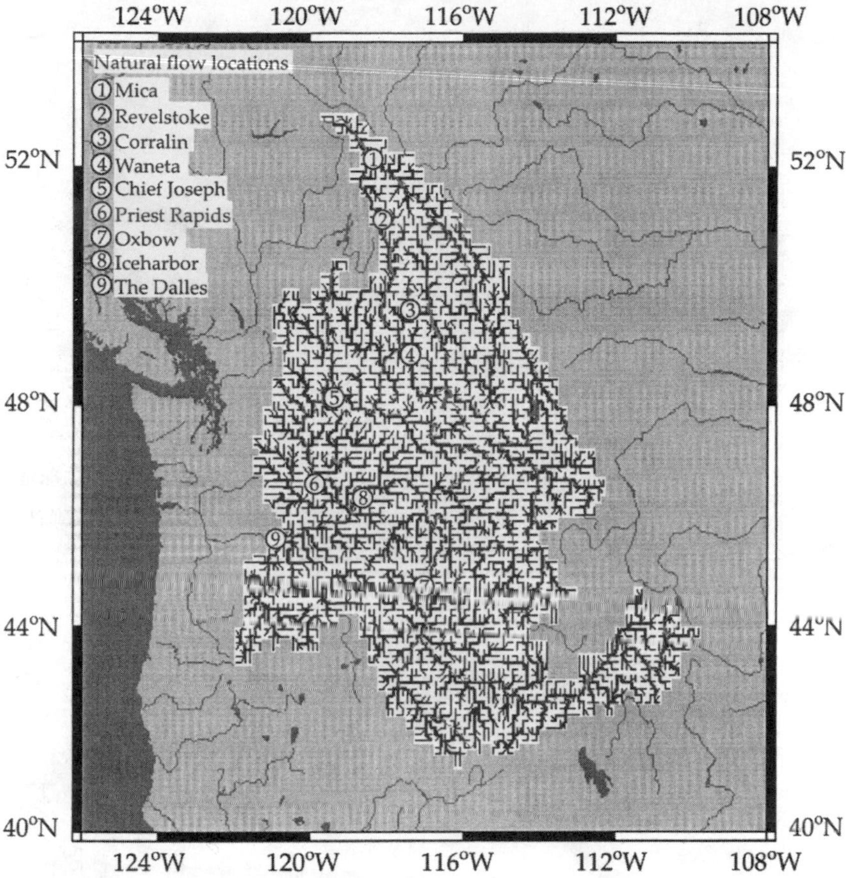

7.5 Channel network for ⅛° resolution VIC model for the Columbia River. SOURCE: Hamlet and Lettenmaier (2000).

using the same simulation and bias-correction tools that were used for January. The results here were less accurate overall than those produced by the January simulations. Although the hydrologic simulations (after bias correction of the climate simulations) were within reasonable bounds, their statistical characteristics were clearly different from those of the historical record. These problems are probably attributable both to the relatively simple design of the bias-correction scheme used for the January experiments and to problems with the spatial distribution and seasonality of precipitation produced by the climate models. If these kinds of problems can be resolved by refining the tools and bias-correction schemes, there is promise of even-

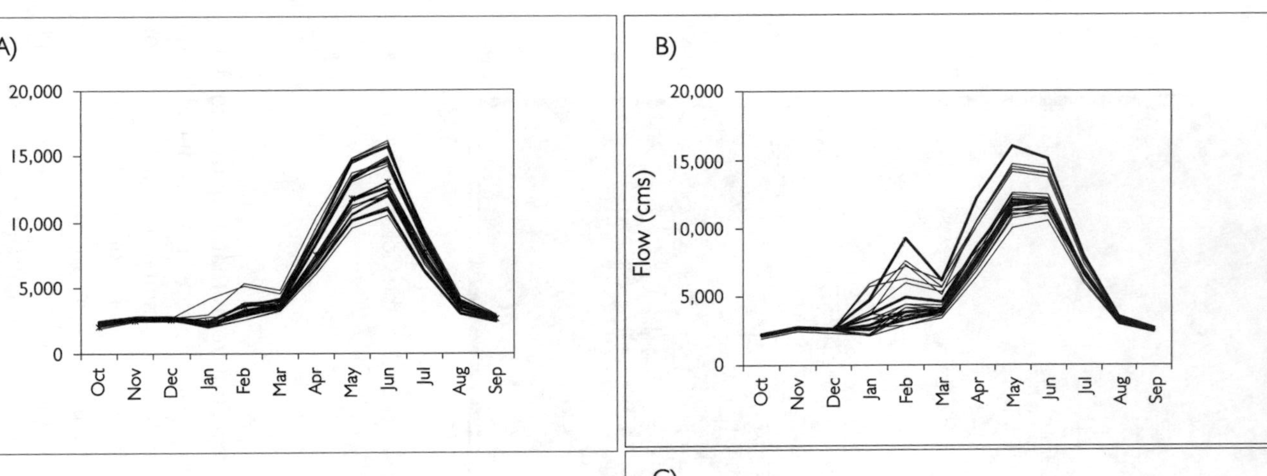

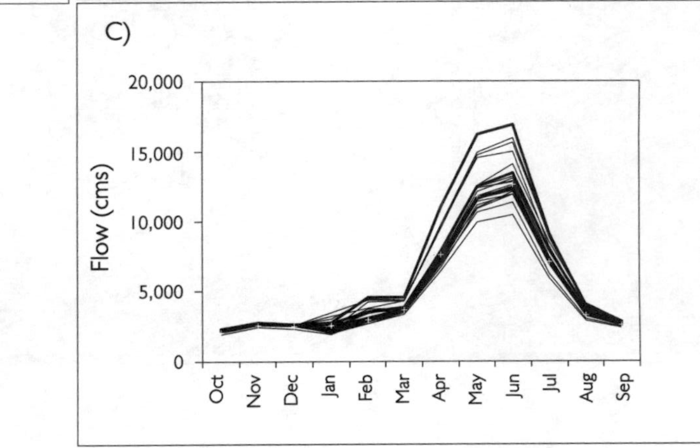

7.6 Effects on Columbia River stream flows of normal SST Januarys for (a) observed Januarys, (b) raw RCM simulated Januarys, and (c) bias-corrected RCM simulated Januarys. SOURCE: Hamlet and Lettenmaier (2000) after Leung et al. (1999). SST = Sea Surface Temperature; RCM = Regional Climate Model.

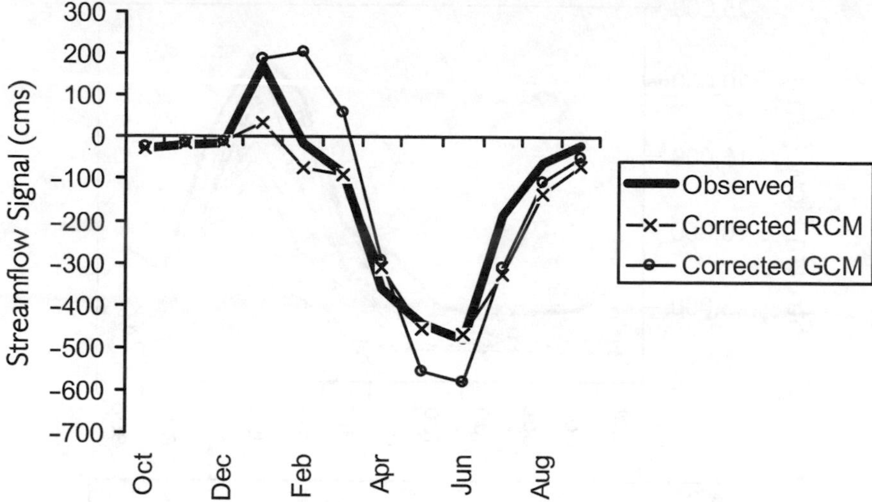

7.7 Monthly streamflow anomalies associated with El Niño Januarys as compared with neutral Januarys for observed data, bias-corrected global climate-model (GCM) simulations of El Niño and ENSO-neutral Januarys, and bias-corrected regional climate-model (RCM) simulations of El Niño and neutral Januarys. SOURCE: Leung et al. (1999). ENSO = El Niño Southern Oscillation.

tually bringing the modeling into an operational context to guide water-management decisions.

SELECTED RESULTS FROM FORECASTING EXPERIMENTS CONDUCTED BY RESAMPLING METHODOLOGY

The results presented here are taken from Hamlet and Lettenmaier (1999). Figure 7.8 shows forecasts of natural flow at The Dalles in the Columbia River produced for water year 1999 in real time. Forecasts for 1999 (as well as 1998 and 2000) were produced on about June 1 preceding the water year (i.e., with about a twelve-month lead time in predicting summer stream flow). Figure 7.8 shows the approximate simulated upper and lower bounds estimated by the highest (high climatology) and lowest (low climatology) monthly simulated flows produced by the hydrologic model for the period 1948–1988, the forecast ensembles based on estimated initial conditions of soil moisture in September preceding the water year and the climate forecast, and the observed naturalized stream flow (dashed line). The year 1998 was assumed to be cold-phase PDO based on the 1997 high flow event and the methods used to predict regime shifts described by Hamlet and Lettenmaier (1999). El

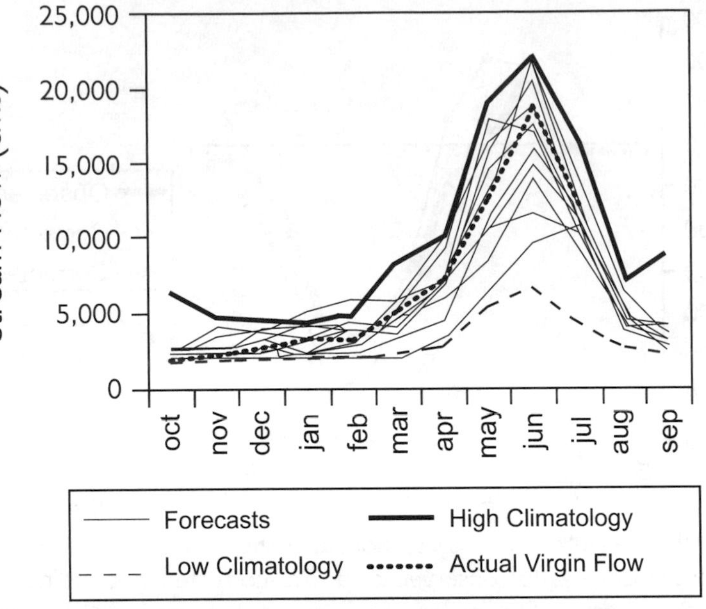

7.8 Real-time ensemble streamflow forecast for the Columbia River at The Dalles for water year 1999 produced in June 1998, plotted with actual stream flows that occurred in 1999.

Niño was forecast for 1998, and 1999 and 2000 were forecast as La Niña. Hamlet and Lettenmaier (1999) also retrospectively forecast ten water years not shown here and discussed potential uses of long-lead forecasts of this type for water-resources management in the Columbia basin.

CONCLUSIONS

Considerable progress has been made in the past several years in producing accurate, long-range ENSO forecasts, in understanding climate variability in the PNW at various timescales, and in developing streamflow forecasting methods that capitalize on this information. It would appear that over the long term, forecasting methods that use climate models as the primary drivers for hydrologic simulation tools would be most effective and useful, especially in view of the potential for rapid changes in global climate. In the short term, however, practical problems with these methods, and the time needed to resolve the problems, may make semiempirical approaches such as the resampling method more effective.

REFERENCES

Abdulla, F. A., D. P. Lettenmaier, E. F. Wood, and J. A. Smith. 1996. Application of a macroscale hydrologic model to estimate the water balance of the Arkansas Red River Basin. *Journal of Geophysical Research* 101(D3) (March): 7449–7549.

Anthes, R. A., E. Y. Hsie, and Y. K. Kuo. 1987. Description of the Penn State/NCAR mesoscale model version 4 (MM4). NCAR Tech. Note NCAR/TN-282+STR, Boulder, CO.

Barnston, A. G., H. M. Van den Dool, S. E. Zebiak, T. P. Barnett, M. Ji, D. R. Rodenhuis, M. A. Cane, A. Leetmaa, N. E. Graham, C. R. Ropelewski, V. E. Kousky, E. A. O'Lenic, and R. E. Livezey. 1994. Long-lead seasonal forecasts—where do we stand? *Bulletin of the American Meteorological Society* 75: 2097–2114.

Battisti, D. S., and E. Sarachik. 1995. Understanding and predicting ENSO. *Review of Geophysics* 33: 1367–1376.

Cayan, D. R., K. T. Redmond, and L. G. Riddle. 1999. ENSO and hydrologic extremes in the western United States. *Journal of Climate* 12: 2881–2893.

Daly, C., R. P. Neilson, and D. L. Phillips. 1994. A statistical-topographic model for mapping climatological precipitation over mountainous terrain. *Journal of Applied Meteorology* 33: 140–158.

Dettinger, M. D., D. R. Cayan, H. F. Diaz, and D. M. Meko. 1998. North-south precipitation patterns in western North America on interannual-to-decadal timescales. *Journal of Climate* 11: 3095–3110.

Grell, G. J., J. Dudhia, and D. R. Stauffer. 1993. A description of the fifth-generation Penn State/NCAR mesoscale model (MM5). NCAR Tech. Note NCAR/TN-398+IA, Boulder, CO.

Hamlet, A. F., and D. P. Lettenmaier. 1999. Columbia River streamflow forecasting based on ENSO and PDO climate signals. *ASCE Journal of Water Resources Planning and Management* 125: 333–341.

———. 2000. Long-range climate forecasting and its use for water management in the Pacific Northwest region of North America. *Journal of Hydroinformatics* 2(3): 163–182.

Ji, M., A. Kumar, and A. Leetmaa. 1994. A multi-season climate forecast system at the National Meteorological Center. *Bulletin of the American Meteorological Society* 75: 569–577.

Kumar, A., and M. P. Hoering. 1995. Prospects and limitations of seasonal atmospheric GCM predictions. *Bulletin of the American Meteorological Society* 76: 335–345.

Kumar, A., A. Leetmaa, and J. Ji. 1994. Simulations of atmospheric variability induced by sea surface temperatures and implications for global warming. *Science* 266: 632–634.

Latif, M., T. P. Barnett, M. A. Cane, M. Flugel, N. E. Graham, H. von Storch, J. S. Xu, and S. E. Zebiak. 1994. A review of ENSO prediction studies. *Climate Dynamics* 9: 167–179.

Lettenmaier, D. P., and D. C. Garen. 1979. Evaluation of Streamflow Forecasting Methods. In *Proceedings, Western Snow Conference.* Sparks, Nevada, April, pp. 48–55.

Leung, L. R., A. F. Hamlet, D. P. Lettenmaier, and A. Kumar. 1999. Simulations of the ENSO hydroclimate signals in the Pacific Northwest Columbia River Basin. *Bulletin of the American Meteorological Society* 80(11): 2313–2329.

Liang, X., D. P. Lettenmaier, E. F. Wood, and S. J. Burges. 1994. A simple hydrologically based model of land-surface water and energy fluxes for general circulation models. *Journal of Geophysical Research* 99: 14415–14428.

Lohmann, D., E. Raschke, B. Nijssen, and D. P. Lettenmaier. 1998. Regional scale hydrology: II. Application of the VIC-2L model to the Weser River, Germany. *Hydrological Sciences—Journal des Sciences Hydrologiques* 43: 143–158.

Mantua, N. J., and S. R. Hare. 2000. Empirical evidence for North Pacific regime shifts in 1977 and 1989. *Progress in Oceanography* 47: 103–145.

Mantua, N., S. Hare, Y. Zhang, J. M. Wallace, and R. Francis. 1997. A Pacific Interdecadal Climate Oscillation with impacts on salmon production. *Bulletin of the American Meteorological Society* 78: 1069–1079.

Mason, S. J., A. M. Joubert, C. Cosijn, and S. J. Crimp. 1996. Review of seasonal forecasting techniques and their applicability to southern Africa. *Water SA* 22: 203–209.

Matheussen, B., R. L. Kirschbaum, I. A. Goodman, G. M. O'Donnell, and D. P. Lettenmaier. 2000. Effects of land cover change on streamflow in the interior Columbia River Basin. *Hydrological Processes* 14: 867–885.

McPhaden, M. J., A. J. Busalacchi, R. Cheney, J. R. Donguy, K. S. Gage, D. Halpern, M. Ji, P. Julian, G. Meyers, G. T. Mitchum, P. P. Niiler, J. Picaut, R. W. Reynolds, N. Smith, and K. Tekeuchi. 1998. The tropical ocean global atmosphere observing system: A decade of progress. *Journal of Geophysical Research* 103: 14169–14240.

Mote, P., M. Holmberg, and N. Mantua. 1999. *Impacts of Climate Variability and Change, Pacific Northwest.* Regional Report to the National Assessment of Climate Variability and Change. JISAO Climate Impacts Group, University of Washington, Seattle, November.

Nijssen, B., D. P. Lettenmaier, X. Liang, S. W. Wetzel, and E. F. Wood. 1997. Streamflow simulation for continental-scale river basins. *Water Resources Research* 33: 711–724.

Trenberth, K. E. 1997. The definition of El Niño. *Bulletin of the American Meteorological Society* 78: 2771–2777.

Trenberth, K. E., G. W. Branstator, D. Karoly, A. Kumar, N. C. Lau, and C. Ropelewski. 1998. Progress during TOGA in understanding and modeling global teleconnections associated with tropical sea-surface temperatures. *Journal of Geophysical Research* 103(C7) (June): 14291–14324.

Yeh, W.W.G., L. Becker, and R. Zettlemoyer. 1982. Worth of inflow forecast for reservoir operation. *Journal of the Water Resources Planning and Management Division* 108: 257–269.

Has Modeling of Water Resources on the Basis of Climate and Hydrology Reached Its Full Potential?

EIGHT

JOHN LABADIE
LUIS GARCIA

This chapter reports the main results of a workshop held in Boulder, Colorado, on June 21–22, 1999. The workshop dealt with the development potential for water-resources modeling based on climate and hydrology. The workshop participants, who represented a range of expertise in the field of water-resources management and water-resources modeling, were asked to consider whether coupled hydrologic and climatic models have substantial undeveloped potential or whether the main causes of variation that could be useful in modeling have already been exploited to the fullest extent possible.

FRAMEWORK FOR EVALUATION OF MODELING POTENTIAL

The workshop participants first created a framework for evaluating the basis for modeling in support of water-resource management (Figure 8.1). With reference specifically to the western United States, the diagram shows a critical coupling between climatic variables that can be described mechanistically in terms of atmospheric dynamics and the development of snowpack, the key determinant of water availability for management purposes in most

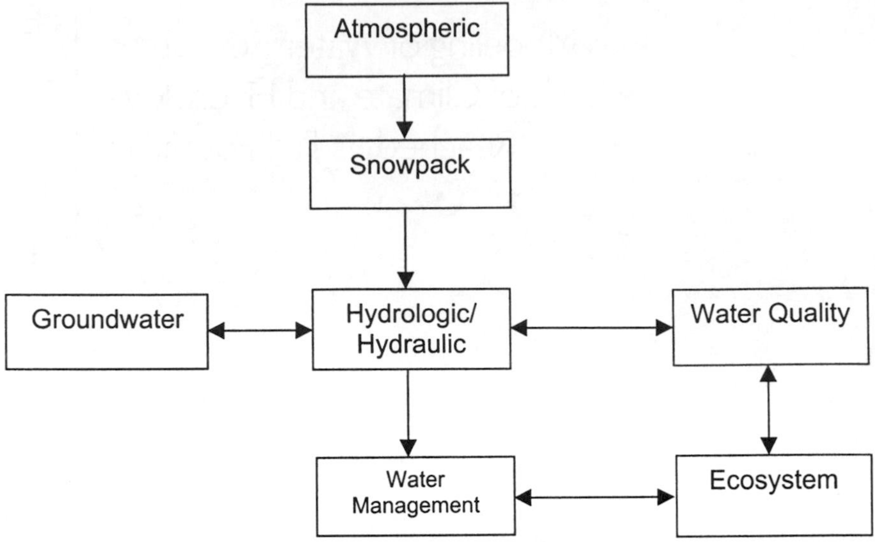

8.1 Framework for evaluation of the status of climate and hydrologic modeling in support of water-resource management in the western United States.

of the western United States. The framework also shows the coupling between snowpack and hydrologic or hydraulic conditions that interface directly with water management. Groundwater also is involved, albeit less directly, and water quality and ecosystem status may also be important considerations interfacing with both hydrologic conditions and water-management priorities.

The workshop participants also identified a set of issues, concerns, and priorities that may limit the use of modeling in support of water-resource management. Table 8.1 gives a summary of these issues. Table 8.1 and Figure 8.1 served as the basis for workshop discussions.

KEY ELEMENTS FOR MODELING IN SUPPORT OF WATER RESOURCES

Snowpack

The modeling of snow accumulation and snowmelt is crucial because approximately 70 percent of annual runoff in the western United States originates from snowmelt (Chapter 2). Although energy-based snowmelt models are well-known at the microscale, little has been done to extend these models to a regional scale. Current work is attempting to use remote sensing and satellite imagery for estimating regional snowpack. Unfortunately,

Table 8.1—The most important considerations that may determine limits on the potential for climatic and hydrologic modeling to serve the needs of water management.

Model accuracy versus precision	Ability to propagate uncertainty, not just mean estimates	Verification and validation of models
Scale issues, both spatial and temporal	Ability to incorporate forecast skill	Incorporation and geographic information system and remote sensing for spatial base development and analysis
Computational efficiency	Good science versus practicality	Inclusion of stakeholders
Data requirements	Robustness and generality	
Ability to model extremes		

extending models to the regional scale raises serious calibration and model verification issues that are not yet resolved.

It may be that even with more complete data on snowpack, highly detailed models are best used for diagnostic purposes rather than for development of ensemble predictions. More simplified, conceptual snowpack and snowmelt models may be more appropriate for development of ensembles. For example, a conceptual snow accumulation model might receive temperature and precipitation as input data and produce estimates of monthly snowfall as output. Primary parameters for such a model may be monthly temperature thresholds above which all precipitation is rain and thresholds below which all precipitation is snow, with ranges between these thresholds assumed to vary in some way. Snowfall estimated from such a model would be added to the snowpack; both snowfall and snowmelt could occur over the same interval. Snowmelt also would be predicted from threshold temperatures (cf. Chapter 5); it would be calculated as a fraction of snow storage, up to a parametric maximum possible fraction. Studies conducted in the western United States have shown that these simple conceptual models are able to explain up to 75 percent of the variability in snowmelt runoff. In summary, undeveloped potential exists for use of detailed, energy-based snowmelt models conjunctively with more conceptual models for ensemble development in support of water-resource management.

Watershed Models of Hydrology

The workshop participants considered hydrologic watershed models and hydraulic flow-routing models to have reached a high proportion of their potential, although further opportunities for improvement remain. Advanced, spatially distributed, physically based models such as MIKE-SHE (Storm and Refsgaard 1996) incorporate the latest in scientific understanding of the hydrologic processes that determine water availability. Further potential exists in linking these models with remote sensing and geographic information system (GIS) technology for preparation of the complex spatial database required by the models. Representation of the spatial variability of soil characteristics for soil moisture estimation is a difficult requirement that could benefit from linkages with remote sensing and GIS. Other underrepresented components of physically based models that could benefit from application of GIS include vegetation, terrain, elevation, and slope.

Unfortunately, the mechanistic realism of present hydrologic models is also a cause of weakness in developing ensemble predictions. Large-scale, regional applications of multiparameter, spatially distributed hydrologic models may become computationally intractable, thus requiring many hours, days, and even weeks of computer execution time—even on advanced computing platforms. In addition, calibration and verification of these models raise serious issues related to degrees of freedom when inadequate amounts and types of data are available. Inadequate data sets may result in many nonunique combinations of variables yielding essentially the same model output.

Workshop participants suggested that a need may exist in hydrologic modeling to reverse the trend toward increasingly mechanistic models and move back to more conceptual models that have the ability to propagate statistical uncertainty in model output. Discussion also centered on possible use of detailed, physically based models for training artificial neural networks or systems based on fuzzy rules for representing regional hydrologic systems. Once trained, these heuristic models could be used efficiently for development of ensembles. Because climatic variations are of such an extensive spatial scale, statistical modeling techniques may prove most valuable in developing ensembles. For example, it may be possible to find strong correlations between temperature variation and runoff response at the regional scale. In effect, macroscale temperature changes associated with long-term global warming may actually simplify the modeling requirements for ensemble development.

Ecosystems and Water Quality

Workshop participants agreed that useful watershed-scale aquatic ecosystem models are not available. The river continuum concept, a unifying set of principles, has never been carried beyond a limited spatial extent. A great challenge remains in predicting ecosystem effects of climate variability or change. Participants agreed that water-quality modeling also remains a difficult area, in part because it is handicapped by lack of data. Current stream water–quality models rely on old technology and are of questionable accuracy. Participants expressed concern about the possible aggravation of salinity problems in regions of extensive irrigated agriculture in the West under conditions of climate change, but little published literature is available on this issue.

Groundwater

Although many scientifically valid numerical groundwater models exist, problems persist in the lack of data needed to calibrate these models. Degrees-of-freedom problems also exist here because of a lack of the correct amounts and types of data necessary for model verification and validation. Interconnected stream-aquifer systems are a particular challenge because of the complexity of their hydraulic connections. Some states in the western United States still fail to recognize this interaction in water law and administration. Workshop participants felt that effects of climate change and variability on groundwater have not been properly addressed and that current patterns of water acquisition (i.e., groundwater versus surface water) may need to be altered in the future. Many parts of the western United States may rely excessively on groundwater resources. Under future climatic scenarios, groundwater resources may decline, resulting in changes in demand patterns, reduced yields, and higher pumping costs. At the regional scale, conceptual, lumped-parameter models may be most valuable; detailed numerical models may be used primarily for diagnostic purposes.

Water Use and Management

Workshop participants stressed the need for predicting future changes in amounts and patterns of water demand, both spatial and temporal, under various global climate-change scenarios. This is particularly important in the western United States, where it is estimated that irrigation accounts for 80 to 90 percent of total consumptive water use. Participants considered it important to use models of evapotranspiration that are able to incorporate information on climatic variability. Methods such as Penman-Monteith, which are also capable of propagating population variances and extreme values, are important in properly assessing the effects of climate change.

Runoff ensembles must be evaluated with respect to the management of water resources. Many of the major river-reservoir systems in the western United States are not managed in a truly integrated fashion but rather rely heavily on predetermined rules designed to capture the combined effects of what is often an overwhelming body of water law, compacts, and agreements. Workshop participants agreed that a great need exists for development of operational models capable of including forecast uncertainty as well as mean estimates and translating that information into uncertainty estimates on the socioeconomic effects of river-system management. Operation of river and reservoir systems must become more flexible and must use sophisticated optimization methods for finding the best among many alternative operational strategies. Heuristic programming methods may be a valuable adjunct for analyzing complex systems that defy application of traditional methods. It is important that administrative and legal issues be fully incorporated into these methodologies, and there is a need for greater efficiency. Participants cited the use of shared vision modeling and decision support systems as particularly valuable tools for incorporating the needs of stakeholders and reaching consensus on climate variability and change.

ACKNOWLEDGMENTS

We gratefully acknowledge the contributions of the workshop participants: Tom Dickinson, Department of Geography, University of Colorado; Marshall Flug, U.S Geological Survey, Fort Collins, Colorado; Dennis Lettenmaier, Department of Civil and Environmental Engineering, University of Washington; Diane McKnight, Institute for Arctic and Alpine Research, University of Colorado; Anne Nolin, National Snow and Ice Data Center, University of Colorado; Diana Perfect, Cooperative Institute for Research in Environmental Sciences, University of Colorado; and John Pitlick, Department of Geography, University of Colorado.

REFERENCES

Storm, B., and A. Refsgaard. 1996. Distributed Physically-Based Modelling of the Entire Land Phase of the Hydrological Cycle. In *Distributed Hydrological Modelling*. M. B. Abbott and J. C. Refsgaard, eds. Kluwer, Dordrecht, Netherlands, pp. 55–69.

PART THREE

Responses of Water Managers to Climate Variability and Climate Change

Water managers are well aware of recent advances in the prediction of climate and of the possibility of change in climate. Just as they have used other new technologies, they are prepared to use climate predictions as a component of management, and some have already begun to do so. A number of barriers to the full assimilation of climate-prediction capabilities into water management remain, however. The main difficulties involve institutional inertia, technology transfer, and the uncertain credibility of predictions. The chapters in this section address each of these issues.

Stakhiv's chapter deals with institutional factors based on his long experience with the U.S. Army Corps of Engineers, a major water manager, but he also treats institutions in general. He lays out guidelines for the extent to which managers should be responsive to hypothetical climate conditions and makes a strong case, based on historical observations, for autonomous adaptation of water management. He argues convincingly that adaptation to contemporary climate variability is the foundation for adaptation to climate change or to increasing climate variability.

Chapters by Fulp and Cech summarize two very different management schemes. The U.S. Bureau of Reclamation's management approach, as described by Fulp, represents practices of a large government agency that must be responsive to an extraordinary diversity of influences extending from federal law and policy to regional requirements and public preferences. Cech's perspective is an interesting contrast in that it portrays the history, development, and current operations of a water provider that operates largely through the private sector, under constraints from Colorado water law and traditional practice.

The final chapter of this series, authored by Wallentine and Matthews, is an assessment of the present utility of climate predictions obtained through a workshop with water managers representing a wide variety of water users. Not surprisingly, water managers seek greater specificity and accuracy in prediction. At the same time, some of the larger and better-funded managers are already incorporating predictions into their standard practices. This chapter shows the mix of receptivity and skepticism that is characteristic of the interface between climate prediction and water management at present.

What Can Water Managers Do About Climate Variability and Change?

NINE

Eugene Stakhiv

Adaptation to climate variability and change will require that all levels of government and all sectors of the economy respond in many ways. Sound management of water resources is central to the efficient functioning of many water-dependent activities—especially transportation (navigation), energy (hydropower), and agriculture. Advances in water management are especially critical for meeting the basic need for drinking water supply of our populace and for sustaining aquatic ecosystems, which would be particularly threatened if the most severe of possible climate scenarios were to materialize (Stakhiv and Major 1997).

The evidence for global climate change has been adequately covered in many publications (e.g., Lettenmaier et al. 1996; Lins et al. 1997). The Intergovernmental Panel on Climate Change (IPCC) reports (1996, 1997), which state that there is a perceptible human influence on climate, should alert us to the possibility that a broad range of potentially adverse and beneficial effects could be caused by a doubling of CO_2 in the atmosphere (see also Chapter 6). The question is, what should water-management organizations such as the U.S. Army Corps of Engineers (the Corps) do to anticipate these changes, and what are the limits to adaptation, given institutional constraints and uncertainties associated with a wide range of seemingly more pertinent exogenous factors such as the socioeconomic structure of the U.S. and global economies?

The thesis of this chapter is that progressive (autonomous) adaptation to hydrologic changes caused by contemporary climate variability and to shifting demands provides a familiar platform for most water managers to use in dealing with climate change. The key issue is not whether water managers are interested in or capable of preparing for climate change in their operational or strategic decision making. Rather, it is more the case that external, policy-driven institutional constraints and conditions are likely to be the larger factors in determining the direction and pace at which adaptation occurs. International attempts at controlling greenhouse-gas emissions are likely to have far greater consequences for energy efficiency and water use than any management actions undertaken directly by the water-resources sector.

As a general proposition, the IPCC itself considers that water resources can be adapted to all but the most severe climate-change scenarios:

> Water management is a continuously adaptive enterprise, responding to changes in demands, hydrological information, technologies, the structure of the economy, and society's perspective on the economy and the environment. This adaptation employs four broad interrelated approaches: new investments for capacity expansion; operation of existing systems for optimal use; maintenance and rehabilitation of systems; and modifications in processes and demands (e.g., conservation, pricing and institutions). These water-management practices, which are intended to serve the present range of climate variability (which in itself is considerable), may also serve to ameliorate the range of perturbations such as droughts that are expected to accompany climate change. However, adaptations come at some social, economic and environmental costs. (IPCC 1996, Vol. II, p. 471)

Indeed, a recent comprehensive comparative analysis of six major water systems and river basins managed by the Corps shows that these systems can be effectively managed for all but the most severe climate-change scenarios (Lettenmaier et al. 1999).

Many regions of the world and urban centers are becoming more stressed in many aspects of water resources—water supply, agricultural irrigation, sanitation, and rural water supply. As a society we must confront these problems, which are daunting even in the absence of climate change. The same mechanisms and approaches are prerequisites for adaptation to climate change. The only issue is whether adaptation can be achieved through a progressive (autonomous) management strategy or will require a special strategy of coordinated, anticipatory measures.

No unique solutions come to mind for water management other than the distinct possibility that significant technological breakthroughs in the next twenty to thirty years can alleviate both energy problems and scarcity of water supply (e.g., cheap energy would enhance the cost-effectiveness of desalination technology, which is energy intensive). Biotechnology will play a major role in future food production. Ausubel (1995) makes a compelling case that trajectories exist for technological innovation and that technical progress in many fields is quantifiable. He believes that the process of decarbonization—decreasing intensity of carbon use for production of energy—has been evident since 1900 but has not been taken into account in the IPCC forecasts.

POLICY IMPETUS

A good deal of measurable, autonomous adaptation is occurring in the water-resources sector. Even with a large amount of waste and inefficiency, total freshwater withdrawals in the United States have been decreasing since their peak of 378 billion gallons per day (bgd) in 1980 (down to 341 bgd in 1995; Solley et al. 1998). Most of the reduction came indirectly as a result of the regulatory requirements of the Clean Water Act and the Safe Drinking Water Act. Legislation has forced industries to recycle water more efficiently, and the rising costs of wastewater treatment have served as an indirect pricing mechanism for municipal users. Also, innovations in irrigation have reduced withdrawals and consumption. All of this was accomplished despite a 16 percent increase in the U.S. population between 1980 and 1995.

Much of the response to increasing water demands must be met through institutional and policy changes that focus on more coherent management of water and related natural resources at the watershed level. These changes in governance will determine the effectiveness of agencies and the private sector in adapting to change and variability. Indeed, institutional and legal changes provide the impetus and means for implementation of reforms at all levels of government and magnify the more limited set of technical adjustments that are the responsibility of agencies. Institutional changes are broad and can be achieved only by federal, state, and local governments because they deal with organization, authority, economic incentives, licensing or permitting, taxes and other economic incentives, and legislation. The cumulative effect of mutually reinforcing initiatives sets the foundation for an effective strategy of operational and design changes. Other significant changes must come directly from the individual in the form of changes in behavior and attitudes that affect the use of resources.

Numerous institutional changes drive public-policy changes, which affect all organizations. The National Environmental Policy Act is a clear example. In the area of water policy, the Water Resources Planning Act of

1965, which formed the U.S. Water Resources Council and river basin commissions, was a milestone. Similarly, the report of the National Water Commission (1973) influenced legislation that reformed the planning and cost-sharing policies of federal water-resources management agencies. Numerous equivalent recent initiatives will have a lasting, positive effect on water-resources management and will set the stage for more responsive adaptive-management strategies, with or without consideration of climate change. These include the report of the Interagency Floodplain Management Review Committee (1994), also known as the Galloway Commission, which has already engendered a broad series of legislative and administrative reforms of floodplain management policies for federal and state agencies. Also, the Western Water Policy Advisory Commission (1998) released its report on a comprehensive approach to water-resources management reforms in the seventeen western states; many of the recommendations will influence policy initiatives. The federal establishment, under the direction of Vice President Gore, issued the Clean Water Action Plan (1998) for restoring and protecting the nation's waters. The National Drought Policy Act, passed by Congress in 1998 (PL 105-199), set up a commission to recommend changes in federal drought management and response policies and to improve coordination among the numerous agencies involved in this endeavor. Congress found an increasing need to emphasize preparedness, mitigation, and risk management rather than crisis management. Many other comparable initiatives in the agricultural, transportation, and energy sectors will result collectively in a set of reinforcing reforms capable of accelerating the pace of autonomous adaptive management. These initiatives and changes will improve the responsiveness of government and private sectors to climate change and further reduce society's vulnerability to climate variability and natural hazards.

Gleick (1998: 10) recognized the marked shift in traditional water-resources management:

> Traditional approaches to water planning, while still firmly entrenched in many water-planning institutions, are beginning to change. Among the factors driving these changes are high costs of construction, tight budgets, deep environmental concerns, new technological advances, and the development of innovative alternative approaches to water management. The search for new solutions is also being pushed along in some places by the changing nature in the demand for water, particularly in North America and western Europe.

Along with major changes in public water-resources management policies, many small economic, legal, financial, and regulatory policy changes over the past thirty-five years have collectively directed or influenced profound changes in the philosophy of water management. Too little credit is given to increased nonfederal cost-sharing provisions and to improvements in benefit-cost analysis and other evaluation procedures that have demonstrably reduced the number of water-development projects, as well as their scale and size, and have resulted in more environmentally benign designs since the mid-1980s—especially since the enactment of the Water Resources Development Act of 1986. These changes are the primary determinants of progressive (autonomous) adaptation. Now a more holistic view of substantive institutional changes is needed to position the nation for more flexible adaptation to social, economic, and environmental changes, as well as those of climate. Water-management agencies and utilities, which are inherently technical in nature and responsive to external mandates, can only seek to optimize the services that they deliver under the specific sets of competing objectives and constraints imposed by parent institutions.

INITIATIVES AND IMPLEMENTATION OF THE U.S. ARMY CORPS OF ENGINEERS

The Corps is an example of an institution that can considerably influence the rate of change and adaptation in management of water resources. Like all other agencies, however, it can only respond to the policy and budgetary directives and constraints imposed by Congress and the Executive Branch. The Corps can exercise greater initiative and responsiveness in defining the pace at which new technologies are introduced, but even that function is limited by budgetary decisions. Despite these constraints, water-resources management is inexorably driven toward more efficient use and management of water resources and a true realization of sustainable development according to the concepts of the President's Council on Sustainable Development (1996).

The greatest potential for improving adaptive management in the water-resources sector lies in institutional settings and policy, its most politically contentious forum. Water-management agencies today have less direct control than ever over water-management decisions, except perhaps in the seventeen U.S. western states where federal government manages much of the surface-water resources and a good deal of the land surface. Generally, the federal government—largely through the Department of the Interior (Bureau of Land Management, Bureau of Indian Affairs, Fish and Wildlife Service, Geological Survey, and Bureau of Reclamation)—may have more influence over, but not direct responsibility for, policies leading to the allocation of water supply and facilitating voluntary transfers of water through a water-market system. Furthermore, the regulatory controls of the Safe Drinking Water Act and the Clean Water Act, both of which are managed by the U.S. Environmental Protection Agency, have played a major role in reducing inefficient water use in the municipal and industrial sectors. Increasingly stringent nonpoint-source pollution controls in U.S. watersheds will continue to impose further restrictions on agricultural water use in the future. It is anticipated that future greenhouse-gas emissions controls will require that the U.S. government undertake a major energy efficiency and technology development initiative to benefit the water-resources sector.

Each agency at every level of government has a role to play in improving the effectiveness of adaptive-management mechanisms geared to meet contemporary demands in the context of climate variability. The sum of these efforts can help U.S. society manage its way through climate variability and change. The key changes that the Corps could significantly effect in its response to global warming would be largely focused in the three technical and administrative approaches listed by the IPCC (1996, Vol. II: 411): (1) planning new investments for expansion of capacity, (2) operating existing systems for optimal use, and (3) maintaining and rehabilitating existing systems.

Even changes in these approaches require considerable congressional approval because many aspects of project planning and design—including major rehabilitation of the existing infrastructure—are based on a strict sequence of approval, procedure, and criteria that dictates cost sharing, evaluation, discount rate, and similar factors. Each potential project goes through several phases of congressional review, direction, and approval—including project feasibility planning, authorization of advanced planning, engineering and design, and appropriation for construction. These sequential controls and criteria profoundly influence the ultimate choice of a project alternative, definition of purposes, and determination of project location and size. Nevertheless, the Corps is at least an active participant in the formulation of project alternatives and the analytical process with its federal, state, and local partners. Some flexibility is possible in choosing and designing more robust and resilient options in anticipation of climate change. The key is how explicitly to introduce future climate scenarios, or at least climate variability, into project planning and design when economic efficiency requirements and cost-sharing concerns drive the selection of alternatives that are neither robust nor resilient (Hashimoto et al. 1982).

Ironically, the introduction of explicit risk and reliability analysis and estimation of benefit-cost ratios as an integral feature of operations for the Corps contributes to "brittle" designs, that is, projects that operate under increasingly narrow margins of reliability. Risk analysis was intended to avoid overdesign and eliminate redundancies, but these are the very features that provide extra margins of reliability (e.g., levee freeboard, which accounts for uncertainties in hydraulic and hydrologic knowledge). On the other hand, the analytical methods developed by the Corps (1992, 1996) for dealing with risk and uncertainty explicitly account for much of the anticipated climate change and variability through procedures based on Monte Carlo simulation. The Corps computes the risk of flooding caused by combinations of hydrologic events, hydrologic parameter uncertainty, uncertainty in reservoir operation, state-discharge relationships, and levee performance. The methodology allows the Corps to investigate the influence of the uncertainties on economic outcomes and benefit-cost justification of various project features, design levels, and optimal size. One can explicitly examine the performance of various alternative designs and project scales in terms of direct measures of robustness, resiliency, and reliability. The Monte Carlo technique involves sampling a complete flood-frequency distribution (log-Pearson III) within the 95 percent confidence limits for return periods beyond 1,000 years. The same is true for the stage-frequency and damage-frequency combinations. The National Research Council (1995) examined the Corps' methodology and concluded that "the new USACE risk and uncertainty procedures are an innovative and timely development. The explicit recognition of modeling uncertainty should result in a better understanding of the uncertainty of flood risk and damage reduction estimates" (p. 161).

PLANNING FOR NEW INVESTMENTS IN EXPANSION OF CAPACITY

Several noteworthy initiatives within the Corps deal with climate change to some extent through a more explicit examination of hydrologic variability and trends and their influence on decisions. The Upper Mississippi River Flow Frequency Study, authorized by Congress in the aftermath of the 1993 flood in the upper basin, deals with the problem of accounting for extreme floods through improved hydrologic forecasting. The principal focus is on developing a better hydraulic model that can predict the river stages for given amounts of precipitation and runoff. The study in part considers whether basic hydrologic methods need to be revised to account for an apparent trend toward increasing precipitation and flood peaks over the past fifty years. The study will have long-term implications for floodplain delineations, levee certifications, and levee design throughout the region. Preliminary results, however, suggest no compelling evidence to deviate from Bulletin 17B (developed by an interagency group in 1982), which gives the basic federal guidelines for performing flood-frequency analysis.

Similarly, a study of the Devils Lake flooding problem relied heavily on a simulation of future probabilities of increased precipitation and runoff into a closed basin as the basis for determining whether to build a pumping station and outlet to the Sheyenne River or continue to add to the system of levees protecting the tow and vital infrastructure. A net investment savings of over $40 million depended on near-term climate-state possibilities and future trends. A study of climate trends, including the influence of the El Niño–Southern Oscillation and other longer-term oscillatory meteorological phenomena (North Atlantic Oscillation, Pacific Decadal Oscillation, and others; Chapter 1), was undertaken by the Corps in its attempt to understand more completely the relationship between historical climate fluctuations and a permanent shift in the means of climate variables as a result of global warming. Such inquiries are beginning to represent a significant portion of the Corps' approach to examining the hydrologic response of large river basins.

One of the most comprehensive reviews of the adequacy of the Corps' planning criteria, procedures, and evaluation principles in light of global warming and climate uncertainty was undertaken by Resources for the Future. A report on that work (Frederick et al. 1997) concluded that the Corps' increased emphasis on risk and uncertainty in all aspects of its planning studies, especially economic justification for new investments, is consistent with adaptation to climate change. Stakhiv and Major (1997) further concluded that the Corps' methodologies, based on the U.S. Water Resources Council's principles and guidelines, complemented the approach advocated by the IPCC—as stated in its *Technical Guidelines for Assessing Climate Change Impacts and Adaptation* (1994)—more closely than the traditional National Environmental Policy Act of 1969 environmental impact assessment guidelines.

The basic planning process used by the Corps promotes the formulation of numerous alternatives that vary by location, technology, and operation, as well as in behavioral, institutional, and regulatory characteristics. Large uncertainties surround each option. Formal methods for analysis of uncertainty and risk are the principal means for determining the reliability, robustness, and resiliency of alternative measures under a variety of conditions, stresses, and future demands. This multiobjective method for formulating and evaluating plans is the backbone of adaptive management.

OPERATION OF EXISTING SYSTEMS FOR OPTIMAL USE

The science of weather forecasting is improving, but forecasts are still too imprecise for reservoir operators (Chapter 12). New forecasting procedures are being developed, but they must be tested and adapted to the needs of practicing water managers. Too much of the theoretical development is in the hands of researchers; too few practical experiments and joint applications are being done with water managers and operators.

One example of a technology that holds great promise for revolutionizing operational water management was recently reported by Georgakakos et al. (1998). Reliable forecasts of reservoir system inflows are essential for efficient reservoir operations because they increase the flexibility for managing and adapting to regional water-management needs. The authors tested and compared the standard rule-based operational forecast with a coupled forecast control system that employs probabilistic ensemble-flow forecasts. The coupled forecast control system explicitly accounts for flexible reservoir management strategies encompassing uncertain forecasts. The sensitivity of reservoir management to climate variability can therefore be explicitly analyzed and thereby minimized. Data from the regulated 14,000 km^2 upper Des Moines River basin were used together with the operational rules of Saylorville Reservoir, which is operated by the U.S. Army Corps of Engineers for flood control, recreation, low-flow augmentation (ecosystems and water quality), and water supply. The study showed that current reservoir management practices cannot effectively accommodate historical climate variability. A substantial gain in resilience to climate variability is possible when the reservoir is operated by a control scheme that uses reliable forecasts and accounts for their uncertainty. The study shows that such coupled decision systems can mitigate adverse effects of climate forcing on regional water resources (Georgakakos et al. 1998). The authors conclude that the failure to incorporate realistic optimization procedures for reservoir operation in climate-change impact studies such as the one developed for the Des Moines River would lead to overly pessimistic results. A similar study is being conducted on improvements in weather and event forecasting capabilities in the Ohio River basin as the prerequisite for improved operation of the water-management system by the Corps. This effort is part of a larger Global Energy and Water Cycle Experiment Continental Scale International Project for which the Mississippi River basin and its major tributaries (Missouri River, Ohio River) have been selected as suitable for coordinated experiments. It is important for operating agencies such as the Corps and the Bureau of Reclamation to participate directly in such studies and technology transfer activities, which will provide the basis for major improvements in water management (National Research Council 1998).

MAINTENANCE AND REHABILITATION OF EXISTING SYSTEMS

Most of the Corps' construction program is directed toward major rehabilitation of the aging U.S. system of locks, dams, and levees. The Corps has developed a risk-analysis method for analyzing each major component of infrastructure. A diverse set of tools and procedures was developed during the past decade for risk-based hydrologic, hydraulic, and economic analysis (Institute for Water Resources 1998). A new generation of risk-based techniques is also being developed for environmental restoration and remediation.

New methods and analytical procedures are important in dealing with forecasting uncertainty, climate variability, and uncertainty in climate change. Widely ranging adaptive-management strategies are routinely formulated during the planning process. Each has different performance characteristics under different operating ranges. The Corps has been developing tools that can be used in analyzing these performance characteristics uniformly. The analyses are important to the Corps' treatment of such problems as dam safety and rehabilitation of hydropower facilities, navigation locks, and levees and the disposal of contaminated dredge material. The performance of each component of infrastructure is tested under a wide range of circumstances that include conditions and events anticipated under many future climate scenarios. Fifteen recently issued technical procedures (1992–1998) provide guidance on risk, uncertainty, and reliability analysis in all aspects of the Corps' planning, design, operations, and maintenance programs.

CONCLUSIONS

Much of what a water-management agency does to adapt to contemporary climate variability also serves as the platform for adapting to prospective climate change. The most effective components of a targeted adaptive-management strategy are likely to be policy and institutional changes that are externally driven, principally as political solutions to social and technical problems. Water-management agencies such as the Corps can become agents of change and adaptation by playing a more active role in transferring technologies associated with climate forecasting. Analysis of risk and uncertainty is the major avenue for explicitly considering the implications of climate variability, change, and uncertainty for all aspects of Corps planning, design, operations, and maintenance.

SUMMARY

Water managers (planners, designers, operators) cannot prevent climate change or variability. Historically, water management has been a process of continuous adaptation to variation of climate and accommodations—through redundancies of engineering design—for uncertainties associated with our lack of understanding about climate. Adaptive management, which involves monitoring and learning from mistakes, has been the foundation of water-resources management since the time of Noah. The societal response to both variability and change is virtually the same—to upgrade and intensify introduction of innovative and cost-effective supply-side and demand-side management measures and to continue to create institutions that are more flexible in adapting to both social and physical changes. Policy initiatives that affect legal and institutional controls on water management are likely to play a much larger role in future adaptation to climate change than are technical and engineering responses. Engineers can design and operate systems in ways that increase robustness and resiliency and reduce vulnerability, but engineering refinements are not useful without institutional arrangements that are reconfigured to ensure that future water-resources services can be provided in a sustainable and equitable manner under a wider range of circumstances.

There are two tiers of change in adaptive management: policy mandates and agency implementation. Many of the changes that will prepare society to deal better with future uncertainty related to climate are already being debated and implemented through changes in policy and institutional reforms that deal with an increasingly complex host of issues and that include such matters as river basin compacts; new partnerships among federal, state, and local entities; nonstructural reduction in flood damage; valuation of water as an economic and environmental commodity; and increasing requirements for environmental protection and aquatic ecosystem restoration. These strategic policy changes will influence future goals, objectives, and responses of water-management agencies. The components of water-resources management directly under the control of or influenced by water managers include adoption of improved methods of hydrologic analysis coupled with risk analysis, improvement of forecasting methods for systemwide analysis, and integrated analysis of multiple watershed needs and outcomes. Future management systems will be more robust and resilient to anticipated climate variability and change and to evolving societal demands.

REFERENCES

Ausubel, J. H. 1995. Technical progress and climate change. *Energy Policy* 23: 411–416.

Clean Water Action Plan: Restoring and Protecting America's Waters. 1998. Interagency Report to Vice President Gore. U.S. Government Printing Office, Washington, DC.

Frederick, K., D. Major, and E. Stakhiv, eds. 1997. *Climate Change and Water Resources Planning Criteria.* Kluwer Academic Publishers, Dordrecht, the Netherlands.

Georgakakos, A., H. Yao, M. Mullusky, and K. Georgakakos. 1998. Impact of climate variability on the operational forecast and management of the Upper Des Moines River Basin. *Water Resources Research* 34: 799–821.

Gleick, P. 1998. *The World's Water 1998–1999: The Biennial Report on Freshwater Resources.* Island, Washington, DC.

Hashimoto, T., J. Stedinger, and D. Loucks. 1982. Reliability, resiliency and vulnerability criteria for water resources system performance evaluation. *Water Resources Research* 16: 14–20.

Institute for Water Resources. 1998. *Tools for Risk-Based Economic Analysis.* Draft Research Report. U.S. Army Corps of Engineers, Fort Belvoir, VA.

Interagency Floodplain Management Review Committee. 1994. *Sharing the Challenge.* U.S. Government Printing Office, Washington, DC.

Intergovernmental Panel on Climate Change. 1994. *IPCC Technical Guidelines for Assessing Climate Change Impacts and Adaptation.* Cambridge University Press, Cambridge.

———. 1996. *Climate Change 1995: Impacts, Adaptations and Mitigations: Contribution of Working Group II to the Second Assessment Report of the IPCC.* Cambridge University Press, Cambridge.

———. 1997. *IPCC Special Report—The Regional Impacts of Climate Change—An Assessment of Vulnerability, Policymakers Summary*. Cambridge University Press, Cambridge.

Lettenmaier, D., G. McCabe, and E. Stakhiv. 1996. Global Climate Change: Effect on Hydrologic Cycle. In *Water Resources Handbook*. L. Mays, ed. McGraw-Hill, New York, pp. 29.3–29.33.

Lettenmaier, D., A. Wood, R. Palmer, E. Wood, and E. Stakhiv. 1999. Water resources implications of global warming: A U.S. regional perspective. *Climate Change* 43: 537–579.

Lins, H., D. Wolock, and G. McCabe. 1997. Scale and modeling issues in water resources planning. *Climate Change* 37: 63–88.

National Research Council. 1995. *Flood Risk Management and the American River Basin—An Evaluation*. Water Science and Technology Board. National Academy Press, Washington, DC.

———. 1998. *GCIP—Global Energy and Water Cycle Experiment (GEWEX) Continental Scale International Project: A Review of Progress and Opportunities*. National Academy Press, Washington, DC.

National Water Commission. 1973. *Water Policies for the Future: Final Report to the President and to the Congress of the United States*. U.S. Government Printing Office, Washington, DC.

President's Council on Sustainable Development. 1996. *Sustainable America: A New Consensus for Prosperity, Opportunity, and a Healthy Environment for the Future*. U.S. Government Printing Office, Washington, DC.

Solley, W., W. R. Pierce, and H. Perlman. 1998. Estimated use of water in the United States in 1995. U.S. Geological Survey Circular 1200, Washington, DC.

Stakhiv, E., and D. Major. 1997. Ecosystem evaluation, climate change and water resources planning. *Climate Change* 37: 103–120.

U.S. Army Corps of Engineers. 1992. *Guidelines for Risk and Uncertainty Analysis in Water Resources Planning, Volumes I and II*. IWR Report 92-R-1, 92-R-2. Institute for Water Resources, Fort Belvoir, VA.

———. 1996. *HEC-FDA Flood Damage Reduction Analysis—User's Manual*. Hydrologic Engineering Center, Davis, CA.

Western Water Policy Advisory Commission. 1998. *Water in the West: Challenge for the Next Century*. National Technical Information Service, Springfield, VA.

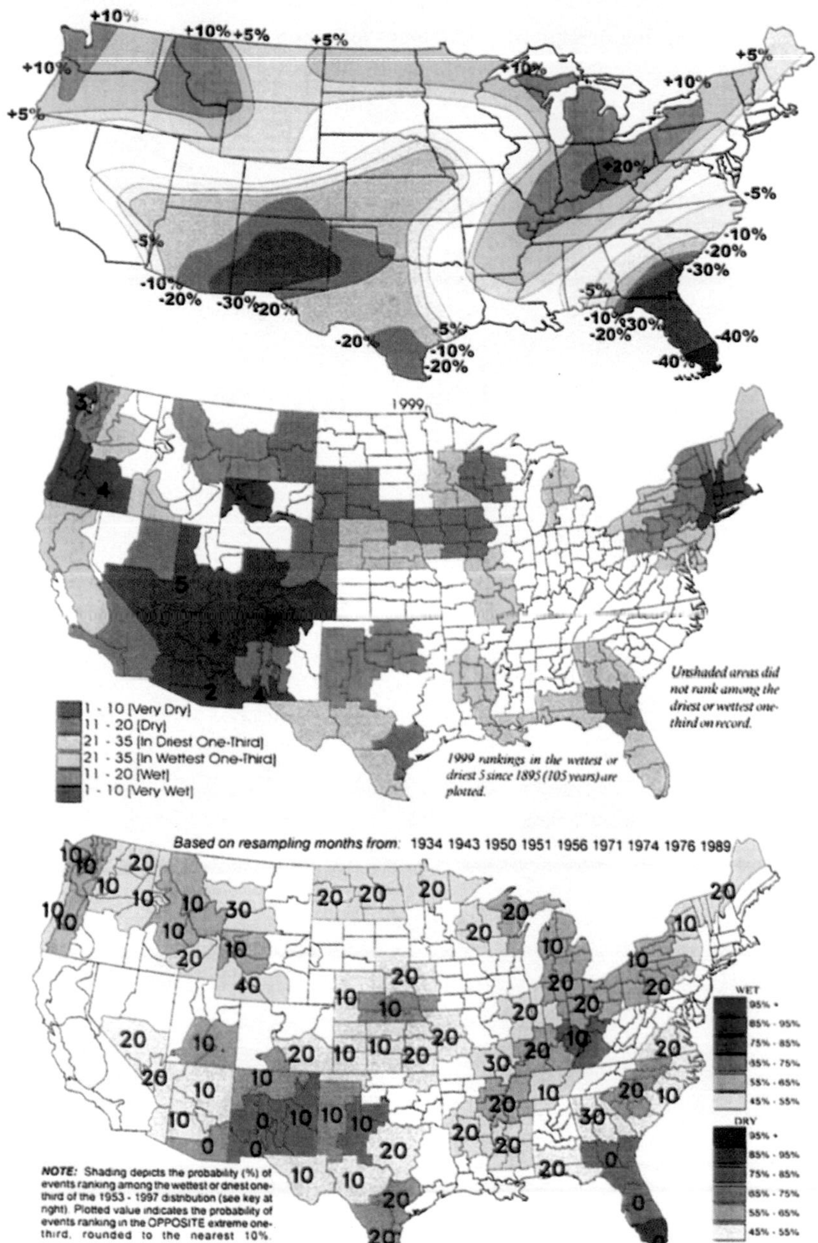

PLATE 1 Official January–March 1999 precipitation forecast issued by the NOAA Climate Prediction Center (TOP PANEL), verification for the same period (MIDDLE PANEL), and statistical estimate of precipitation distribution for this period obtained from nine previous analogous years (BOTTOM PANEL).

Spring Precipitation Extremes following El Niño Winters

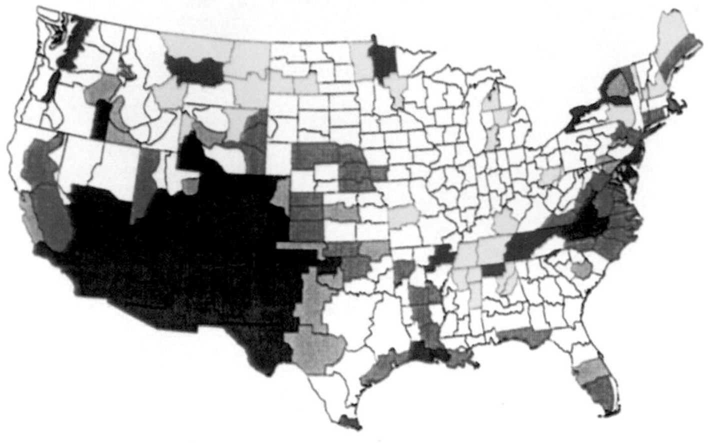

Spring Precipitation Extremes following La Niña Winters

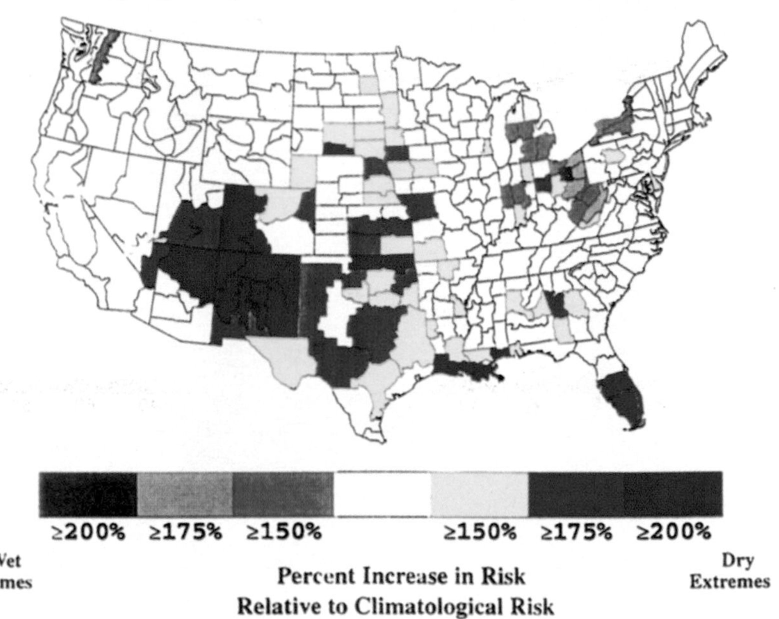

PLATE 2 Risk estimate expressing ratio (as percent) of either very wet or very dry springs following El Niño winters (TOP PANEL) or La Niña winters (BOTTOM PANEL); a value of 200 percent would correspond to a doubling of risk. Note the increased risk of very wet springs following El Niño winters through much of the Southwest and, conversely, the increased risk of very dry springs in the same region following winter La Niña conditions.

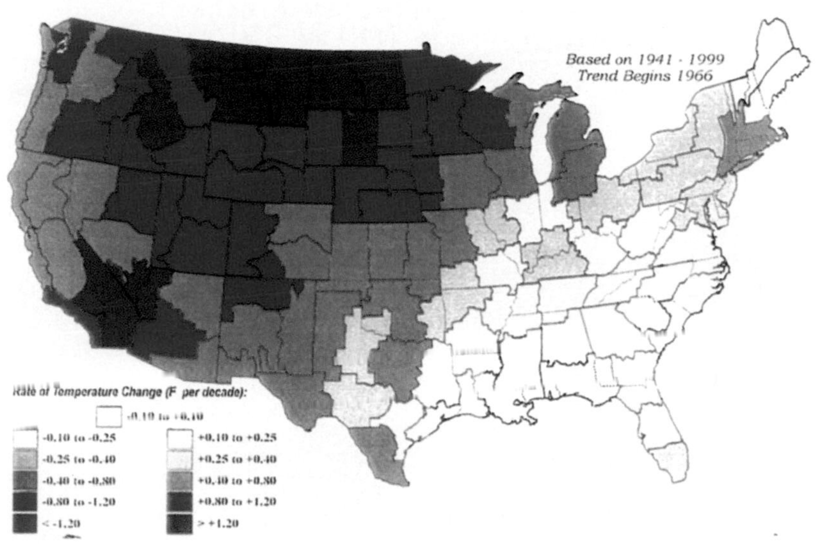

PLATE 3 Rate of change in temperature for the period beginning in 1966, relative to the interval 1941–1999. Red indicates warming; warming exceeds 1°F per decade over portions of the far northern plains.

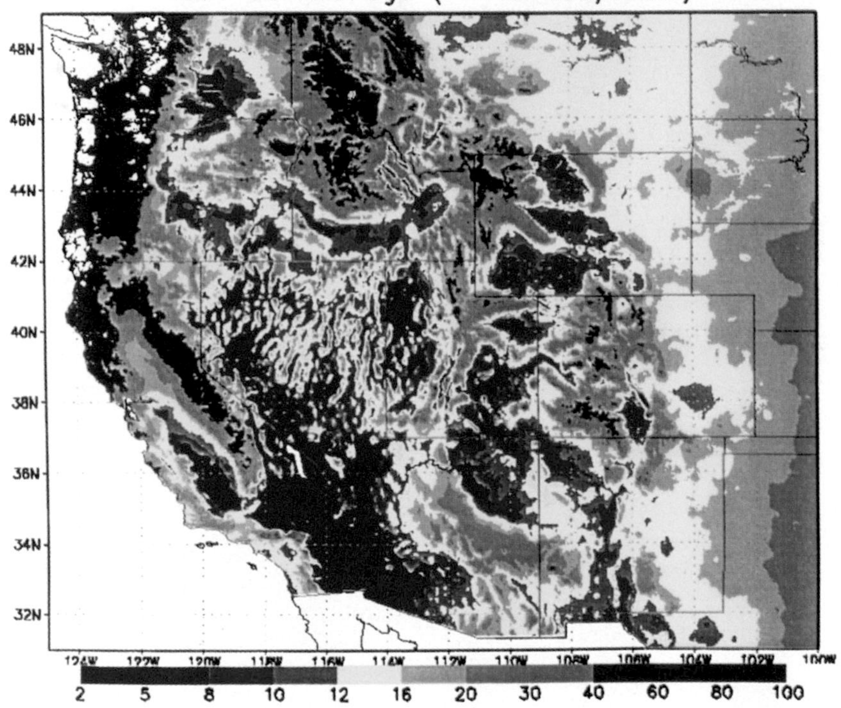

PLATE 4 Average annual precipitation map, 1961–1990, for the western United States based on the Precipitation Regression on Independent Slopes Method (PRISM) output from Oregon State University.

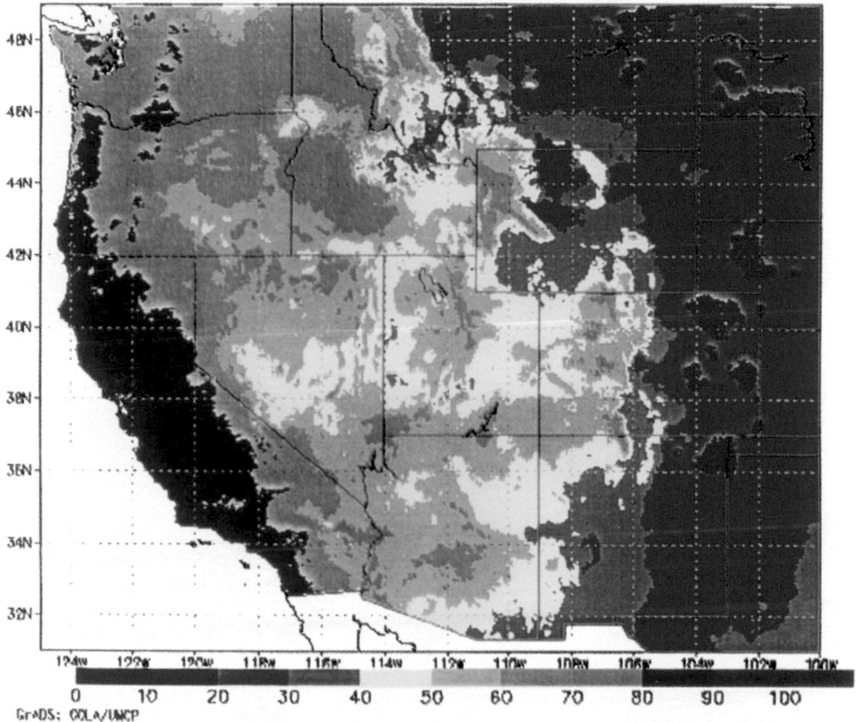

PLATE 5 Percentage of 1961–1990 average annual precipitation falling in the winter months of October–March, based on PRISM output from Oregon State University.

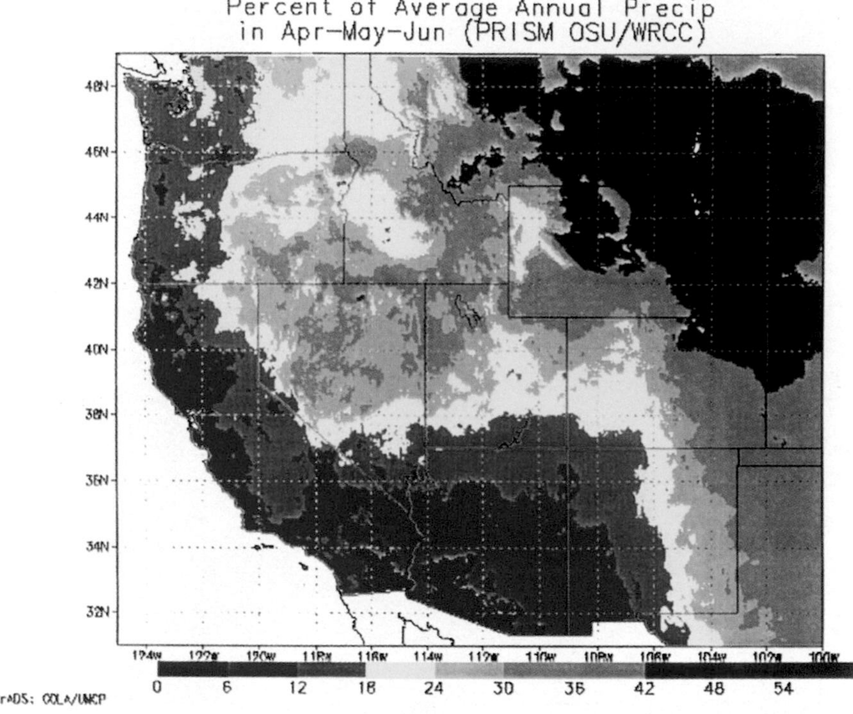

PLATE 6 Percentage of 1961–1990 average annual precipitation falling in the late spring months of April–June, based on PRISM output from Oregon State University.

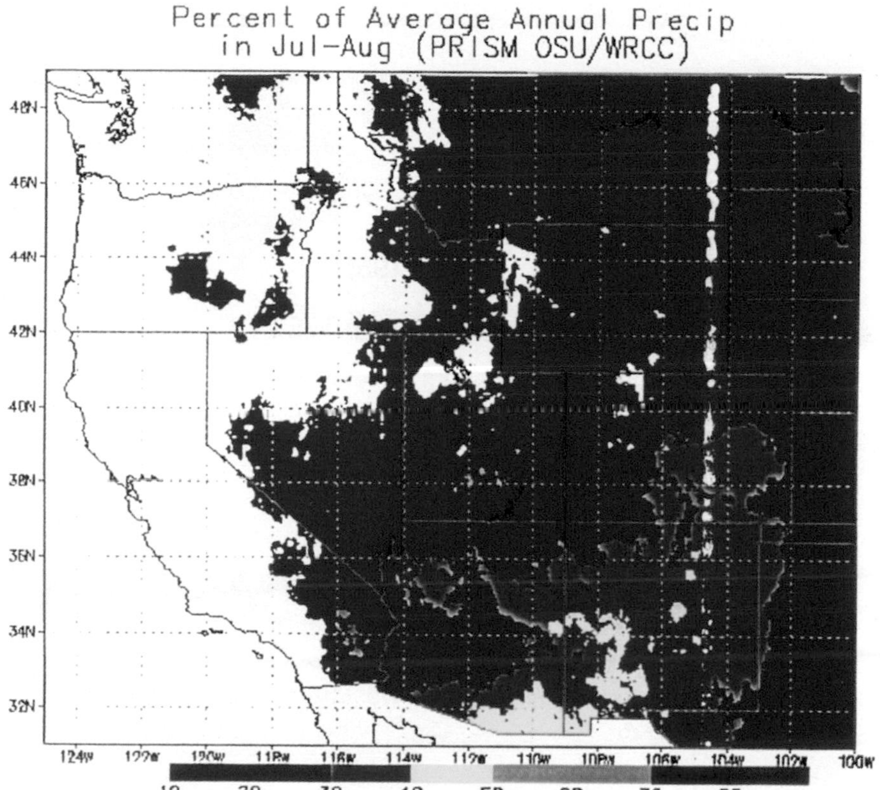

PLATE 7 Percentage of 1961–1990 average annual precipitation falling in the summer monsoon months of July and August, based on PRISM output from Oregon State University.

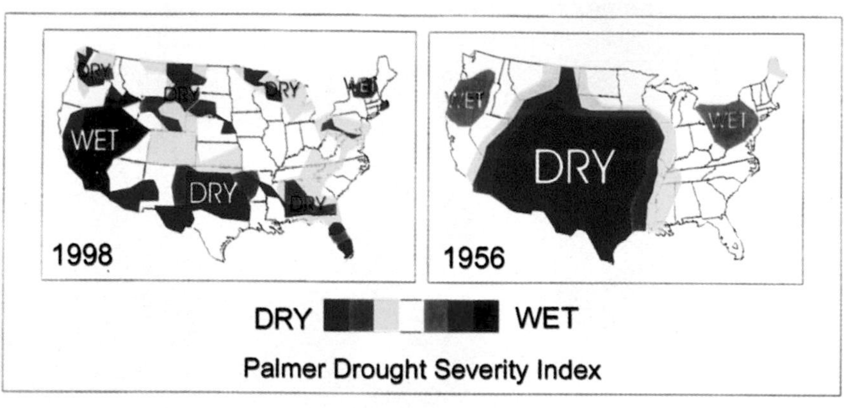

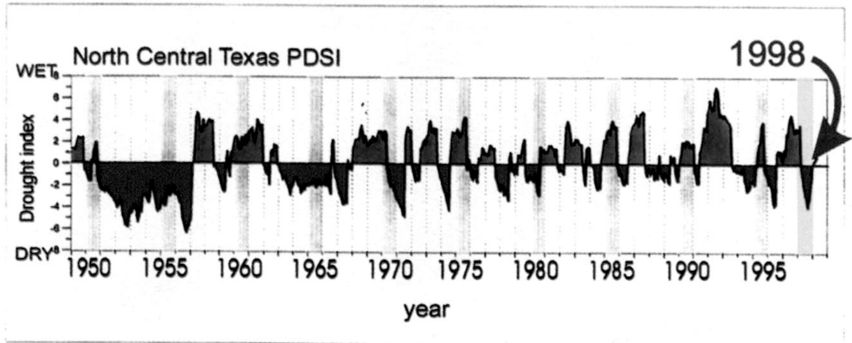

PLATE 8 Comparison of the Palmer Drought Severity Index (PDSI) in 1998 and in 1956, the worst year of the 1950s drought (TOP) and PDSI index values for north-central Texas, 1949–1998 (BOTTOM). DATA SOURCES: http://www.ngdc.noaa.gov/paleo/drought.html (TOP) and http://www.ncdc.noaa.gov/onlineprod/drought/main.html (BOTTOM).

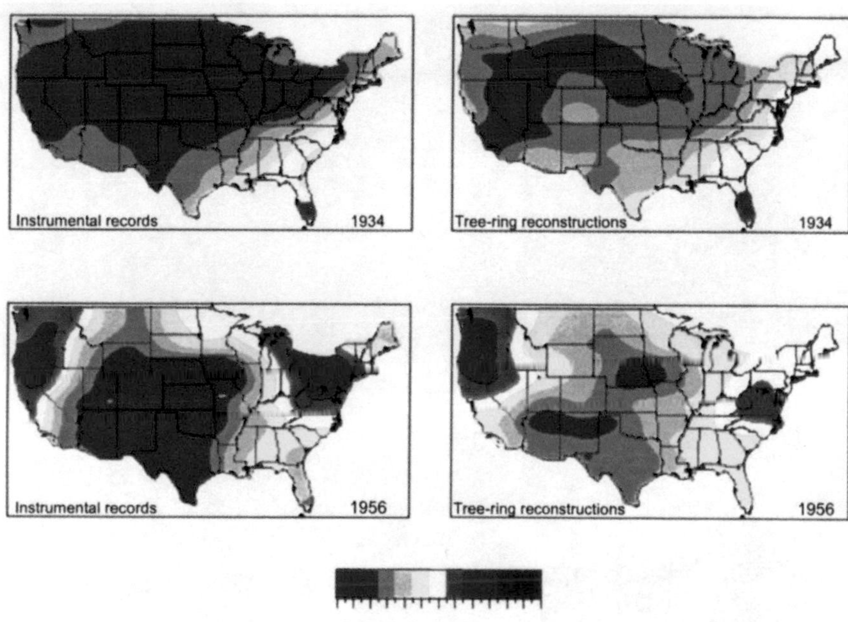

PLATE 9 The two left panels show patterns of drought from observed climate records for 1934 and 1956. The two right panels show tree-ring reconstructions of drought for the same two years. From http://www.ngdc.noaa.gov/paleo/drought.html.

Reconstructed Drought, selected years

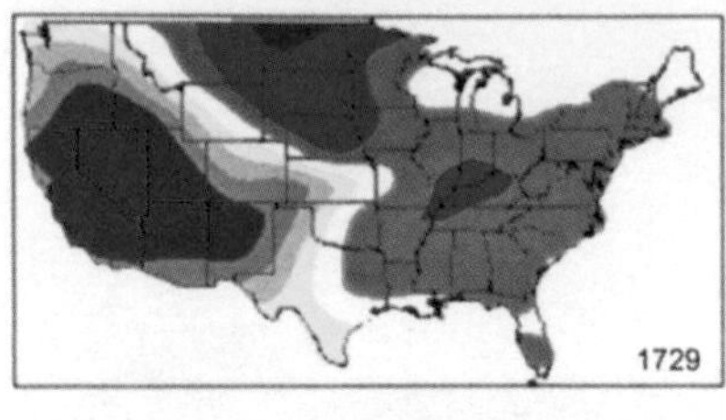

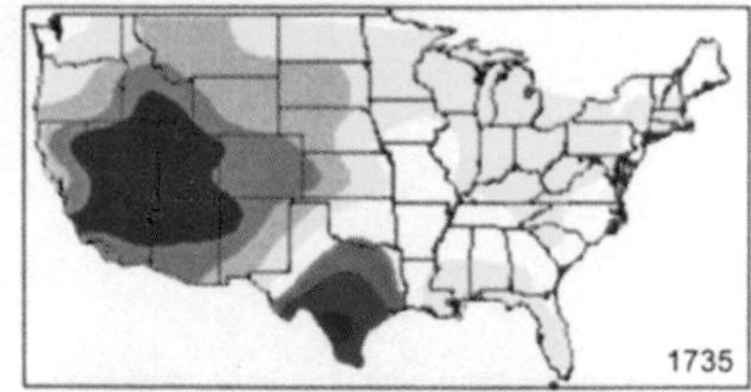

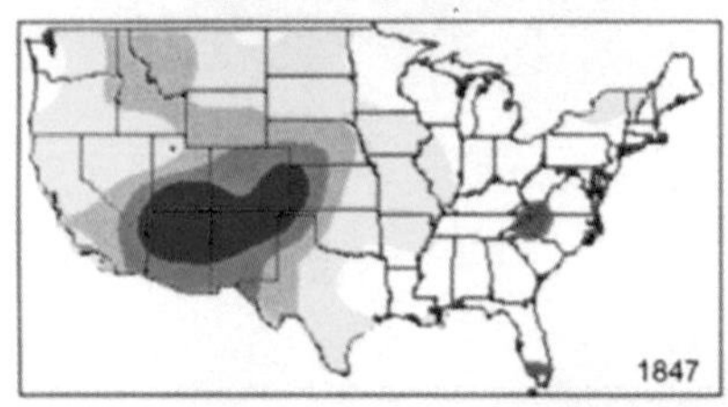

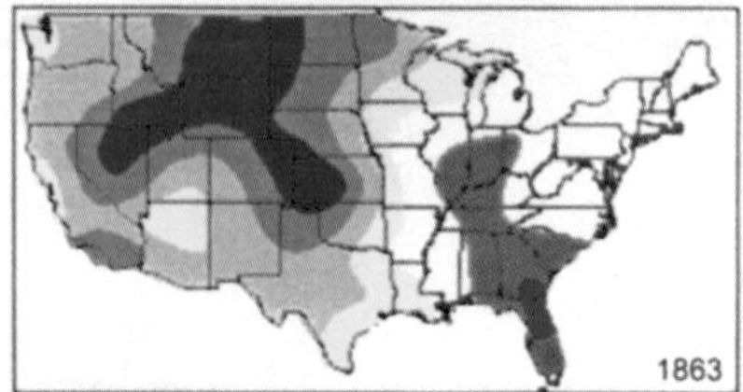

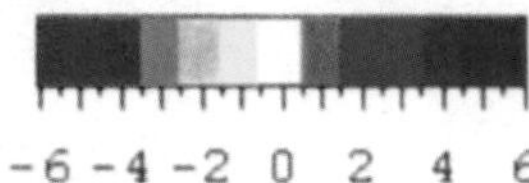

PLATE 10 Patterns of drought reconstructed from tree rings for the four worst drought years in droughts that lasted two to six years. From http://www.ngdc.noaa.gov/paleo/drought.html.

PLATES 11a and 11b (ON PAGES 11–12 OF COLOR SECTION) Hovmuller diagrams of hydrologic forecast simulations for the six variables illustrated in Figure 5.2. The left column shows the root mean square error (RMSE) for the forecasts based on climatology, and the middle column illustrates the RMSE for the simulation using forecasts from the National Center for Environmental Prediction's (NCEP) atmospheric model. The right column shows the improvement in forecast skill over climatology (middle column subtracted from left column). In the first two columns, blue shows lower errors and red shows higher errors; in the last column, blue shows NCEP more accurate than climatology, and red is the reverse.

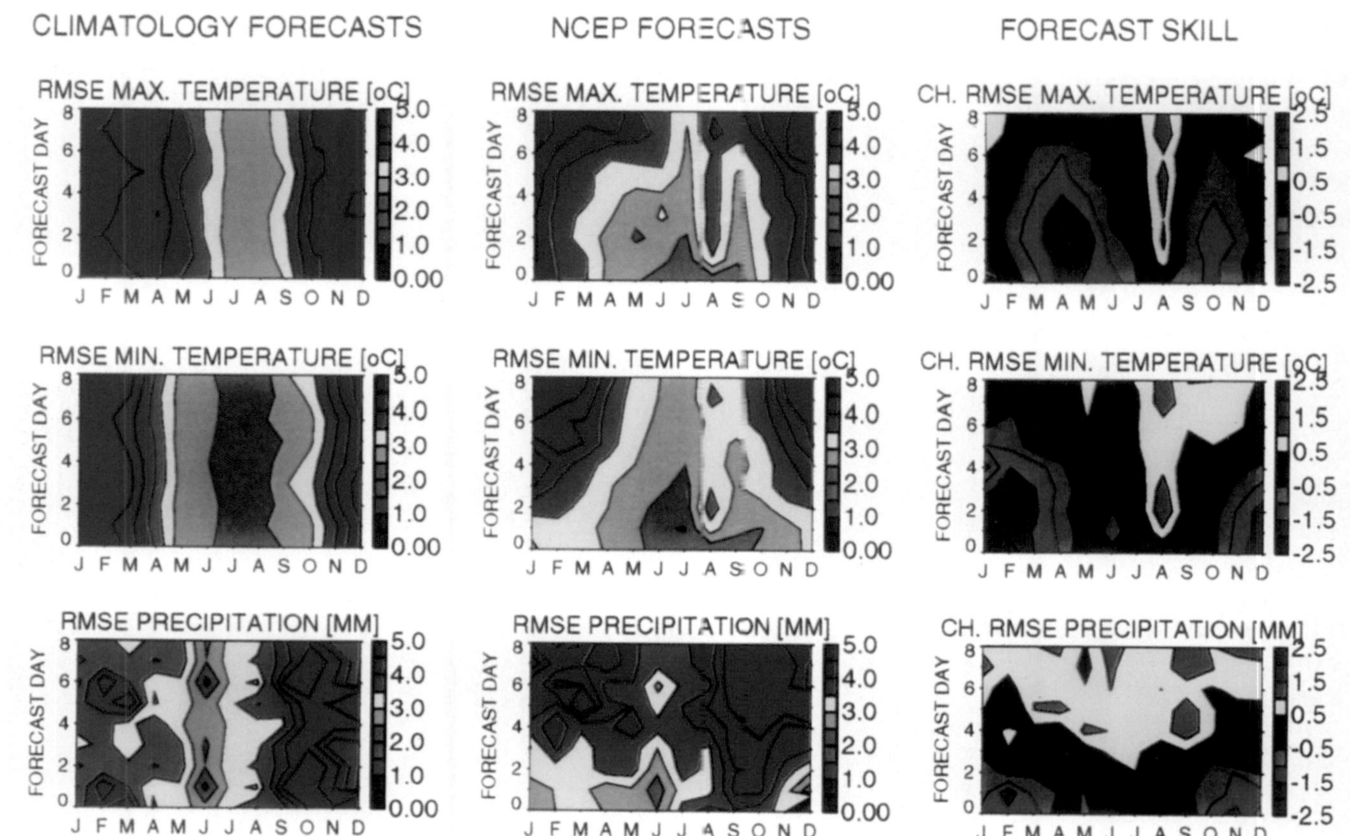
CLIMATOLOGY FORECASTS
NCEP FORECASTS
FORECAST SKILL
RMSE MAX. TEMPERATURE [oC]
RMSE MIN. TEMPERATURE [oC]
RMSE PRECIPITATION [MM]
CH. RMSE MAX. TEMPERATURE [oC]
CH. RMSE MIN. TEMPERATURE [oC]
CH. RMSE PRECIPITATION [MM]
FORECAST DAY
J F M A M J J A S O N D

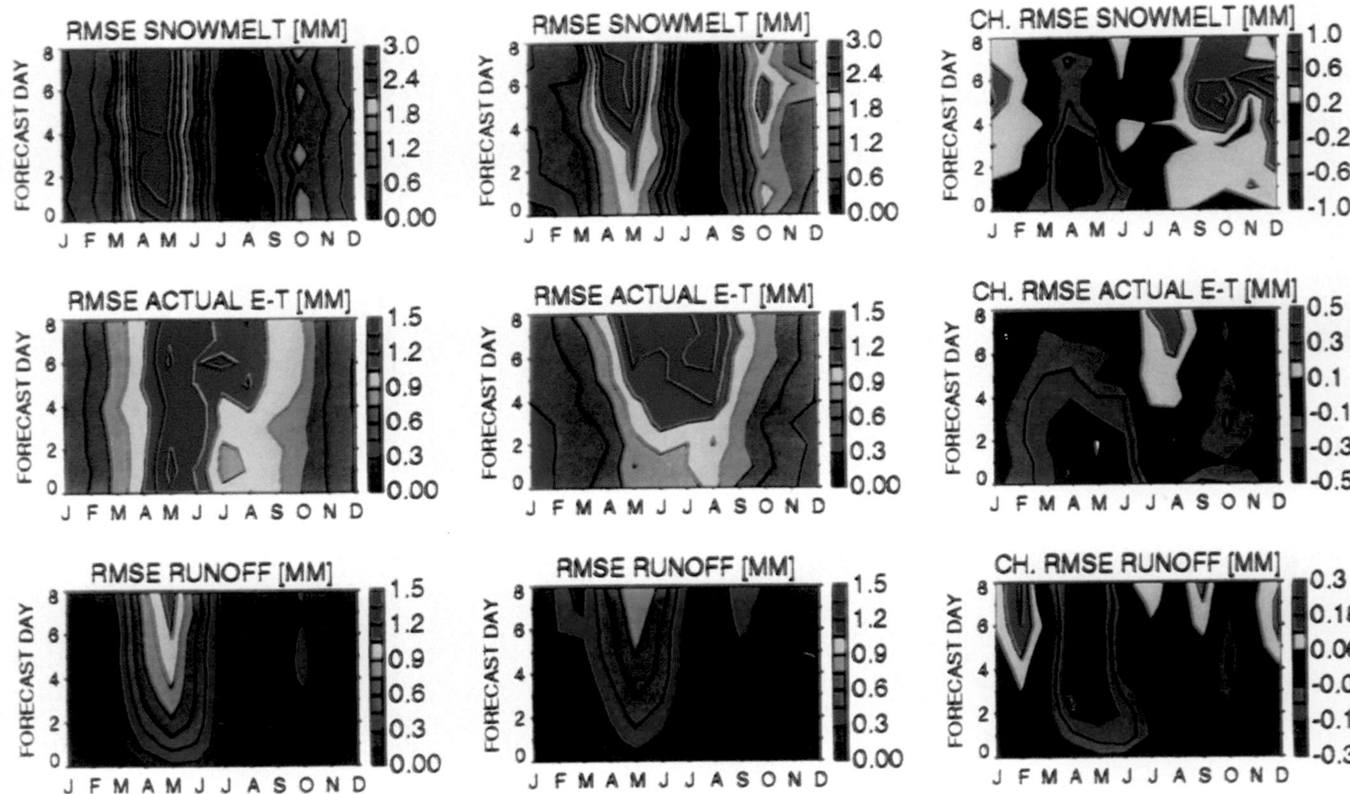
RMSE SNOWMELT [MM]
RMSE SNOWMELT [MM]
CH. RMSE SNOWMELT [MM]
RMSE ACTUAL E-T [MM]
RMSE ACTUAL E-T [MM]
CH. RMSE ACTUAL E-T [MM]
RMSE RUNOFF [MM]
RMSE RUNOFF [MM]
CH. RMSE RUNOFF [MM]
FORECAST DAY
J F M A M J J A S O N D
3.0 2.4 1.8 1.2 0.6 0.00
1.5 1.2 0.9 0.6 0.3 0.00
1.0 0.6 0.2 -0.2 -0.6 -1.0
0.5 0.3 0.1 -0.1 -0.3 -0.5
0.3 0.18 0.06 -0.06 -0.18 -0.3

Management of Colorado River Resources

TEN

TERRANCE J. FULP

The Colorado River basin drains over 240,000 square miles (647,500 square kilometers) of the western United States. Its main stem originates in Colorado just west of Boulder; its four major tributaries are the Green, Gunnison, San Juan, and Gila Rivers. The basin covers portions of seven states (Wyoming, Colorado, Utah, New Mexico, Arizona, California, and Nevada) and drains into the Gulf of California in Mexico.

The Colorado River is the major source of water for consumptive use (about 12.5 million acre feet per year, or 1.54×10^{10} m^3 per year) in the seven states. Ten major reservoirs lie within the basin, with a combined storage capacity exceeding 60 million acre feet (7.40×10^{10} m^3). The average annual natural inflow at Lees Ferry, Arizona, is about 15 million acre feet (1.89×10^{10} m^3) and has ranged from 3 million acre feet (0.67×10^{10} m^3) to 24.5 million acre feet (3.08×10^{10} m^3) since 1906 (Figure 10.1). Natural flow is defined as the observed flow corrected for upstream consumptive use and reservoir regulation. The total storage capacity is over four times the average annual natural flow at Lees Ferry.

As in most watersheds, the primary management objectives on the Colorado River often conflict with each other and are incommensurable; no single

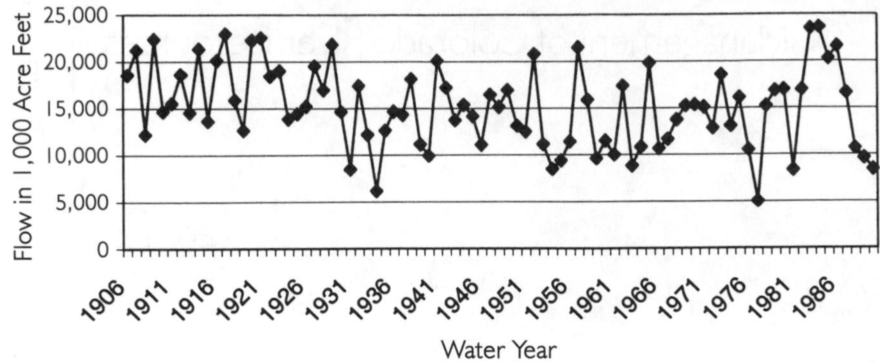

10.1 Natural flows at Lees Ferry, Arizona.

value system can be used to express and compare the objectives. Although much work has been done on multiobjective optimization and decision analysis—particularly with regard to generating a set of nondominated solutions, or Pareto optimality (Koopmans 1951)—it remains difficult to apply strictly mathematical, multiobjective optimization methods to the management of Colorado River resources.

The goal of Colorado River management is to find an equitable balance of objectives. Equitable management treats each objective and stakeholder fairly and creates a plan that is acceptable to all stakeholders (Goicoechea et al. 1982). To make equitable decisions, management must consider legal and political constraints, use sound technical information, and involve stakeholders in the decision-making process from start to finish.

FACTORS THAT AFFECT MANAGEMENT

Legal and political constraints may include any of the following: (1) legislation passed by local, state, or federal governments; (2) agreements (usually in the form of compacts) between states and local governments; (3) contracts between the U.S. Bureau of Reclamation (USBR) and water and power customers; (4) international treaties; (5) court decisions on the local, state, and federal levels; and (6) administrative decisions at the local, state, and federal levels. The Colorado River is subject to all of these kinds of constraints. When taken together, they form the "Law of the River" (Nathanson 1980). Examples include (1) the Colorado River Compact of 1922; (2) the Boulder Canyon Project Act of 1928; (3) the Seven Party Water Agreement; (4) water and power contracts of 1930 and later; (5) the Mexico Water Treaty of 1945; (6) the Upper Colorado River Basin Compact of 1948; (7) the Colorado River Storage Project Act of 1956; (8) the Supreme Court Decree, *Ari-*

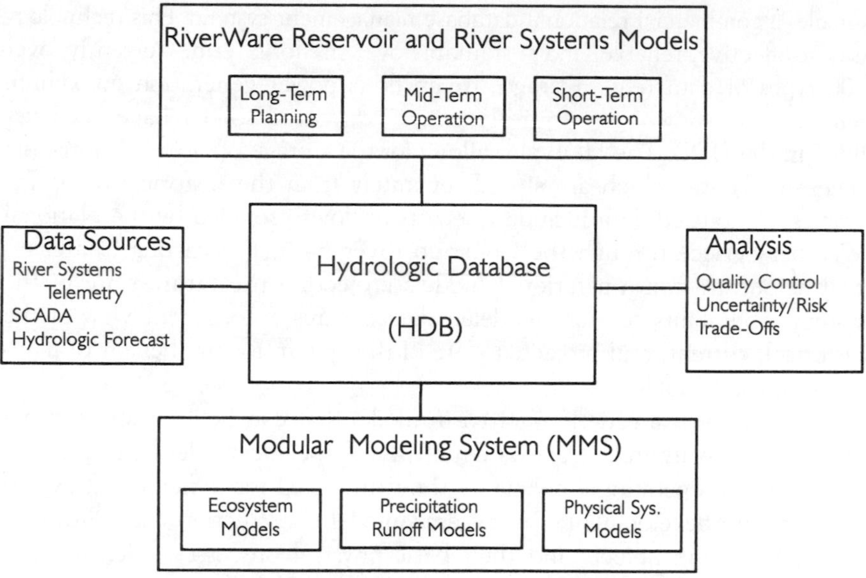

10.2 Decision support system for the Colorado River.

zona v. California, of 1964; (9) the Colorado River Basin Project Act of 1968; (10) the National Environmental Policy Act of 1970; (11) the Endangered Species Act of 1973; (12) the Colorado River Basin Salinity Control Act of 1974; and (13) the Grand Canyon Protection Act of 1992.

Any management plan for the Colorado River must lie within the legal framework but also must be viable from the political perspective—that is, people must consent to use the plan (USBR 1996a). One way to ensure this is through strong involvement of stakeholders, as discussed later.

Sound technical information is the second component of a successful management plan. For the Colorado River the USBR, through a joint research program with the U.S. Geological Survey, has created the Watershed and River Systems Management Program, a generic decision support system (DSS). The DSS, although still being developed, has been implemented on the Colorado River. Its major components include the Hydrologic Database (HDB), the RiverWare reservoir and river-system model, the Modular Modeling System (MMS), and various data collection, query, display, and reporting functions (Figure 10.2).

The HDB is designed for the storage and efficient retrieval of hydrologic time-series data, physical attributes, statistical information, and other types of data pertinent to water management. It has been implemented by use of

Oracle, a commercial relational database management system. This technology is cost-effective, reliable, and sustainable over the long term. Currently, over 100 types of data (e.g., average stream flow, power generation, maximum temperature, snow-water equivalent, total dissolved solids) have been defined in the HDB. The HDB also allows for the storage of modeled (forecast) time-series data, which are stored separately from the historical data. Examples of modeled data include forecasts of flows provided by the National Weather Service through the Colorado River Basin Forecasting Center, as well as output from operational models projecting the future state of the system. The ability to store modeled data ensures a consistent view of the historical, current, and projected state of the system for the benefit of managers and stakeholders.

RiverWare is a generic river-basin model that can be used for both operations and planning. It allows detailed, site-specific modeling of physical processes and operational policies without the need for difficult and expensive software development. RiverWare's model construction capabilities include river-basin objects and their basic physical processes (such as mass balance), as well as libraries of specific algorithms for other processes such as river routing, evaporation, and hydropower generation. Time-step size ranges from one hour to one year, and there is no limit on the time range of a model run (Zagona et al. 2001). RiverWare has three solvers: pure simulation, which is driven by user inputs; rule-based simulation, which is driven by prioritized operating policies as provided by the user; and linear, preemptive goal programming optimization, which is driven by a set of user-supplied, prioritized goals. The policies for the second and third solvers are created in languages specific to RiverWare by use of graphical editors. The policies are saved as data and can be modified without recompiling the software.

The MMS is an integrated model that can be used to simulate a variety of processes involving water, energy, and biogeochemistry. The primary use of MMS is in the development and analysis of physically based Precipitation Runoff Modeling Systems (PRMS) for individual watersheds. The PRMS component of the MMS provides geographic information system capability that can be used in describing and analyzing the spatial distribution of the hydrologic variables needed for the prediction of runoff. Various physical processes are available, and users can develop their own modules as needed (Leavesley et al. 1996). To date, PRMS models have been developed for the Gunnison (Ryan 1996) and San Juan basins and are being tested.

Additional applications are being integrated into the DSS as needed. For example, the USBR has used software from the Environmental Systems Research Institute to develop a map-based query capability for the HDB (Ryan 1996). Other software of various kinds provides additional functionality with low development and maintenance costs. The USBR has used one such tool

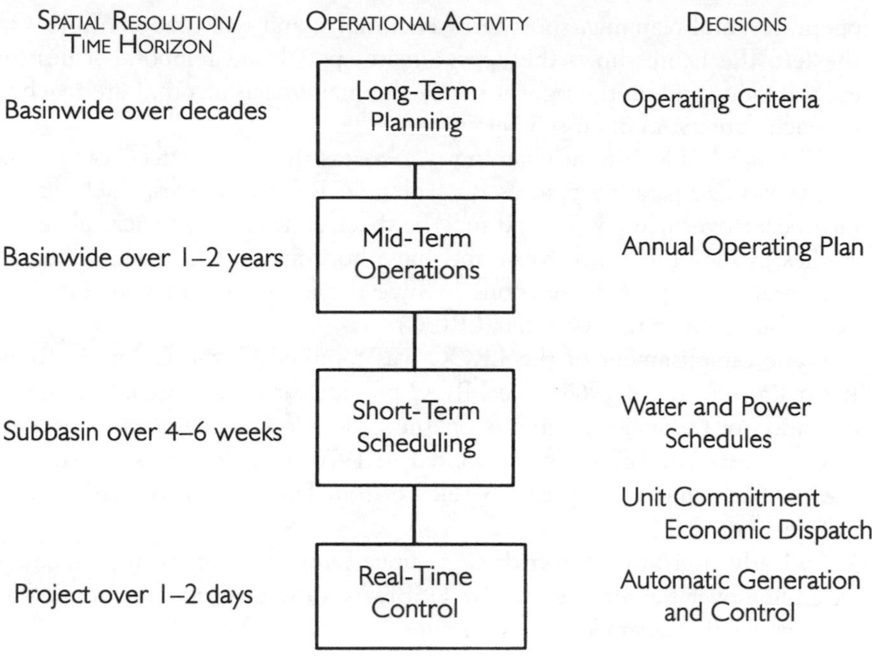

10.3 Decision hierarchy used on the Colorado River

(a query and reporting package from Brio Technologies) to develop a partially customized, ad hoc query and update capability.

Involvement of stakeholders is, from the viewpoint of the USBR (1996b), a means of encouraging and incorporating public input into the agency's decision making. Such a process ensures that all stakeholders are consulted, that the process is open and fair, and that communication is clear, complete, and consistent.

For the Colorado River the USBR employs a variety of methods for involving stakeholders: (1) periodic meetings of formal groups of stakeholders, (2) formal notification and comments (e.g., through the Federal Register), (3) informal stakeholder meetings, (4) accurate and timely dissemination of technical information, and (5) education and training. Involvement of stakeholders is becoming a part of USBR's daily business routine.

APPLICATION AND MANAGEMENT PRINCIPLES FOR THE COLORADO RIVER

On the Colorado River the USBR uses a hierarchical decision process (Figure 10.3) that to date has produced the desired balance of objectives. The major activities are depicted in the center (long-term planning, midterm operations and planning, short-term operations, and real-time control). On the left, the figure shows the approximate spatial and temporal scales for each activity, and on the right it shows the major decisions that are reached for each activity, as discussed below.

The goal of long-term planning is to assess the future effects of current and proposed operating policies on Colorado River resources. Such effects may occur over many years and may be diverse. Therefore typical planning studies include the entire basin and have horizons on the order of fifty to sixty years. The primary decisions involve the establishment and review of Long-Range Operating Criteria (LROC).

The establishment of the LROC was required by the Colorado River Basin Project Act of 1968. The LROC provide for coordinated, long-range operation of the major reservoirs on the Colorado consistent with the Law of the River. The LROC, first adopted in 1970, include provisions such as the establishment of a minimum release from Lake Powell of 8.23 million acre feet per year.

Clearly, a great deal of hydrologic uncertainty is inherent in predictions extending over the long term. The USBR uses the historical record (1906–current) for the Colorado River to estimate various future hydrologic possibilities and then models multiple scenarios that lead to estimates of uncertainty for the variables of interest. Such modeling provides technical information for decision makers.

The Colorado River Basin Project Act also requires that the LROC be established with the involvement of the seven basin states and that a formal review occur at least every five years. Since 1990, USBR has involved stakeholders in these reviews through a formal process.

The goal of midterm planning and operations is to ensure that forecast reservoir operations are consistent with the LROC and that the relatively large uncertainties in hydrologic forecasts are recognized appropriately on a year-to-year basis. Typical midterm studies have horizons on the order of one to two years and span the entire basin. The primary decision is the establishment and update of the Annual Operating Plan (AOP).

The AOP process was established as part of the LROC in 1970. It includes the publication of an annual report describing the actual operation of the previous year and the projected plan of operation for the current year. As stated in the LROC, "The projected plan of operation shall include rules and quantities as may be necessary and consistent with the criteria contained herein, and shall reflect appropriate consideration of the uses of the reservoirs for all purposes" (USBR 1970: 1). In this way the USBR ensures that the operation is consistent with legal and political constraints.

Because reservoir storage and hydrologic forecasts vary greatly from year to year, the USBR uses a monthly operations model to plan the operation for the coming year. At the beginning of the water year (October 1), three different hydrologic forecasts bracket operations for the AOP (10th, 50th, and 90th percentiles of historical flow to Lake Powell). The plan is then updated each month throughout the year as the hydrologic forecast changes and actual inflows are recorded.

In development of the AOP, involvement of stakeholders occurs through the Colorado River Management Work Group—a formal organization made up of representatives of the seven basin states, other federal agencies, various environmental groups, and other interested parties and agencies. During the operating year, monthly reports are disseminated to the group.

The goal of short-term operations is to ensure that the forecast reservoir operations set by the AOP include due consideration of uncertainty in the inflow and water and power demand forecasts on a short-term basis. Typical short-term studies have horizons of four to eight weeks and apply to particular subbasins (e.g., the lower Colorado). The primary decisions involve the establishment and update of operating schedules on a daily time step for each watershed within the basin.

If the USBR meets the monthly targets established by the AOP (e.g., monthly target elevations and releases from reservoirs), operations are consistent with the LROC and also allow flexibility within the month to meet changing water and power demands. The USBR uses models of daily operations to assess the trade-offs between various short-term demands within the monthly targets. At the beginning of each day, observations are loaded into the daily model along with the operations projected on the previous day. Changes in demand schedules and new projections are then made. In the Upper basin, the USBR is currently testing PRMS models that produce a forecast of the inflow on a daily basis.

Involvement of stakeholders in short-term operations occurs on an ad hoc basis. Typically, stakeholders (particularly water users) communicate with the USBR each day with regard to their changing schedules. Meetings occur as needed to discuss special operations such as reduced flows for plant maintenance. Daily reports also are disseminated, primarily by electronic means.

The goal of real-time scheduling and control is to ensure that daily schedules are met, even when demands are changing rapidly. Typical schedules are produced for the current and the next day every hour. The primary decisions involve the establishment and update of operating schedules on an hourly time step for each facility within the basin. Consistency with the hierarchy of decision criteria ensures operation within legal and political constraints. Some additional regulations are imposed at this level, however, with regard to power-plant operations and are beyond the focus of this chapter.

Involvement of stakeholders in real-time operations is patterned after short-term operations in that many stakeholders communicate with USBR each day and even each hour with regard to their changing schedules. At Hoover Dam, signals are received for demands at four-second intervals. Periodic meetings are also held concerning power-plant performance as well as costs of operation and maintenance, which determine the cost of power.

CONCLUSIONS AND FUTURE CONSIDERATIONS

Although management objectives for water systems in the West often compete and conflict with one another, we can achieve an equitable balance of those objectives by considering legal and political constraints, using sound technical information, and involving stakeholders in the decision-making process from start to finish. The USBR hierarchical decision-making process on the Colorado River has produced such a balance, but maintaining that balance will become more difficult as increased pressures are placed on the system in the future. These pressures are derived from historical overallocation of water to consumptive uses, possibly decreased water supplies caused by climatic changes (e.g., Chapter 6), increased demands for water to meet environmental objectives, and changing public attitudes. Several questions arise from consideration of these pressures.

We can expect that water will be allocated by law to meet environmental objectives in the future. This has already occurred in some basins (e.g., the Truckee River in California/Nevada) where water has been purchased for environmental purposes (DOI 1996). A more rigorous enforcement of the concept of beneficial use also may be applied in basins where water is overallocated (Chapter 14).

Recent advances by agencies within the National Oceanographic and Atmospheric Administration give hope for better forecasts of both climate and weather (Chapter 1). As we increase our understanding and quantification of the uncertainties in forecasts, we will be able to improve water-management decision making. Improved projections of water and power demands also will facilitate better operational projections.

Although the USBR's stakeholder involvement process has been successful, it may require further development, particularly with regard to environmental issues. More formal analysis of the effects of operational changes (through environmental assessments, environmental impact statements, and Endangered Species Act consultations) will be done. Also, the USBR will likely see adaptive-management strategies become more standard practice as the agency attempts to increase its understanding of the physical systems and their response to USBR actions. Management of water resources, particularly in the West, will continue to be a difficult task. The USBR will continue to improve its ability to balance competing objectives.

SUMMARY

Management of water resources, particularly in the western United States, has the primary goal of achieving an equitable balance among competing objectives. On the Colorado River these objectives include, but are not limited to, flood control, consumptive use and recreation, ecosystem maintenance, recovery and preservation of endangered species, and hydropower. One definition of an equitable balance is a plan with which all stakeholders can agree. To create such a plan, the manager must use sound technical information, consider legal and political constraints, and involve stakeholders in the decision-making process from start to finish. On the Colorado River the U.S. Bureau of Reclamation uses a hierarchical decision-making process to accomplish these objectives and has to date produced the desired balance of objectives.

ACKNOWLEDGMENTS

Funding for the research, development, and implementation of the decision support system on the Colorado River was provided by USBR's Research and Technology Transfer program and the Upper and Lower Colorado Regional Offices.

REFERENCES

Department of the Interior (DOI). 1996. *Truckee River Water Quality Settlement Agreement.* DOI, Carson City, NV.

Goicoechea, A., D. Hansen, and L. Duckstein. 1982. *Multi-objective Decision Analysis With Engineering and Business Applications.* Wiley, New York.

Koopmans, T. C. 1951. *Activity Analysis of Production and Allocation.* Cowles Commission for Research in Economics, Monograph 13. Wiley, New York.

Leavesley, G. H., P. J. Restrepo, S. L. Markstrom, M. Dixon, and L. G. Stannard. 1996. *The Modular Modeling System (MMS)—User's Manual.* U.S. Geological Survey Open File Report 96-151, Denver.

Nathanson, M. N. 1980. *Updating the Hoover Dam Documents.* U.S. Department of the Interior, Bureau of Reclamation. U.S. Government Printing Office, Denver.

Ryan, T. P. 1996. *Development and Application of a Physically Based, Distributed Parameter, Rainfall Runoff Model in the Gunnison River Basin.* Global Climate Change Response Program Report. U.S. Bureau of Reclamation, Denver.

U.S. Bureau of Reclamation (USBR). 1970. *Criteria for Coordinated Long-Range Operation of Colorado River Reservoirs Pursuant to the Colorado River Basin Project Act of September 30, 1968 (P.L. 90-537).* Published in M. N. Nathanson. 1980. *Updating the Hoover Dam Documents.* U.S. Department of the Interior, Bureau of Reclamation. U.S. Government Printing Office, Denver, Appendix VII, p. VII-5.

———. 1991. *Hydropower 2002: Reclamation's Energy Initiative.* U.S. Bureau of Reclamation, Denver.

———. 1996a. *How to Get Things Done: Decision Process Guidebook.* U.S. Bureau of Reclamation, Denver Technical Center, Denver.

———. 1996b. *Public Involvement Zone.* U.S. Bureau of Reclamation, Mid-Pacific Region Public Affairs Office, Sacramento, CA.

Zagona, E., T. Fulp, R. Shane, T. Magee, and H. Goranflo. 2001. RiverWare: A generalized tool for complex river basin modeling. *Journal of the American Water Resources Association* 37: 913–929.

Water Development and Management Along the South Platte River of Colorado

ELEVEN

TOM CECH

"Here is a land where life is written in water."

These words of Thomas Hornsby Ferril, Colorado's poet laureate in 1940, are etched in the rotunda of the State Capitol Building in Denver. Colorado's pioneers carved their water legacy in the earth when they developed an intricate web of irrigation ditches across the state. Today, water providers continue that tradition by tapping into nature's riches to provide dependable water supplies for a wide range of uses.

Colorado irrigators have struggled with the state's climate for centuries. Although extensive irrigation did not begin in Colorado until the mid-1800s, ancient cliff dwellers in the arid southwestern part of the state developed irrigation based on small dams made of rocks and brush to harvest water from rainstorms and melting snow. In 1787, Colorado's first irrigation ditch was constructed near Pueblo. It was short-lived, however, and the ditch was abandoned. Other irrigation occurred at Bent's Fort along the Arkansas River in 1832 and at early trading posts such as Fort Lupton (created in 1837) and Fort St. Vrain (1838) along the South Platte River. All were abandoned, however, when the fur trade ended. In 1852 the San Luis People's Ditch was developed in the San Luis Valley; it has the distinction of being the oldest ditch in continuous use in Colorado (Radosevich et al. 1976).

A boom hit Colorado in 1859 in the colors of gold and green. Most pioneers hoped to get rich from gold and silver mining, but others saw that enormous wealth lay within the fertile soil along the river bottoms. David Wall of Indiana dug the first ditch in the South Platte River drainage in 1859 near Golden. He was a former California gold miner who knew miners craved fresh vegetables. Wall irrigated 2 acres of garden vegetables and cleared $2,000 his first year. That fall others followed Wall's lead and built small irrigation ditches along Bear Creek and Boulder Creek. Ditch construction occurred along the St. Vrain Creek near Longmont and the Cache la Poudre River near Windsor in 1860. In 1861 the first diversion was made from the Big Thompson River. Just three years from the start of the gold rush in Colorado, irrigation ditches were diverting water from all principal streams in the upper South Platte River basin.

In the 1860s most ditches were built by individuals or small partnerships. These systems were short, narrow, and primarily confined to river bottoms. Construction techniques were crude, but some farmers were aided by surveyors from the mining camps. Other potential irrigators simply started at a stream bank and plowed a furrow in the earth, moving just ahead of the flowing stream of water with their team of horses and plow.

The Union Colony irrigation ditches were the first large canals built by community effort in Colorado. The first residents of the Colony, later called Greeley, arrived on April 18, 1870. Water from the first irrigation ditch, a river-bottom ditch called the Greeley Number Three, reached town on June 10, 1870. Each member of the Colony paid an initiation fee in exchange for a farm plot, water rights, and a town lot. B. H. Eaton, who became Colorado's governor in 1885, was instrumental in helping Union Colony settlers construct their first ditch (Norris and Lee 1990). General Robert Cameron also helped establish the Union Colony, then moved on to help the new settlers at the Agricultural Colony near Fort Collins.

David Boyd, a University of Michigan graduate, was hired by the Greeley Number Three Ditch Company as its first ditch rider. His job was to ride the ditch daily to allocate water to the various headgates. He wrote (Boyd 1987: 61–62):

> It was ill luck, so far as peace of mind is concerned, but good luck so far as valuable experience goes, to be put in charge of the Number Three when water was let into it on June 10th. It was duty not only to ride the ditch and plug the gopher holes, but also to see that the water got around to the two hundred more or less patches of land in and around town, which, as dry as ash heaps, had been planted and were awaiting the water to germinate the seeds.

Construction techniques left a great deal to be desired on the Union Colony's second irrigation ditch, the Greeley Number Two. A miner named McDonald, who had built ditches for placer mining operations in the mountains, was hired to bring water to the higher ground north of Greeley. No specific construction methods existed for construction of long ditches connecting the foothills to benchlands miles away. Ditch builders of the time thought erosion would quickly deepen a new ditch to the desired depth. McDonald constructed a ditch 10 miles long and 8 feet wide but only 15 inches deep; he expected the flowing water to deepen the ditch by erosion. The fall of the ditch bottom was judged with a carpenter's level. Lack of equipment and experience defeated the project, however. Along the bluffs of the Cache la Poudre River, the bends of the Number Two were too sharp. In some places the water did not erode the ditch bottom as expected, but erosion was excessive on hill slopes. Gardens and shade trees imported from Illinois died of thirst.

During the 1870s, ditch construction was generally financed by cooperative efforts. Most river-bottom ditches had been constructed by individuals, and new systems were needed to lead water to higher benchlands. In some cases fields to be irrigated were over 50 miles from the nearest streams. English investors provided significant capital to develop these long irrigation ditches in northeastern Colorado.

Even though irrigation development in Colorado had taken on an international flavor, ditch construction was still slow, backbreaking work. On the Larimer and Weld Ditch north of Greeley, for example, ten yoke of oxen pulled a double moldboard plow to tear through the tough prairie sod. Men with picks and shovels walked behind the plows to further loosen tough sagebrush roots. These crews were followed by men directing teams of horses or mules that excavated the ditch bed with slip scrapers. The embankment on the low side of the ditch was usually constructed as a wagon road for ditch riders to use as they inspected the ditch on a daily basis.

During the wet years of 1872–1873, water allocation was not a problem between ditches along the Cache la Poudre River, but the summer of 1874 was hot and dry. Water was scarce, and water allocation had not yet been formally adopted in the South Platte River basin. Irrigators generally followed the English common-law doctrine of riparian water rights. Simply stated, the owner of a property could divert unlimited amounts of water from a stream crossing the property. The upstream irrigators at Fort Collins followed this eastern practice and nearly dried up the Poudre River. Tempers flared when Greeley irrigators rode to Fort Collins and saw what was happening to their water. Threats were made to tear out the junior upstream ditches to protect their own senior downstream water rights. Fort Collins irrigators did not agree with this plan, and a meeting was quickly called at the Eaton school.

The meeting was lively, to say the least. General Cameron and B. H. Eaton tried to keep everyone calm. They proposed to appoint some disinterested person for that year to divide water according to the greatest need. The idea was not widely accepted, and the Greeley delegates "hurled defiance in hot and unseemly language" (Boyd 1987: 120).

Fortunately, Eaton and Cameron were able to quiet the mob, but no solution was found. As luck would have it, heavy rains the next few days reduced tensions. Soon after, irrigators along the Cache la Poudre River adopted the principles of prior appropriation ("first in time, first in right"). This system had been used in the goldfields of California to reduce bloodshed in water disputes. Because many irrigators in the South Platte basin were former gold miners, it was natural that they would adopt a similar standard in their disputes over irrigation water. The water rights along the Cache la Poudre River soon were recorded with the state courts. In 1876 Weld County irrigators were instrumental in having the "Colorado Doctrine" written into the state constitution: "The right to divert the unappropriated waters of any natural stream to beneficial uses shall never be denied. Priority of appropriation shall give the better right as between those using water for the same purpose" (Article XVI, Section 6).

The adoption of the Colorado Doctrine created the concept of "first in time, first in right" for the beneficial use of water. This system of prior appropriation is still used today in Colorado and has been adopted by most other western states.

Foreign capital helped construct numerous reservoirs, such as Riverside, Prewitt, and Jumbo, which represented the second generation of water development in Colorado. Ditch construction slowed, however, during the 1890s. Then in 1908, farmers in the Hudson-Keenesburg area northeast of Denver decided to move away from dryland farming to irrigated agriculture. The Henrylyn Irrigation District was formed and authorized the sale of bonds to finance construction of a ditch that would begin north of Denver on the South Platte River and extend to the Prospect Valley. The bonds levied indebtedness of $40 per acre against all lands in the district, and the Bank of France purchased the bonds for approximately $3 million. When the bonds came due in 1918, the franc had dropped to 10 percent of its original value because of a recession in France. That was good news for the farmers in the Henrylyn District; the bonds were bought back from the Bank of France for 10 cents on the dollar.

In later years large irrigation projects were funded by the federal government through the Reclamation Act of 1902, which created a revolving fund through the Bureau of Reclamation to develop additional irrigation in the West. The Colorado–Big Thompson (C-BT) project (1938–1959) is one of the largest Bureau of Reclamation projects in the state. It diverts water from the West Slope near Granby and delivers it through the Alva B. Adams tunnel to the Big Thompson River basin above Estes Park. Approximately 310,000 acre-foot units of irrigation, municipal, and industrial water can be delivered annually to eastern Colorado through the C-BT system (Tyler 1992).

The next wave of development involved groundwater. Initially, wells were dug by hand near streams and were used for domestic supply and livestock. In the South Platte basin the first wells were dug in the early 1900s. Later, after development of mechanical drilling methods, well installation boomed. During the droughts of the 1930s and 1950s, the South Platte River slowed to a trickle and was completely dry in many places. Crops died, and irrigators looked to the vast underground supply of water to save their crops in future droughts.

Thousands of wells were drilled to provide a supplemental water supply for crops. Later, with the invention of the center pivot in Nebraska, uplands too steep to flood irrigate were taken into production. This conjunctive use of surface and groundwater provided a reliable source of irrigation water in the South Platte basin. Use of alluvial groundwater caused tension, however, between ditch and reservoir irrigators and the owners of these wells, which had junior priority. Ditch and reservoir irrigators had developed an elaborate legal and management system for allocating scarce surface-water supplies during dry summer months. Surface irrigators knew they had to be "in priority" before they could divert water from a stream. All of that changed, however, with the installation of wells. Groundwater irrigators could simply push a button to pump alluvial groundwater. Ditch and reservoir owners were convinced that this tributary groundwater, if located near a stream, fed the stream during the summer months. When groundwater was pumped, they knew it had a negative effect on stream flows and on their senior water rights.

Colorado's legislature agreed with this argument. After extensive engineering studies and lobbying by surface-water owners, the Water Rights Determination and Administration Act of 1969 was adopted. The act stated that all tributary irrigation wells had to follow the priority system of water allocation that was developed for ditches and reservoirs. In effect, a groundwater user who drilled a well in 1949, as a result of the act, had a priority date of 1949. That was not good, considering that most ditches had priority dates in the 1870s and reservoirs had decrees dating to the early 1900s. In addition, well users were required to cease use of their wells during the dry months of July and August when the wells were not in priority. Thus the 10,000 irrigation wells in the state proved to be less useful than anticipated.

The problem of inadequate water for wells was solved through augmentation, which involves creation of new water supplies sufficient to allow well users to pump their wells during the summer without injuring downstream senior appropriators (Radosevich et al. 1976). Well owners who pump from alluvial aquifers are required by law either to belong to an augmentation plan such as the Groundwater Appropriators of the South Platte (GASP) in Fort Morgan or the Central Colorado Water Conservancy District (Central) in Greeley or to have an individual augmentation plan. Well owners who are members of Central or GASP pay an annual fee to receive augmentation coverage. This money is used to purchase or rent water that is delivered to, and thus augments, stream flows. In this way downstream senior diverters are protected from injury caused by pumping.

Central has a variety of sources of augmentation water. Most common is the purchase of water rights from a ditch or reservoir company. Central can only claim the historical, consumptive use of water shares from such a company. This means that any water historically returned to the South Platte River through recharge or surface return flow cannot be claimed for augmentation. This protects downstream senior water users that have historically relied on these return flows to the river.

Central also rents reusable municipal effluent from cities along the Front Range and has developed groundwater recharge projects where South Platte River water is diverted during the spring and fall into recharge basins or dry creek beds. This water will return to the South Platte River at a later date and can be claimed as an augmentation credit.

Finally, Central has developed numerous lined gravel pits to hold augmentation water. Mined pits are lined with a bentonite slurry wall that forms an underground curtain around the perimeter of the gravel pit from the land surface down to bedrock. The bentonite curtain prevents water from escaping. Water is delivered into these gravel-pit lakes through adjacent irrigation ditches by pumps or by gravity from the river.

Water education is another aspect of water management that is very important in Colorado. Too many residents are unaware of the importance of water conservation, history, and management. Most do not know that we live in a semiarid climate, that water rights are used to allocate water during times of scarcity, and that water rights can be bought and sold. In addition, growing conflicts among wildlife, recreation, and irrigation continue to increase. Therefore Central has sponsored numerous water-education activities, including a water curriculum for grades preschool through twelve, installation of groundwater monitoring wells at schools in the region, and creation of Children's Water Festivals for fifth graders and their teachers (Central Colorado Water Conservancy District 1995).

Central's Children's Water Festival is held each year at Aims Community College in Greeley; over 1,500 students attend. Activities include classroom sessions on water history, water rights, water quality, water and art, water conservation, and water and music. Students can also learn more about water from displays and presentations in the Exhibit Hall. Children's Water Festivals have been extremely successful in Colorado; sixteen were held in 2000. Over 100,000 students have attended these festivals since 1991.

Dry climate, variable precipitation, and drought have caused Colorado to develop elaborate laws and management plans to protect water rights. Groundwater has been developed extensively in the South Platte River basin to provide supplemental water supplies during water shortages. We must continue to prepare for drought and wisely manage our water resources. If we are successful, the rich water heritage described by Thomas Hornsby Ferril will always be with us.

REFERENCES

Boyd, D. 1987. *A History: Greeley and the Union Colony of Colorado.* Kendall, Greeley, CO.

Central Colorado Water Conservancy District. 1995. *Colorado WaterWise Curriculum.* Greeley, CO.

Norris, J. E., and G. Lee. 1990. *Written in Water—The Life of Benjamin Harrison Eaton.* Swallow and Ohio University Press, Athens.

Radosevich, G. E., D. Allerdice, C. Kirkwood, and K. C. Nobe. 1976. *Evolution and Administration of Colorado Water Law: 1876–1976.* Water Resources Publications, Littleton, CO.

Tyler, D. 1992. *The Last Water Hole in the West.* University Press of Colorado, Niwot.

Can Climate Predictions Be of Practical Use in Western Water Management?

TWELVE

C. Booth Wallentine
Dave Matthews

Water managers are uniquely qualified to evaluate the utility of current and proposed climatic predictions that can be related to availability of and demand for water. A workshop for water managers in Boulder, Colorado, on June 21–22, 1999, provided an unusual opportunity to gather information from water managers about the current and future uses of predictions. The outcome of the workshop for water managers is summarized here through their responses to a series of questions concerning current, anticipated, and ideal means by which climatic predictions and their hydrologic and water-demand correlates can be used. The workshop participants gave a diversity of responses, particularly with regard to the differing needs of various categories of water users, but also showed some common needs and perspectives on the current usefulness of predictions.

WHAT INFORMATION DO WATER MANAGERS NEED?

The most general need of water managers is for an improved understanding of the basis for forecasts as a means of more confidently interpreting them. Key concepts include forecast procedures and synoptic patterns related to forecasts of runoff, precipitation, temperature, wind, surface-energy budgets at timescales from days to years, and verification data showing the quality of performance.

Examples of specific needs include summer demand forecasts of consumptive use by crops, such as those of the Colorado Water Conservancy District (CWCD) in Glenwood Springs. Experience of the CWCD has shown that demand forecasts are sensitive to short-term variations in cloud cover and temperature; water releases for irrigation on a cloudy day, for example, could differ 10 percent from those on a sunny day. Demand forecasts based on evapotranspiration (ET)—a function of precipitation, relative humidity, temperature, and soil moisture—are important. Another important consideration is travel time from release of water at a reservoir to its arrival at the irrigation-diversion points, which varies from hours to days. Travel time determines the forecast lead times that water managers need in daily operations. For example, at some locations on the Colorado River, three-day travel times require a seventy-two-hour forecast.

The Denver Water Department (DWD) noted that for its needs related to domestic supply, forecasts for the period of snowmelt (May to July) are most critical and that probability, verification, and consistency are essential. The DWD's forecasts currently are provided by extended streamflow predictions (ESP) for the Colorado River based on historical hydrographs.

One can weight forecasts by probabilities associated with El Niño–Southern Oscillation (ENSO) or Pacific Decadal Oscillation as determined by historical patterns (Chapter 1). During drought, forecasts of surcharge for reservoir flood-control pools are helpful. ET demand forecasts with 2- to 4-day lead times help managers plan releases and may enhance water conservation. In contrast, flood control requires medium-term (15–30 days) to short-term (1–5 days) forecasts that include the timing and relative magnitude of storms. Confidence levels for longer-lead forecasts help water managers decide if they should use a forecast in planning and daily operations. A balance between user needs for irrigation and risks of flooding, damage to property, and loss of life requires timely decisions based on forecasts and status of reservoir systems.

Accuracy for a 1-day forecast is about 85 percent; for a 30-day forecast it is about 55–60 percent (Tom Potter, personal communication, 1999). Forecast skill varies considerably by location, ENSO signal strength, and season. For example, the skills of forecasts differ seasonally. In October, sea-surface temperature patterns are well defined and have higher persistence. In March, however, systems are in transition; thus uncertainty is high. Managers need to know the accuracy of water-supply predictions. Conditional probability may be useful in this regard. For example, if a forecast shows 50 percent probability of very dry weather, a manager may hedge conservatively. Managers may need consultants to translate the forecasts into useful terms. For example, during the strong La Niña spring of 2000 in Colorado, the DWD on March 1, 1999, forecast very dry conditions—but from the end of March to mid-April moisture was 175 percent of normal.

Coupled atmospheric-hydrologic models at high resolution also are useful in forecasting. These models can provide objective estimates of temporal and spatial distributions of precipitation and runoff. For example, the U.S. Bureau of Reclamation requires quantitative uncertainty estimates along with streamflow and demand forecasts (Chapter 10). Such forecasts become critical in periods of drought and flood.

Irrigation companies may require accurate and timely forecasts of wind, relative humidity, temperature, and precipitation—all of which affect soil moisture. Many ranchers use 5- to 12-day radio weather forecasts. Livestock producers need forecasts, especially during calving and lambing season; temperature, wind, snow, and seasonal water supply are all important.

Decisions must often achieve a balance among conflicting demands (Chapter 9). For example, water during the irrigation season can be used in support of fish, recreation, or crops; and managers must consider environmental needs for minimum stream flows. For example, managers of the upper Colorado River need six-month forecasts as a basis for planning in support of nonconsumptive uses. The forecasts are used in setting conservation-pool storage in the reservoirs for environmental purposes, irrigation, and reserve storage capable of mitigating flood damage. These multiple demands reinforce the need for specialized forecasts with probabilities, quantitative estimates of flow, and confidence intervals.

HOW DO RESOURCE MANAGERS USE INFORMATION?

Needs for information vary among users as follows: water-management agencies—supply and demand forecasting; fire management agencies—regional and local staffing; ski resorts—feasibility of snow making; agriculture—planting, irrigation scheduling, harvesting, wind damage; endangered fishes—seasonally augmented flows; recreation—maintenance of flows for rafting, fishing, and skiing. One basic water-management policy is to fill storage pools without spilling water. Storage leads to reduced flows from the end of June through July, when recreational users may desire high flows. Accurate forecasts of snowmelt and runoff are one means of relaxing storage requirements to suit nonconsumptive uses such as recreation. Today's emphasis is on the paying customer with water rights, but forecasting can improve possibilities for accommodating other kinds of uses.

Long-term watershed management includes maintenance of water quality. For example, feedlot designs are based on historical 25-year records and 24-hour peak runoff forecasts. Forecasters should use relationships between

ENSO and runoff to design wastewater treatment plants. Regulatory agencies and other managers need to be ready to use these types of weather and climate forecasts.

Agricultural credit is based on expectations of profit, which are based in part on consideration of climatic data from the past 25 years. More accurate projections of the effects of climate variability on crops may be beneficial here.

Water-supply planning takes a 50- or 100-year look ahead. Thus climate projections should be considered. One example is Douglas County, Colorado, where water planners have attempted to estimate the extended life of aquifers. Millions of dollars in infrastructural investment may be at risk if such estimates fail to account correctly for climate variability. In most cases climatology is the best source for projections of climate. The possibility of global change in climate must also be considered.

Municipal water users and snow removal teams use long-range planning to estimate staffing and formulate city budgets. Seasonal to interannual budgeting for the costs of roadway applications is critical in the planning process of local communities, hence the value of long-range forecasts.

Flood prevention and management also deal with land-use patterns, including long-range planning of floodplain development and weekly management intended to avoid uncontrolled spills. It is important to identify where long-range climate is expected to generate significantly more precipitation and higher stream flows than in the recent past.

WHEN AND HOW OFTEN IS PLANNING NEEDED?

Water-resource managers often require a fully integrated temporal suite of forecasts that includes daily, weekly, biweekly, monthly, and seasonal forecasts as well as multiyear trends. The water-year forecast (from October 1 to September 30) is especially important. This forecast should be revised monthly or even biweekly under some circumstances. In the West, spring runoff forecasts are often critical. Forecasts of the peak flow and hydrograph from April to July are required, and reforecasts should be revised each day from May through peak runoff.

During summer, demand forecasting of water use is needed with 5- to 10-day lead times for agriculture. Forecasts of demand are valuable for 2 to 4 days, depending on travel time between reservoirs and irrigation diversions. Demand is a function of soil moisture as well as precipitation, ET, temperature, radiation, relative humidity, and wind.

From September to January, predictions of stream flows for ecosystem management (quantity and quality) are needed. Extended streamflow predictions for the fall must use climate forecasts rather than historical means.

During winter, the water-supply forecast for mountain snowpack is critical for planning reservoir operations and maximizing use of conservation pools while maintaining adequate flood-control space. In addition, recreation (snow for skiing) and safety (winter storm warning) forecasts are important for planning snow-making strategies, preparing for snow removal and sanding of highways, and building soil moisture, which is used in water-supply forecasts.

HOW CAN FORECASTS BEST BE USED?

In the West, spatial variation may occur at the scale of a valley, a tributary watershed, a river basin, or an aquifer. Detailed forecasts should reflect these differences, as well as differences related to elevation, exposure, and other factors that affect weather (Chapter 2).

Agricultural applications include the long-range forecast of precipitation, which would improve decisions regarding the type of crop to plant and time of planting. Also, forecasts of wind, moisture, and temperature can be used in determining plant spacing. Thus information about the relative strength of winds for a given growing season would be useful for crop management and for estimating ET losses.

WHAT CLIMATE FORECASTS WILL MEET THE NEEDS OF WATER MANAGERS?

Long-term (5- to 10-year) planning of the transition from agricultural to urban use of water requires forecasts of water supply and demand. In the West, snowpack forecasts are the most critical for judging seasonal water supplies. The Southwest monsoon moisture also is critical in Arizona, New Mexico, southern California, and Nevada from July through September. Most areas require accurate snowpack forecasts from April through May and runoff forecasts from May through July. The forecasts should include precipitation, temperature, wind, and solar irradiance, which control the amount and timing of the melt and ablation of snowpack and peak runoff. From June through September, water-demand forecasts of snowpack become important. These require estimates of ET as estimated from temperature, humidity, precipitation, soil moisture, wind, and solar irradiance. Energy demand is critical during the summer and can be forecast from temperature, cloud cover, insolation, and humidity.

ARE CURRENT CLIMATE FORECASTS ACCURATE ENOUGH?

Even though significant progress has been made, forecasts still lack sufficient accuracy to be useful for most purposes. Further progress might come from informed collaboration between providers and users. Education of managers and the public regarding the physical basis of forecasting is needed. Presently, water managers do not use long-range forecasts because they have little confidence in them. Forecasters must build trust by providing users with a better understanding of forecast strengths and weaknesses. Some forecast periods with strong global forcing such as El Niño and La Niña are well suited to prediction, as evidenced in 1998, whereas variation with other conditions is less predictable.

The relative accuracy of forecasting and the confidence of forecasters in their products need to be conveyed to users of forecasts. For example, National Weather Service (NWS) forecasters use a skill index to indicate confidence, but engineering users prefer more conventional confidence intervals. Probability percentiles need to be more precise and should provide higher spatial and temporal resolution. For planning of major water-management facilities, forecasts may be more useful than historical data, but the comparative value of these two approaches is still unclear.

DO USERS UNDERSTAND THE UNCERTAINTY IN FORECASTS?

Improvements in understanding uncertainty would include: (1) better information on uncertainty (probabilities) presented explicitly with the forecasts; (2) description of sources of variation (e.g., ENSO) in uncertainty; (3) workshops and communications that describe the basis of forecasts, probabilities, and relative confidence in the forecast and enhance interpretation for specific users; and (4) extension of interpretation for climate forecasts. An example of the role of private-sector forecast support comes from the state of Colorado, which contracts with a private forecaster capable of providing customized services.

DO WATER MANAGERS CURRENTLY TAKE ACCOUNT OF UNCERTAINTY?

Formal operating criteria are being tested for the American River of California by use of ensemble forecasts to drive the National Weather Service River Forecast System in collaboration with the River Forecast Center, the Hydrologic Research Center, and the U.S. Bureau of Reclamation's (USBR) Dam Safety Program. This plan will incorporate uncertainty and risk-based information designed to avoid flood-control releases and thus will affect operations of Folsom Dam in the Central Valley Project. Only the largest water-management organizations can afford to employ this type of risk management. Examples of these agencies include the DWD, the USBR, and the California Department of Water Resources.

HOW WILL WATER MANAGERS USE FORECASTS WHEN UNCERTAINTY IS INCLUDED IN PREDICTIONS?

Most water managers use some form of risk management in their daily operations. For example, managers of the Colorado River use probability forecasts for 10 percent, 50 percent, and 90 percent exceedance based on climate records, and irrigators use probabilities of 60 percent for drought and 40 percent for normal precipitation when planning planting strategies. Even so, wider understanding of uncertainty is needed.

An example of probable future efforts to use uncertainty estimates is the RiverWare modeling framework for the Colorado, Rio Grande, and Yakima basins. Temperature-control devices at Flaming Gorge and forecasts of temperature and rate of heating are being tested as a means of managing temperatures at the confluence of the Green and Yampa Rivers, where a critical habitat for endangered species exists. These tests will use data from medium-range forecasting model ensemble forecasts for 5- to 10-day periods.

The NWS River Forecast System Advanced Hydrologic Prediction System will use information from nested global climate models to provide regional-scale weather forecasts that may enhance streamflow forecasts over the next decade. The ensemble global climate forecasts will likely provide probability information.

WHAT IS THE IDEAL CLIMATE-FORECAST INFORMATION FOR WATER MANAGERS?

The goal over the next decade should be to improve the mean accuracy of major streamflow forecasts to 90–95 percent and to provide probabilities or confidence limits for 1-, 3-, and 6-month forecasts. Also needed is forecasting of daily flows for the current water year; forecasts should be updated bimonthly at critical points. Streamflow forecasts should give explicit predictions of snowpack water content that incorporate surface-energy budgets and precipitation.

Many meteorologists do not believe that accuracies of 90–95 percent can be achieved. The addition of probabilities or confidence intervals to forecasts of stream flow may, however, lead to the successful use and practical application of extended-range forecasts.

Water-demand forecasts that incorporate evapotranspiration, hydropower, and consumptive use are needed with daily to seasonal lead times. One specific goal would be to improve accuracy of forecasts for temperature, wind, precipitation, evapotranspiration, and cloud cover by 15 percent over the next decade.

CONCLUSION

Climatological forecasting can be of practical use in western water management, but only through a partnership among water operations engineers, managers, forecast providers, and the research community. The partnership should build trust through understanding.

ACKNOWLEDGMENTS

The workshop leaders wish to acknowledge the contributions of workshop participants David Brandon, Paul Gleason, Chris Goemans, Kenneth Kolm, John Leese, Gregory McCabe, Roger Pielke Jr., Thomas Potter, Anton Seimon, Cathy Tate, Marc Waage, John Wiener, and Robert Wilby.

PART FOUR

Perspectives on Society, Institutions, and Water

The framework within which human adaptation to climate variability must occur in the West consists most visibly of water-management institutions, as described in Part 3. A less visible but no less important part of the framework consists of institutions and societal forces that are not directly connected to water management but nevertheless influence it powerfully. These include demographic trends, land-use management, water and environmental law, economic and commercial systems for exchange and distribution of water, and a network of organizations that either make or influence policy in ways that will affect distribution and use of water. The four chapters in this section deal with these influences on society's ultimate adaptability to climate variability as seen through variability in water supply.

It is often said that the West is becoming more populous and more urban. Implications of these trends are not always obvious, however. Certainly, waters pass from agricultural to municipal use as population increases. The rate at which these changes are occurring and their ultimate consequences for water distribution are not clear and require careful analysis. Travis, in the opening chapter of this section, describes trends in demography and land use and explains their probable consequences for use and distribution of water.

The chapter by Getches is a masterful and scholarly treatment of the control of water through legal and paralegal mechanisms. His thesis, which is startling yet obvious as he explains it, is that the western states have scarcely used the legal control over water that has been available to them since the earliest phases of water development. The states' reticence has left the federal government with a much larger role in controlling water than was originally anticipated. Environmental legislation at the federal level has added to the influence of the federal government over water at the state level, but Getches shows how the federal government has steered away from command and control and toward collaboration, thus helping watershed groups and informal associations fill the policy gap left open by states' passivity in matters of water policy. He is critical of unmodified prior appropriation principles in the modern setting on grounds that this system is too rigid and thus will not allow the kinds of adaptation that would be desirable in response to climate variability, for example. He shows a diversity of adaptations by individual states to modern conditions, including water scarcity, environmental regulations, and demands for greater public participation. He argues that true structural adaptation will occur only in response to crisis but that states and water-based interests would be wise to plan for opportunities that will come inevitably with the next water crisis. This chapter is thoughtful, provocative, and an excellent reference as well.

Loomis, Koteen, and Hurd deal with economic and institutional strategies, thus complementing the work by Getches. They describe trends in water use and show how the anticipated costs and benefits to agriculture, municipal supply, and other uses would change in response to changes in mean precipitation or mean temperature. They use California extensively, as does Getches, in showing the effectiveness of various experimental strategies for improving the efficiency of water use and introducing flexibility presently absent in many western states. The blockage of even temporary water transfers between uses—an obvious cause of inflexibility—is well illustrated by these authors, and remedies are proposed.

Miller and Gloss, reporting the results of a working group, analyze current capacities of water-management institutions and society at large to respond to variability in water supply. They emphasize the role of planning but also show its complexity as viewed through an extensive and diffuse hierarchy that in aggregate decides the use and distribution of water.

THIRTEEN

A Changing Geography: Growth, Land Use, and Water in the Interior West

William R. Travis

The Interior West stands out in any geographic analysis of U.S. regions for a number of reasons. Physically, the region encompasses the most mountainous terrain in the contiguous United States, from the broad swath of the Rocky Mountains to the multiple ranges of the Basin and Range province. The landscape is also politically atypical because roughly half of the land is in federal ownership. Socially, portions of the West are among the fastest-growing regions of the country, especially in the last decade. The eleven western states grew by 10.2 million people in the 1990s, or 20 percent, compared with a national rate of growth of 13.2 percent. This growth during the 1990s continues a historical trend that put the West ahead of national growth rates for four consecutive decades. Five western states—Nevada, Arizona, Utah, Colorado, and Idaho—were the fastest growing in the United States. These five states grew from 10.8 million to 14.9 million residents (4.1 million additional people, or a 37 percent increase—almost three times the national rate) during the decade 1990–2000.

Demographic growth in the West has been accompanied by a regional economic transition in which the vast majority of new jobs and income are derived from the services and professional sectors, not from resource extraction, agriculture, or manufacturing (Power 1996). In this way the West tracks the rest of the national economy, although its transition to services industries is especially notable in a region with roots in resource extraction.

Regional shifts in demography and economic development interact with natural resources, including water, in a variety of ways. In particular, resources are redefined in the new economy; water is shifted from agricultural to urban uses but also becomes a critical element of species conservation. Open space becomes more valuable in a rapidly developing region, raising an obvious tension: the forces causing growth and increasing the value of open lands are simultaneously obliterating open space. This assessment briefly examines such regional demographic and economic shifts and their effects on resources, especially as related to changing land-use patterns and their interaction with western water resources.

DEVELOPMENT AND LAND USE

The West has long been viewed as a rural region, stronghold of the miner, the cowboy, and the oil roughneck. Yet most westerners—82 percent in 2000—live in metropolitan areas (the national average is 78 percent). This has long been true; even during the early gold rushes and in the days of the big cattle drives, cities dominated the economic and social landscape of the West—a fact rarely portrayed in the region's much-told agrarian creation myth but well documented by historians (White 1991; Abbott 1993).

The concentration of westerners in cities means that the region is actually less urbanized in terms of land dedicated to urban uses. Although we lack consistent data on land use, land use can be estimated. Of the 760 million acres in the eleven western states, less than half, or about 325 million acres, could feasibly be developed; the rest is mostly public land. Even this buildable half includes lands with significant limitations, including large tribal holdings and physically undesirable areas such as playas. But how much land is actually developed? This can be assessed either indirectly through demographic data or directly through detailed land-use measurements.

Simply calculating the housing density in census blocks provides one measure of development. By this approach, only around 4.4 million acres of the 325 million acres of potentially developed area were built up in 1990 at urban and suburban densities (i.e., more than two housing units per acre). Another 12 million were developed in a low-density suburban pattern (that is, from 0.5 to 10 acres per house). Another 22 million acres were developed at exurban densities (one unit per 10 to 40 acres), and the remainder can be classified as rural, with less than one house per 40 acres. Together, development of at least suburban density covered 5.6 percent of buildable land in the eleven western states, whereas development of at least exurban density accounted for roughly 10 percent (38.4 million acres) of that land.

In an attempt to measure land use directly, the U.S. Department of Agriculture has tried to measure the status of all private lands every five years through the National Resources Inventory. A combination of on-the-ground point measurements and aerial photo assessment indicated that about 4.2 percent of nonfederal land in the eleven western states was "built up" (USDA 2000 <http://soils.usda.gov/soil_survey/main.htm>).

By several measures, developed area (densities above rural) has grown faster than population in the West for several decades (Riebsame 1997). This is a sign of increasing sprawl as residents progressively lower density settings. Development occurs mostly on agricultural land. For example, urban, suburban, and exurban development in places such as the Colorado Front Range and Utah's Wasatch Front is taken entirely from agricultural land because any private land in this area not developed or taken by some other intensive use (e.g., mines, landfills) is used for agriculture or at least is counted as agricultural land for census and tax purposes. Conversion of land in this way has evoked an agricultural land-protection movement (American Farmland Trust 1997). Yet even during the 1990s, a decade of rapid growth, the actual amount of land converted from agriculture constitutes a very small fraction of the total base.

WATER AND WESTERN DEVELOPMENT

Water guided early western development in many ways. Certainly, its availability shaped early land uses. Prehistoric population concentrations, some of which later became towns and cities, were near water sources (e.g., along major waterways such as the Missouri and Columbia Rivers, at springs such as those at Flagstaff, or along the shores of freshwater lakes such as Walker Lake, Nevada). Early Euro-American incursions into the region, such as those along the Santa Fe and Oregon Trails, either followed watercourses or traversed from one water source to another (Beck and Haase 1989: 32–33). And early Euro-American settlements—such as the Spanish colonization of the upper Rio Grande, Mormon settlement in the Salt Lake Valley, gold and silver mining camps in the Rocky Mountains, and the first permanent towns along the Rocky Mountain Front Range—were all on streams or rivers (Beck and Haase 1989; Wyckoff 1999).

Even some settlements that seem hydrologically illogical today had water at their base; for example, Las Vegas began as a stop on the Salt Lake–to–Los Angeles freight trail instituted by Mormon leader Brigham Young. He needed access to the Pacific from the inland State of Deseret, and Las Vegas—about halfway along the trail—offered several springs and ponds. Those springs are inadequate for southern Nevada water demands today, but they anchored the primal Vegas strip.

Lack of water left its mark on the cultural landscape, too. Trails avoided larger deserts and playas in Utah, Nevada, and California (e.g., the Great Salt Lake desert), although various shortcuts and cutoffs did entice some immigrants across these arid landscapes—sometimes with disastrous results. The reader of historian Patricia Limerick's stories of travelers caught without water might want to have a glass of water on hand (Limerick 1989).

Even true deserts could not thwart early development. Prehistoric cultures of the Southwest built urbanlike developments in dry landscapes, most notably the Chacoans and other Ancestral Puebloans in the San Juan basin (Lekson 1993) and the Hohokam around what is now Phoenix (Lekson 1993; Wescoat 1990). Lekson (1993, 1999) argues that both Chacoans and other Ancestral Puebloans developed and receded in a irregular pattern but that the Hohokam prospered longer and remained more stable because of their extensive irrigation works.

The irony of modern western development is not lost on water-resource analysts: the very driest part of the nation is the fastest growing. Indeed the two driest large cities in the United States—Phoenix and Las Vegas—are the fastest-growing metropolitan areas in the country. Any quaint ideas about how aridity would retard regional development have rightly been banished. Most of the 61.4 million people living in the eleven western states are served by centralized water systems that reach into local and distant watersheds to store and move water. Water is moved from mountain catchments to urban distribution systems for essentially all western cities in both wet and dry climates; this includes Seattle, San Francisco, Phoenix, Denver, Los Angeles, and smaller cities such as Cheyenne, Wyoming (population 73,000), which reaches two mountain ranges away for some of its supply. In all of these cases local watersheds and aquifers are insufficient, yet lack of geographically proximate water has not limited urban development. Instead of water driving land use (or any integrated planning of the two), land use drives water development and affects the entire water geography of the West.

Because growing cities need more water, water used in agriculture shifts increasingly to municipal and industrial uses. But most water in the western states is still used for agriculture (from 85% of all water use in California to 97% in Idaho). Moreover, the much discussed transfer of water from agriculture to cities does not appear to be associated with a decrease in irrigated area (Table 13.1). In fact, reported irrigated land area increased in the 1990s, a decade of rapid urban growth in the West. This is illustrated by a closer look at the South Platte basin in Colorado.

THE SOUTH PLATTE: IRRIGATED FIELDS AND SUBURBS

The South Platte basin near the Colorado Front Range is the site of one of the largest transbasin water diversion and irrigation projects in the country and one of the West's largest and fastest-growing urban corridors. Basin

Table 13.1—Irrigated land in the eleven western states, 1992 and 1997 (in millions of acres).

State	*1992*	*1997*
Arizona	0.96	1.01
California	7.57	8.71
Colorado	3.17	3.43
Idaho	3.26	3.49
Montana	1.98	1.99
Nevada	0.56	0.76
New Mexico	0.74	0.80
Oregon	1.62	1.71
Utah	1.14	1.21
Washington	1.64	1.71
Wyoming	1.46	1.71
Total	24.1	26.53

Source: U.S. Department of Agriculture and U.S. Census Bureau <http://www.census.gov/epcd/www/.html.>

population grew by about 600,000 people in the 1990s, or 25 percent (Figure 13.1)—almost all of that growth occurred in the Denver metro area, including the single fastest-growing county in the United States during most of the decade (Douglas County south of Denver). The basin is expected to grow by at least another million people in the next fifteen to twenty years. Virtually all regional economic growth is in services, technology, and construction. Natural-resource industries such as agriculture, logging, and mining are flat or decreasing in terms of number of jobs and share of the regional economy.

Economic and demographic growth means that the South Platte basin is marked by residential, commercial, and infrastructural land uses replacing agriculture, natural areas, and other open spaces. This footprint of development is expanding rapidly along with population growth; but land-use data are very poor, and we have only a rough idea of land use in any given region. Still, an approximation of developed land and trends in development can be ascertained from census data on housing density. The housing density for all census blocks in the South Platte basin is shown in Table 13.2, along with projected development to 2020 based on population projections and a land-use model developed by David Theobald at Colorado State University (see Theobald and Hobbs 1998). Several aspects of basin development are worth noting. First, a quarter of the basin is in public ownership or is otherwise unbuildable, although this proportion is smaller than those in many parts of the West because much of the South Platte basin lies in the Great Plains, where most land is private. Of the private land that could be developed, most (73%) is still rural. Nevertheless, a fifth of the developable land in the basin is developed at exurban, suburban, or urban densities. This proportion may increase to approximately 26 percent by 2020 at current growth rates.

The Colorado Agriculture Commission (1998 <http://www.ag.state.co.us/resource/documents/98LouisianaAgLeadPgmtalk.html>) estimates that the state loses 90,000 acres of farmland per year, mostly along the Front Range. But as shown later, evidence indicates that irrigated farmland is stable, even growing, in parts of the basin.

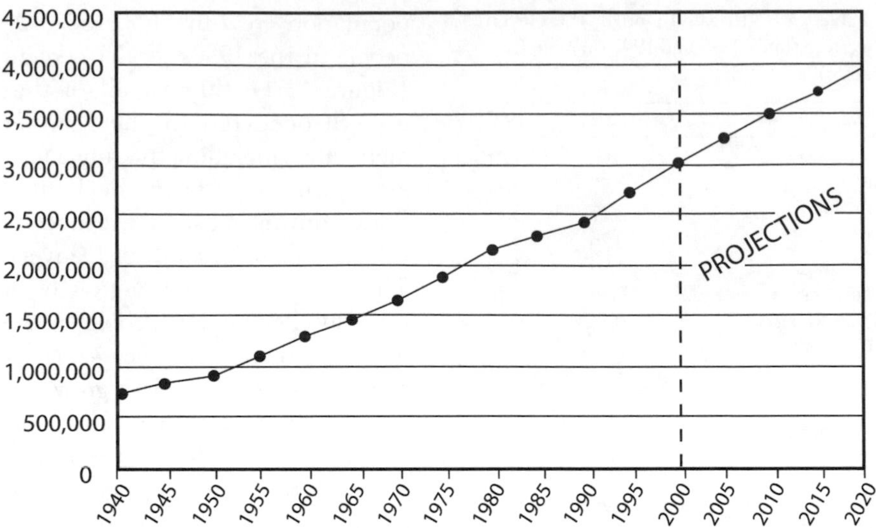

13.1 Population of the South Platte basin. SOURCE: http://sciencepolicy.colorado.edu/wwa/index.html.

Land use and water use interact in several ways, although the two are rarely analyzed in an integrated manner. Instead, water and land are treated separately in both research and practice. This separation is driven in part by the very different policy regimes for land and water. Water is managed by owners and users under a state water-rights system administered by the state engineer and adjudicated—at least in Colorado—by water courts (Nichols et al. 2001). Land is managed and developed by owners and users with very limited regulation by local government. The main role of government in land development is to encourage development and to provide services such as roads, schools, and water lines.

Municipal water use in the basin is increasing with population growth, but because a large proportion of increased urban supply comes from previously developed agricultural supplies, urban growth does not necessarily mean a concomitant growth in new supply. On the other hand, statistics do not necessarily show the decline in agricultural land that one would expect in an urbanizing area where water is shifting to cities.

Agricultural users accounted for 93 percent of total water extractions in Colorado in 1990. In the South Platte River basin, however, agriculture accounted for only 80.2 percent of water use, and municipalities used 12.9 percent in the late 1980s (Litke and Appel 1989)—a proportion that has

Table 13.2—Developed area in the South Platte River basin (acres).

Category	*1960*	*1990*	*2020*
Vacant	24,8296	27,127	27,127
Rural (< 1 unit per 80 acres)	7,820,757	6,473,852	5,674,481
Ranchette (< 1 unit per 40 acres)	345,782	734,250	791,432
Exurban (< 1 unit per 10 acres)	288,462	1,197,948	1,795,646
Suburban (< 1 unit per 2 acres)	55,264	163,667	256,591
Urban (> 1 unit per 2 acres)	76,828	238,545	290,113
Unbuildable (public land, too steep, open water)	3,269,128	3,269,128	3,269,128

certainly increased. Still, irrigated agriculture consumes most of the water, and the South Platte retains a large irrigated area; indeed, most of Colorado's irrigated cropland lies within a triangle formed by Boulder, Fort Collins, and Greeley and along the South Platte main stem (Smith et al. 1996).

INTERACTION OF AGRICULTURAL AND URBAN DEVELOPMENT

The relationships between land use and water use are less clear than often argued because of complications in the transition from agricultural to urban development. According to some studies and conventional wisdom, the South Platte basin is losing agriculture as subdivisions replace farms and cities acquire water supplies. Smith et al. (1996) found that between 1978 and 1992, the Front Range lost 8 percent of irrigated agricultural land; and the Colorado Agriculture commissioner reports that around 270,000 acres per year of agricultural land were converted to other uses between 1992 and 1997 (Ament 2000). But even an initial assessment shows a more complicated picture. Although the transition from resource to service economies is reflected in urban water transfers and declining farmland, urban development does not necessarily obliterate as much cropland—especially not irrigated land—as one might expect.

The fastest-growing county in the nation during the 1990s showed a relatively stable base of irrigated land (Figure 13.2). The complexity of the relationship between urban and agricultural water and land use is illustrated by the Colorado–Big Thompson (C-BT) project. As the largest water project in the basin and the third-largest irrigation project in the western United States, it is evolving from an agricultural system into an urban water system (Figure 13.3a). Yet actual deliveries (Figure 13.3b) and irrigated area (Figure 13.3c) in the C-BT service area suggest that either alternative water is available or greater efficiencies allow irrigators to continue cropping similar acreages while selling water to cities.

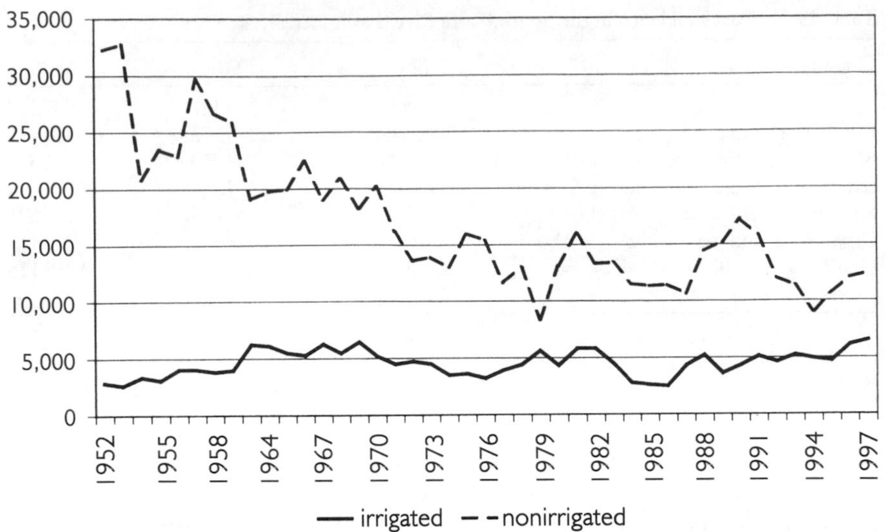

13.2 Irrigated and nonirrigated cropland, Douglas County, Colorado (acres).

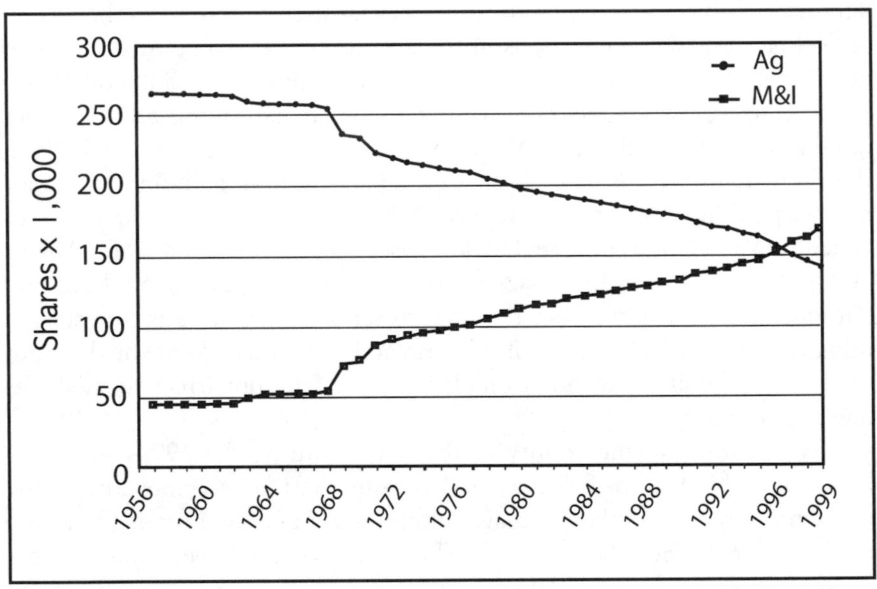

13.3a Water use in the Colorado–Big Thompson project: distribution of water shares between agricultural and municipal owners, 1957–1999. SOURCE: Northern Colorado Water Conservancy District, personal communication, 1999.

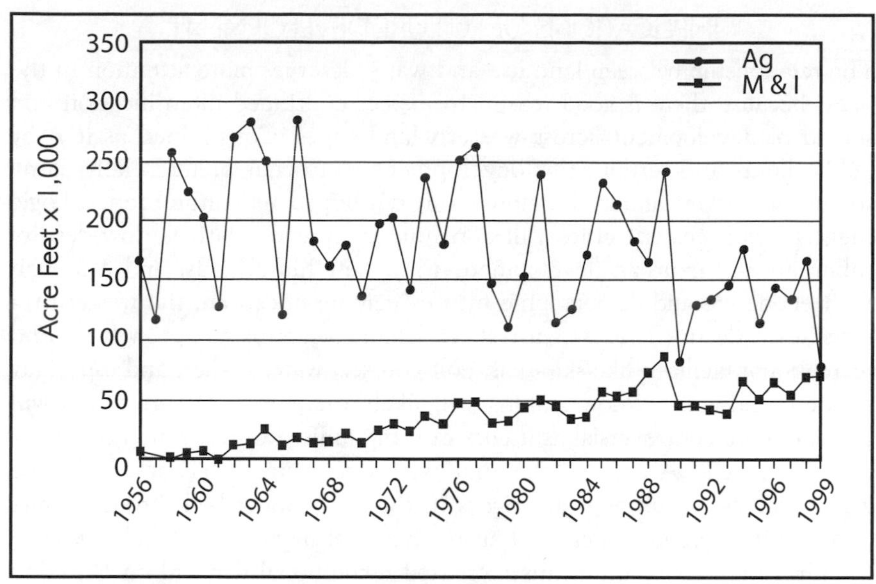

13.3b Water use in the Colorado–Big Thompson project: actual deliveries of water, 1956–1999. SOURCE: Northern Colorado Water Conservancy District, personal communication, 1999.

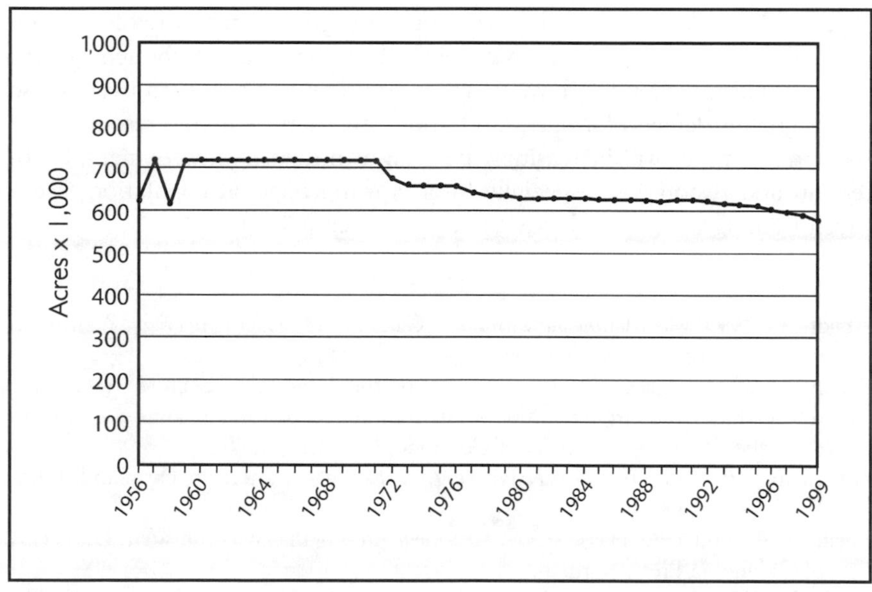

13.3c Water use in the Colorado–Big Thompson project: irrigated area. SOURCE: Northern Colorado Water Conservancy District, personal communication, 1999.

IMPLICATIONS OF FUTURE DEVELOPMENT

The relationship between land use and water deserves more attention in the West because there is good reason to expect continued intensification and spread of development across western landscapes. Constrained as it is by public lands and terrain, this development will occur preferentially along streams and rivers and in the more easily developed agricultural zones. Logic suggests that western cities will continue to spread relatively low-density suburban and exurban development into their hinterlands; and although their economic and demographic futures may be uncertain, the western resorts seem destined for at least several more years of fast growth around recreational facilities like ski areas, golf courses, water bodies, and especially striking landscapes. Snow making will likely increase at ski areas and will become more controversial as it cuts into the limited winter runoff of mountain watersheds. A critic of the Vail ski resort claims it has acted like the western cities, acquiring water rights in the Eagle watershed through a mix of attractive purchase offers and aggressive legal maneuvers (Glick 2001).

Growing population means increased recreational demand on land and water. The West and its waters have been the destination of long-distance tourists since before the first stagecoaches entered Yellowstone National Park. Today, however, most outdoor recreation user days are generated by visitors who travel relatively short distances from nearby urban areas (USFWS 1993, 1997). Thus urban and rural population growth in an area is a major cause of increases in recreational demand in that area. Although growth requires that more land and water switch to urban and suburban uses, it also brings greater demand for open land and water as recreational resources.

Finally, water will increasingly be seen as an aid to environmental protection and restoration, especially by a growing regional population disengaged from agriculture and extractive industries (Chapter 14).

REFERENCES

Abbott, C. 1993. *The Metropolitan Frontier: Cities in the Modern American West.* University of Arizona Press, Tucson.

Ament, D. 2000. Impacts of Water Transfers on the Agricultural Sector. Presentation to the Water and Growth in the West Conference. Natural Resources Law Center, School of Law, University of Colorado, Boulder, June 9.

American Farmland Trust. 1997. *Farming on the Edge.* American Farmland Trust, Washington, DC.

Beck, W. A., and Y. D. Haase. 1989. *Historical Atlas of the American West.* University of Oklahoma Press, Norman.

Glick, D. 2001. *Powder Burn: Arson, Money, and Mystery on Vail Mountain.* Public Affairs, New York.

Lekson, S. H. 1993. Chaco, Hohokam, and Mimbres: The Southwest in the 11th and 12th centuries. *Expedition*: 35: 44–52.

———. 1999. *The Chaco Meridian: Centers of Political Power in the Ancient Southwest.* AltaMira, Walnut Creek, CA.

Limerick, P. N. 1989. *Desert Passages: Encounters With the American Deserts.* University Press of Colorado, Niwot.

Litke, D. W., and C. A. Appel. 1989. *Estimated Water Use in Colorado, 1985.* U.S. Geological Survey, Reston, VA.

Nichols, P. D., M. K. Murphy, and D. S. Kenney. 2001. *Water and Growth in Colorado: A Review of Legal and Policy Issues.* Natural Resources Law Center, University of Colorado, Boulder.

Power, T. M. 1996. *Lost Landscapes and Failed Economies.* Island, Washington, DC.

Riebsame, W. E., ed. 1997. *Atlas of the New West.* W. W. Norton, New York.

Smith, D. H., and twelve others. 1996. *Irrigation Water Conservation: Opportunities and Limitations in Colorado.* Colorado Water Resources Research Institute, Fort Collins.

Theobald, D. M., and N. T. Hobbs. 1998. Forecasting rural land use change: A comparison of regression- and spatial transition–based models. *Geographical and Environmental Modelling* 2: 57–74.

U.S. Fish and Wildlife Service (USFWS). 1993. *1991 National Survey of Fishing, Hunting and Wildlife-Associated Recreation.* U.S. Fish and Wildlife Service, Washington, DC.

———. 1997. *1996 National Survey of Fishing, Hunting and Wildlife-Associated Recreation.* Publication FAW-96 NAT. U.S. Fish and Wildlife Service, Washington, DC.

Wescoat, J. L., Jr. 1990. Challenging the Desert. In *The Making of the American Landscape.* M. P. Conzen, ed. Unwin Hyman, Boston, pp. 186–203.

White, R. 1991. *It's Your Misfortune and None of My Own: A New History of the American West.* University of Oklahoma Press, Norman.

Wyckoff, W. 1999. *Creating Colorado: The Making of a Western American Landscape, 1860–1940.* Yale University Press, New Haven, CT.

FOURTEEN

Constraints of Law and Policy on the Management of Western Water

DAVID H. GETCHES

A quiet revolution has been taking place in western water policy. By the 1980s, western water law seemed outmoded, and states promised reform to satisfy changing public demands and values. The 1990s brought change, but it came largely through initiatives of the federal government and local communities. Meanwhile, state water law and policy evolved haltingly at best.

During the last quarter of the twentieth century, urbanization and society's aroused environmental consciousness put new demands on an old, unyielding system. The water distribution regime was dominated by unadorned notions of legal rights rooted in prior appropriation and by physical facilities, consisting mostly of federal dams and reservoirs, astride the West's waterways. These utilitarian laws and water projects left a legacy of environmental insult, and their inertia impeded dynamic new expressions of public values. New attempts by water developers to meet expanding demand were continually frustrated as they collided with a suite of federal environmental protection legislation passed in the 1970s. When federal subsidies for water projects collapsed in response to public sentiment and economic constraints, the states met change with outrage, arguing that the dual impediments of regulation and a shift away from subsidized development undermined their control over their future.

Some leaders in the West called for the states to take a pivotal role in responding to emerging public values concerning water. This was a bright new recognition of how the West was changing and how the response of states must change, as well as an attempt to prevent federal domination. The agenda for a new western water policy included greater efficiency and conservation, conjunctive use of groundwater and surface water, protection of stream flows, more comprehensive planning, and inclusive public participation at the level closest to the resource.

This chapter surveys the developments in state water law and policy during the 1990s. Although significant changes have occurred in the way important water decisions were made, the most impressive innovations were largely the products of federal regulatory pressure and flexibility and of locally based problem-solving efforts. The past teaches us that state water law furnished an administrative framework with little consciously articulated policy; the framework yielded whenever there was a promise of federal largess.

Western water law and policy are likely to evolve further in response to new public demands. Major stresses on the system—demographics, competing policy demands, climate variability or change—will create crises. Experts, including state officials, may use nontraditional, place-specific problem-solving efforts as laboratories for incubating systemic reform.

A HISTORY OF WESTERN WATER LAW

When most of the country and virtually all of the West were federal land, the federal establishment could have dominated the allocation and control of water rights, but that would have been infeasible because the federal presence in the West was thin. Just as the federal government tolerated and later encouraged settlers—who were little more than trespassers on the public land—to seek out and develop minerals, it also let them take and use water for mining, agricultural, and domestic purposes.[1]

In the era of western expansion, national economic and social policy favored development through mining and agriculture and the establishment of communities by settlers. Thus water was an instrument for realizing policy.[2] If the settlers in the new territories and states could avoid or resolve disputes over water by dividing rights to use water among themselves, the federal government was pleased to step aside. The states did so by allowing individuals to "appropriate" water, even on federal land, for any "beneficial use." Congress allowed this, first by its silence on the subject and later by use of oblique language in mining and homesteading laws.[3] The Supreme Court ultimately held that it was the states' prerogative to allocate water on public lands by any system they chose.[4] But the seminal *Rio Grande Irrigation Com-*

pany case had made it clear that state-authorized water use must not interfere with the rights of the government to protect the flow of water and that state authority over water can be superseded by the exercise of federal powers over commerce and public land.[5]

Congress and the Supreme Court honored, at least in word, the idea that states deserved deference in allocating water according to their own laws. But virtually every iteration of the policy of deference was accompanied by references to *Rio Grande,* which subordinated state control to federal supremacy.[6] Early cases made it clear that federal preemption of state water law need not even occur by express legislation; it could be accomplished simply by setting aside federal land for purposes that required water.[7] Nevertheless, the federal government rarely interfered directly with state-granted water rights. Rhetoric in discussions of the federal role in water matters was dominated by references to "state primacy," which conveniently ignored the reality of federal supremacy.[8]

The myth of state control has always been precarious, depending as it does on congressional forbearance in the exercise of federal preemption. Federal legislation routinely recited that it shall not "be construed as affecting or intended to affect or to in any way interfere with the laws of any States or Territory relating to the control, appropriation, use or distribution of water."[9] But even when a federal law has appeared to preserve state water laws, the overriding purpose of the federal law has controlled the use of water. The Supreme Court in case after case has held that such disclaimers mean nothing in the face of a federal purpose that conflicts with state water rights.[10] Thus federal programs ranging from dam building in the early twentieth century to the environmental regulations that were imposed later can inhibit or preclude the operation of state water law and state-granted water rights whenever a conflict exists between the state legal system and the federal purpose.[11] The expansion of federal environmental laws late in the century and the increasing scarcity of water for all uses in a growing West made conflicts more frequent and sometimes bitter.

Water policy was straightforward in the early days of the West. It first dealt with water allocation and later with water development, distribution, and planning. Accordingly, water institutions provided mechanisms to make and enforce allocation decisions and to develop and deliver water. State legal regimes committed water to the kinds of private uses (primarily mining and agriculture) that prevailed early in the twentieth century. The dams and other structures that were built, mostly with federal dollars, fixed distributional patterns and further tied water to particular uses and places. State appropriation doctrine and federal water development created an inflexible system that satisfied narrow purposes.

The Traditional State Role: Creation and Protection of Private Rights to Use Water

States in the West typically adopted the prior appropriation doctrine, a simple system based on the notion of first-in-time, first-in-right. The allocation function required little institutional support because prior appropriation was conceived as a self-help system. Beginning in the mid-nineteenth century, appropriators claimed rights to quantities of water that they would develop and use beneficially, and the appropriator who was first in time had the first right to use water. There were no local institutions to make adjudicative or administrative decisions.

As rivers became fully appropriated early in the twentieth century and competition for water intensified, the prior appropriation system eventually needed more institutional support. In this laissez-faire system, records of water rights were inadequate and scattered. Before long, states felt the need to standardize and centralize their records.[12] By the twentieth century all western states had expanded the function of water agencies to ensure that water was not used unlawfully. Water institutions recognized that private entities could get property rights to water, even though water was essentially a public resource, and that the law protected these rights against interference by others. Water agencies served the users and mediated conflicting claims.[13] In an era when key issues could be refereed by an official who enforced simple priorities among holders of rights, institutions were unencumbered by considerations of multiple issues and interests.

Water rights under the prior appropriation doctrine were generally granted with no reciprocal responsibility except to avoid harm to other holders of water rights. Nearly all states recognized that water was a public resource, and some state laws and institutions encompassed a limited consideration of the "public interest," a concept whose meaning varied from state to state.[14] This discretion allowed the agencies to confront and make difficult decisions about the quantities of water needed for new uses and whether a new or proposed change in use was consistent with the public interest.[15] At least until recently, however, state engineers rarely made a serious and comprehensive inquiry into the public interest; they simply determined whether the water would be put to a beneficial use. Environmental, social, and economic effects of a water use have rarely been considered in most states and have almost never resulted in legal limitations on the scope or type of water rights or uses permitted by a state.

Legal institutions did evolve to meet the needs of economic development, however. In arid parts of the West, dams and canals were needed to transport and store water. Because water was sometimes appropriated far in excess of average annual flows, uninterrupted use of water in years of drought required storage. In the old West, construction of dams and ditches was initially a task for individual appropriators. But water users often lacked sufficient capital to build the facilities for use of limited water sources, and collective use of resources was uncommon among independent-minded water users. Even when private irrigation enterprises were formed, they often failed for lack of capital and because there was no legal means to force potential beneficiaries to participate.[16] California responded to this problem with a law that gave irrigation districts power to impose taxes on landowners. Failures continued nevertheless, in part because of economic inefficiencies and lack of expertise. Although states tried several other approaches, entities for collective irrigation in the arid West—other than in early Spanish, Indian, and Mormon communities—generally did not succeed.[17]

The U.S. Congress established the federal Reclamation program in 1902. At first the program dealt directly with individual irrigators and attempted to collect repayment for project facilities from them. Eventually, Congress recognized special water districts as the only entities with which it would deal. Today these locally controlled water districts serve diverse purposes. State law created these entities outside the framework of state government to handle some of the most important issues of water development, supply, and management at a regional or local level. They generally enjoy quasi-governmental status and independence from state control.[18]

The Federal Role: Subsidized Water Development

For most of the twentieth century, the U.S. Bureau of Reclamation was the largest single provider of capital in the form of subsidized loans and grants for major water development in the West.[19] Major decisions about size and design of facilities were left to engineers; management and distribution of water were left to the special districts that dealt directly with the federal agency.

The federal government's entry into water development eclipsed state water law; it subordinated state laws and policies to federal dam-building programs. Competition for large federally funded water facilities was keen, and the timing and siting of dams became politically charged. The primary water-policy objective of western politicians for most of the twentieth century was to capture federal assistance for water projects. Development of public works is an attractive arena for politicians because it gives them a mission with tangible results, ostensibly for the benefit of their constituents. Although political rhetoric referred to state control of water resources, states and their leaders seemed willing to sacrifice legal and institutional control for construction of facilities that established physical control of water, even if it meant subordination of the states to federal laws and policies.

Although the states passed enabling legislation for districts to be organized so they could operate the federal projects, the districts were locally controlled, enjoyed considerable autonomy under state laws, and typically took control of all water rights in the area to be served by a project.[20] Besides requiring subordination of the old state-granted individual water rights of landowners who would be served, federal project authorizations sometimes superseded state water law. These measures allowed federal control over sufficient water sources to justify federal projects.

Historically, the intent of most state water planning was to attract federal dollars by satisfying federal guidelines that required plans.[21] Many state plans were used simply to identify and expedite federal water-development projects.[22] Because planning was supply-oriented, it focused on where and how to build but showed little regard for environmental or social consequences or even for the economics of the projects. Only a few state plans did more than justify federal projects.

A 1965 federal law attempted to promote better, more integrated planning at the river basin level.[23] States resisted, although many complied insofar as necessary to remain eligible for federal support for water projects. When the hope of federal funding waned in the 1980s, most of the motivation for state water planning was gone. Finally, the federal effort to promote a more comprehensive approach at the river basin level was abandoned in 1981.[24]

The federal government eventually required fuller consideration of economic feasibility and environmental impacts for federal water projects, but only after funding for federal projects had declined. It was not until the 1970s that the National Environmental Policy Act required federal agencies to prepare an environmental assessment or environmental impact statement before undertaking a major federal action.[25] By 1983, when the government required balancing costs and benefits and a consideration of environmental consequences,[26] most big federal projects were completed or under way.[27] Thus federal projects generally were constructed without concern for their economic, social, or environmental consequences. Purely natural values were generally neglected. If fish and wildlife, recreation, and flood control were considered at all, they were incidental to agricultural, municipal, and industrial purposes.[28]

The Rise of Environmental Consciousness and Economic Rationality

The first half of the twentieth century was an era of rapid population movement, expansion of agriculture, and growth of new cities. The accompanying water development put an indelible mark on the western landscape. Irrigating arid lands and supplying cities far from rivers were not simple tasks because they required transporting water across long expanses of dry land and storing enough water to make up for erratic flows from season to season and year to year. Today more than a million water storage reservoirs dot the West. The U.S. Bureau of Reclamation constructed 355 of the largest reservoirs plus 16,000 miles of canals.[29] The U.S. Army Corps of Engineers (the Corps) expanded its role as a navigation protection agency to include flood control and water supply. Although the Corps worked mostly in the East and Southwest, it took on new responsibilities, especially in the Pacific Northwest and the Missouri River basin. Today Reclamation reservoirs store over 119 million acre feet, and Corps reservoirs in the West store 103 million acre feet of water.[30]

From an engineering standpoint, U.S. water development is a remarkable achievement. Furthermore, it produced impressive economic rewards. Flood-control benefits have been significant, and the power generated by Bureau of Reclamation hydroelectric facilities in the West sells for over $700 million per year. The anticipated benefits to agriculture often have fallen short of expectations, however, because the soils and climates are largely unsuited for intensive irrigation.[31] Notwithstanding subsidies of up to 90 percent of project costs, many farmers could not produce enough to repay the balance.[32] Nevertheless, crop production aided by irrigation projects is valued at $9 billion per year.[33]

It is now widely understood that the unintended consequences of water projects were enormous and far-reaching:

> Dams have flooded valleys and displaced farmers and communities, blocked or disrupted fish migrations, reduced naturally occurring flood frequencies and magnitudes, disrupted natural temperature fluctuations, altered low flows . . . reduced sediment and nutrient loads, changed channel-sediment characteristics . . . narrowed and shrunk river channels, changed channel patterns and eliminated flood plains.[34]

Moreover, a recent report concluded that water development is the single most significant threat to survival of species.[35]

Although the environmental consequences were not assessed and discussed in advance of project development as they would be today, by the 1970s most observers recognized the adverse effects of water projects. Environmentalists had long decried the damage caused by dams and had won historic battles in the 1950s and 1960s by stopping three major dams planned for the Colorado River.[36] These early victories helped to define the environmental movement, but the official government policy still favored subsidized water development, even after Congress had passed an impressive array of environmental protection legislation in the early 1970s.[37]

In 1977, Jimmy Carter became president, and federal water policy began an inexorable change. For westerners interested in water issues, the hallmark of Carter's presidency was the infamous "hit list" of thirty-three water projects of the Bureau of Reclamation and Corps of Engineers that were targeted for elimination from the federal budget.[38] Congress had already approved construction of these elaborate systems of dams, irrigation systems, and hydropower facilities in the West, and they had taken on an aura of birthright for the states. When the new administration announced that the projects were doomed by intolerable environmental impacts or unacceptable benefit-cost ratios, there was a shrill outcry from western politicians and development interests. Colorado governor Richard Lamm wrote:

> However well-intentioned it may have been, the Carter hit list was a study in federal arrogance. Its assumptions were questionable, its conclusions were faulty; it was riddled with anti-western prejudice and wrapped in anti-western ignorance. It reflected no understanding of western conditions, of western people, of the nature of their lives, or of the relentless crushing aridity that shapes their land and everything in it.[39]

Lamm was joined in his unvarnished criticism of the Carter hit list by his fellow western governors.

Carter's message was clear and well-founded. Because the federal projects could not be economically justified, and those that were proposed would cause environmental problems at least as serious as those caused by their predecessors, they must not be built. The Reclamation program was touted as one that paid its own way, but most projects were heavily subsidized and could not be justified based on benefit-cost analyses. Although the Reclamation Act's purpose of aiding growth in the West had been realized, it had come at a considerable cost to the rest of the nation. The costs were not only monetary but were measured also in inefficient use of land and water. The Reclamation program raised questions of interregional equity as crop production was promoted by federal expenditures in the irrigated areas of the West rather than in other more fertile and humid regions. Damage to natural systems was enormous. The public was becoming deeply concerned about major environmental crises, and many of the most troubling ones were traceable to federal water projects.

Carter was new to Washington, and the hit list was a political blunder, but it is bound to be viewed more favorably in the future than it was by contemporary observers. Although Congress retained a few of the projects, Carter was adamant, and the thrust of the policy was maintained. The Reagan and Bush administrations then embraced policies that had the same result. Federal water projects that did not meet certain criteria and lacked significant state or local funding were basically dead.[40] History will rightly mark Carter's decision as signaling the end of the era of federal dam building in the United States.[41]

The same national environmental consciousness that emboldened the Carter administration to produce the hit list had inspired the wave of lawmaking in the early 1970s that accounts for most federal environmental laws. These laws created a new kind of federal presence. In the preceding half century the federal government had profoundly influenced the distribution and use of western water. But with the demise of water project funding and a looming regulatory presence, the federal government was left holding all sticks and no carrots.

It should be clear that "state primacy" was never much more than a shibboleth of western politicians. Most important decisions and responsibilities were ceded to or assumed by the federal government or by special districts. Western water law was simply a framework for allocating rights in the first instance and was so unembellished by rules and so inflexible that it did not well serve modern needs or help solve water problems.

THE MOVEMENT TO REFORM WESTERN WATER POLICY

Western water law as manifested in prior appropriation law served the purposes of an earlier era of western settlement and utilitarian goals, but it did not keep pace with evolving social and economic values. As of 1990, the state role remained essentially administrative. Federally subsidized water development had assisted in distributing water more widely, but it had become obvious that this policy had created major environmental problems. The deficiencies in water policies were recognized in reports and studies beginning in the 1950s.[47] By the 1990s the literature in the field constituted a chorus of reform ideas and recommendations with remarkable similarities.[43]

Public consciousness of the need for reform expanded, but calls for change met systemic inflexibility. The public concern for environmental quality manifested in federal laws and opinion polls was not reflected in state water law. Some growing western cities and industries could see the advantages of supporting the new ideas as a means of alleviating supply problems and facilitating solutions to environmental conflicts, however, and they supported the idea that reform was needed.[44]

The states were jealous of the primacy they had always asserted in western water matters. Aiming to forestall an increased federal regulatory presence, they proposed to make changes in their own systems sufficient to satisfy an expanding public demand for protection of the environmental and social values that are implicated by water use in the West while also responding to the demands of a growing urban population. The states also had begun to press the federal government to provide financial support for negotiated settlements of many Native American water-rights claims that had long cast doubt on the utility and security of much of the West's water rights.

The Western Governors' Association, led especially by Babbitt of Arizona and Lamm of Colorado, recognized that times had changed and that state policies either had to change or become irrelevant.[45] The trends were clear. If dams were going to be built, it would have to be with state or private funds. Even if states could raise the money to build their own dams—and this was a time of declining state revenues—they would face significant regulatory obstacles.

Although some governors had criticized the abrupt shift in the federal role, they recognized the inexorable change in water policy that was necessary to address evolving social values. The progressive governors consequently accepted that water scarcity could no longer be satisfied with expanding supplies and that demand for water had to be controlled. Also, they recognized that the unspoiled resources defining the West's essential character had to be protected. Finally, they perceived that states had an opportunity to supplant the need for widening federal regulation.

The governors took leadership in a growing movement for reform of western water policy. The movement included the Western Governors' Association (WGA) and eventually the Western States Water Council (WSWC), an association of water officials and professionals from the western states. The WGA members passed formal resolutions committing themselves and their states to a vision of change. The new coalition of western state water experts and political leaders proposed improved water conservation, protection of stream flows, integration of quality and quantity management, and use of markets and pricing mechanisms to make water allocation more efficient in serving the West's needs. In addition, they began to accept the idea that water decision making needed to include a variety of interests. These were not really new ideas, since many were traceable to reports of governmental and academic groups over the preceding thirty years. Yet it was rather revolutionary for state officials to embrace them. No one doubted that the prior appropriation doctrine had served reasonably well to promote development in the old West, but it was becoming apparent that without reforms, the doctrine was too simple to meet the complex and multiple demands on water law in the new West.

In 1986 a pathbreaking report commissioned by the WGA called for governors to take the lead in promoting water conservation by working with the U.S. Department of the Interior, water users, and others.[46] It made the case that demand management could help satisfy water needs and still maintain quality of life. Specific reforms discussed in the report included salvage and transfer of water and substitution of alternative supplies of water for senior users through exchanges and other measures. The report urged water marketing as a means of improving efficiency but recognized the need for governmental protection of public, community, and environmental values. Another WGA report elaborated on the need for efficiency-enhancing laws and policies and for federal agencies to be more active in developing information to facilitate water markets and in protecting environmental values of water.[47]

The WGA, even as its membership and leadership changed, continued to support water-policy reform. Together with the WSWC, the WGA sponsored three workshops on western water management in Park City, Utah, in 1991, bringing together a diverse group of water experts who developed a manifesto for western water management known as the "Park City Principles."

The Park City Principles boldly insisted that the states should play the pivotal role in bringing about major changes in western water policy. The accepted principles would incorporate in state laws and policies a recognition of diverse interests in water-resource values, a holistic approach to water problems that recognizes crosscutting issues and impacts, and a policy framework responsive to economic, social, and environmental considerations. This would enable broader state and basin participation in federal policy development and administration, and the federal agency role in local problem solving would become one of participation rather than prescription. Along with the introduction of more negotiations and marketing principles, these measures would minimize the need for command-and-control regulation. It is interesting that recommendations for reform usually did not press the linkage between land use and water demand or between land use and its effects on water resources.[48]

The constituency for changing water laws and policies, however, was limited, and proposals for nearly any change in western water law were controversial. In the past, state laws had incorporated departures from the strict doctrine of prior appropriation in the course of providing incentives for development and creating elaborate statutory programs for permitting.[49] But when it came to reforms that would result in shared decision making for allocation and use of water, some observers expressed suspicion and resisted.[50]

Although the governors and others were out in front and were sometimes at odds with many traditional voices in water policy, by the beginning of the 1990s, reform seemed inevitable and less ominous. The Park City Principles, as well as the conclusions of other reports of the WGA and WSWC, coincided with many of the ideas expressed by earlier task forces as well as the reform-oriented scholarship of the era.

State Law Reforms

The 1990s began with the states expressing their intentions to adapt water laws and administration to the West's rapidly changing needs. Because state politicians apparently embraced the essential reform proposals, there was reason to expect new legislation and more aggressive administrative initiatives that would be tested in judicial decisions. The stage seemed set for action that would encourage greater efficiency and conservation through mechanisms such as statutes promoting reuse and salvage of water and requiring low-flow plumbing fixtures. Groundwater law had long needed renovation in many states to provide for its conjunctive management with surface water.[51] The public was insisting on more effective protection of streams and promotion of other elements of the public interest, and state political leaders could see that creating innovative state programs to respond to these demands was a key to replacing federal controls. This motive seems plausible. When states have gone beyond, but not conflicted with, federal programs in attempting to protect public values, the Supreme Court has upheld both state requirements for releases of water from a federally licensed dam to protect water quality[52] and state water rights permit limitations on filling a federal reservoir to protect fish, wildlife, and recreation below dams.[53] States had begun efforts to improve state water planning to make it more comprehensive and inclusive.[54] Easing barriers to water transfers and marketing was also on the states' agenda.[55]

Political and legal developments at the state level fell short of their potential. The movement for reform of state law that propelled us into the 1990s was not the dominant force in changing water policy. Instead, federal laws and programs combined with the engagement of people and groups outside traditional water institutions to produce the most profound effects on water policy in that decade.

Changes in state policy were mostly ad hoc and not part of the deliberate process of modernization that had begun in the 1980s. As evidenced by California, where the most interesting legislation and court decisions of the 1990s are found, almost all changes were provoked by some threat or crisis. In California, water demands created by urban growth and the threat of enforcement of federal environmental regulations created enormous pressures.

Water policy in the 1990s showed a decline in the importance of traditional water institutions, particularly at the state level. The federal government became the catalyst in the evolution of western water policy and today remains the single most powerful force for change.

To the extent that nonfederal entities have been involved in water-policy reform, the most innovative efforts have bypassed state water institutions and have led to the establishment of new entities. These "outside-the-box" initiatives have created "islands" of policy change. The federal government is an essential participant in these initiatives through its substantial regulatory authority.

State systems risk obsolescence as some of the most important water issues in the West are being entrusted to unconventional federally and locally driven processes. The decline in the influence of state water institutions is ironic not only because of the time-honored myth of state primacy in water but because at the beginning of the 1990s the states seemed ready to take leadership in promoting change. A summary of state legislation and court decisions concerning water policy shows the rather modest results produced from 1989 to 1999.[56]

Conservation and Efficiency

The inefficient use of water in agriculture is notorious, and the documented benefits of conserving water through modification of agricultural use are great.[57] Cities also may save water through pricing methods and improved plumbing. Yet after the western governors identified an agenda for legal changes, most states did little to promote water-use efficiency. This is nevertheless the area where the most state reform has taken place through legislative and judicial measures that promote water conservation. For example, Kansas passed a law requiring water-conservation plans of all water users,[58] and Texas now requires a state water-conservation plan.[59] The Washington Supreme Court has held that when rights are adjudicated, they can be limited to the amount used over time in a reasonably efficient manner.[60] That is to say, the quantity of a water right can be diminished from the amount stated in a permit or court decree to the amount used with reasonable efficiency, even if the full amount of the paper right has been used over time.

A few states passed laws encouraging salvage of wasted irrigation water. Montana's law allows appropriators who conserve water to make use of the amount that they consume ("salvaged water").[61] Washington authorized counties to create boards to undertake improvements in water conservation and efficiency and to pursue redistribution and transfer of salvaged water.[62] Oregon liberalized its definition of salvaged water.[63] New Mexico stopped short of allowing transfers of salvaged water but declared that conserved water will not be forfeited for nonuse.[64] Arizona and Colorado passed laws promoting low-flow plumbing fixtures.[65] Nevada enabled Clark County to regulate the artificial lakes that have proliferated at golf courses and casinos around Las Vegas.[66]

California's efforts stand out in comparison to those of other states. It produced the most significant legislation in the West concerning efficiency, freeing up enough water to supply millions of people. Reuse has become a necessity in a region growing well beyond the capacity of its water resources. The understandable distaste that people may have for drinking water that they might have flushed down the toilet the day before has been overcome by the scientific reality that water can be recycled without sacrificing health or aesthetics. California lawmakers passed new laws almost every year in the 1990s to encourage recycling.[67] They have required standards for gray water—bath and wash water—to be set for residences.[68] Another law declares the use of potable water to be waste if reclaimed water is available.[69] And districts in southern California have designed programs that generate hundreds of thousands of acre feet of water per year through reclaiming wastewater and using it to replenish groundwater.[70]

Groundwater

States made some minor advances in laws dealing with groundwater. Most significant was the long-awaited legislation in Nebraska recognizing that water pumped out of wells that are hydrologically connected with a stream must be managed conjunctively with the water from the stream.[71]

Texas law has always held to the unrealistic idea that a landowner has absolute ownership of underlying groundwater, thus ignoring the hydrologic reality that pumping can affect surface flows and rights to use them.[72] Allowing pumpers the right to virtually unregulated pumping from beneath their property has resulted in odd and inequitable results in disputes between surface and groundwater users; the Texas Supreme Court has even called the rule "harsh and outmoded."[73] But until a federal court order put the Edwards Aquifer under federal management, the legislature did not correct the situation.[74] Depletion of the Edwards Aquifer, which supplies San Antonio, caused flows to decline in springs that feed the Guadalupe River. The springs are the habitat of eight federally listed threatened or endangered species, and the city's water use created a serious conflict with the federal Endangered Species Act.[75] Groundwater withdrawals from the Edwards Aquifer are now regulated, and pumpers must account for the effects of withdrawals on surface waters.[76] The state supreme court upheld the constitutionality of this special regulatory system for the Edwards Aquifer.[77] In the absence of additional legislation, however, the same court later declined to adopt a reasonable use rule for other aquifers.[78] In 1998 the Texas legislature required the state agency charged with applications for new surface-water rights to consider the effects of the proposed use on groundwater and groundwater recharge.[79] This approach appears to leave gaps in conjunctive management outside the Edwards Aquifer for old surface uses that affect groundwater and for all groundwater uses that affect surface uses.

In Nevada the legislature set up a new program directing a local water authority in the Las Vegas area to establish a groundwater-management program designed to protect long-term supply and prevent contamination.[80] In 1999 the state engineer was empowered to revoke groundwater permits temporarily in overdrafted areas.[81]

Conjunctive management can cause practical problems for senior users with old, inefficient means of diversion (either surface diversions or shallow wells) that become ineffective after modern, efficient wells are installed by juniors. Even if enough water is still available to satisfy senior rights, seniors may argue that they should not be required to build efficient diversion works to replace their old methods of diversion. Idaho has addressed this problem by enabling special districts to organize so that junior groundwater users will develop and pay for mitigation plans to alleviate the effects on senior rights.[82]

Most other state groundwater legislation in the 1990s dealt only with the details of existing groundwater statutes. During an era of growth pressures and enlightened management, one might have expected states to address the use of aquifers for storage and legal protection for aquifer recharge. Idaho has taken steps to encourage recharge by recognizing that underground storage is a beneficial use.[83] New Mexico also has legislated promotion of aquifer recharge,[84] and California has encouraged recharge with reclaimed water. Other legislatures have been inactive in this area.

Maintenance of Instream Flows

The most serious and immediately noticeable environmental impact of water development and use is the depletion of stream flows. To deal with their problems of loss of fish and wildlife habitat and recreational opportunities, almost all states had created instream flow protection programs by the end of the 1980s.[85] Nebraska, one of the states lacking regulation of instream flows, enacted a law during the 1990s, and the courts upheld its constitutionality.[86] New Mexico has struggled with the issue but failed to enact a statute, leaving it the only western state without some type of instream flow law.[87]

A few states improved their programs. Utah broadened its existing instream flow law to allow the Division of Parks as well as the Department of Wildlife to acquire existing appropriations and convert them to instream flow rights.[88] Montana fought for years over whether to allow experimental leasing of water rights for instream flows and finally authorized it on five streams.[89] In Washington a court held that the Department of Ecology could condition groundwater permits on maintenance of specific instream flows.[90]

Most developments during the 1990s, however, illustrated state resistance to instream flow protection, and some states retreated from their earlier efforts. Under considerable pressure, Idaho enacted legislation in 1991 and 1995 to authorize use of water for protection of fish.[91] Idaho had failed to respond to demands for better protection of endangered salmon and recognition of Indian treaty rights to water for sustaining fish.[92] One means of making more water available for salmon habitat in the Snake River was to release water from federal Bureau of Reclamation reservoirs. There is little doubt that the government has legal authority to release water.[93] The state objected, however, arguing that although the water was not legally committed to any particular irrigators, it should not be released for salmon recovery because it would then no longer be available for agriculture. The federal agency procured the state's consent to the releases, and the secretary promised that, to the extent the releases would interfere with any deliveries to people who had contracts for water, it would "lease" the contractors' rights.

Under federal pressure, the state legislature enacted two temporary laws giving the government "permission" to do what it probably could have done anyway. Idaho also passed a bill in 1999 to protect the Payette River from future hydropower development.[94]

In Wyoming the supreme court reversed a special master's finding that the Wind River tribes should be able to dedicate some of the water rights allocated to them in a general stream adjudication to instream flows. The court upheld the state engineer's refusal to enforce tribal instream flow permits where it would take water quantified for future tribal agricultural uses and use it to maintain flows to the detriment of other irrigators.[95]

Colorado has one of the oldest and ostensibly most flexible laws for establishing instream flow rights.[96] The Colorado Water Conservation Board, which is in charge of administering the law, and the legislature and courts have removed some of the statute's flexibility and effectiveness.[97] On the other hand, although the legislature declared that only the Conservation Board could hold instream flow rights, the state supreme court effectively allowed the city of Fort Collins to protect instream flows by a creative interpretation of existing law.[98] The court found that Fort Collins's channelizing dam at the beginning of the stretch of stream and its boat chute at the end of the stretch of stream the city hoped to protect were equivalent to diversion dams and therefore were actually effecting an appropriation by diversion and not establishing an instream flow right.

Colorado state courts have rejected most federal attempts to assert instream flow rights under the federal reserved rights doctrine.[99] The Idaho court adjudicating water rights in the Snake River system likewise denied federal claims for flowing water in national forests but found that the purposes of federally designated wilderness areas necessitated reservation of the unappropriated waters within and above streams flowing into those areas.[100] The latter development caused the state Department of Water Resources to declare a moratorium on all pending water-rights applications, including those that would not necessarily be affected by the federal claims.[101]

Public Interest

For many years, virtually all states have required some type of public-interest review of all new appropriations.[102] These laws ostensibly reflected the principle that unappropriated water is a public resource; thus a grant of private rights to use it should be consistent with the public welfare. Application of the laws was uneven among states, however, and was rarely forceful enough to protect social and environmental values. Moreover, most states lacked clear standards for evaluating the public interest and needed better procedures for public involvement in the process. Some public-interest review was no more than a cursory determination by the state engineer.[103]

Apart from the public-interest requirement specific to water allocation, some states have required environmental impact reports (EIR) on major developments, including water projects and water allocation decisions.[104] The EIR process provides a forum for raising public interest factors in particularly controversial cases.[105]

During the 1990s, states gave little attention to improving public-interest consideration of water decisions. A few states extended their public-interest requirement to changes and transfers of existing rights, recognizing that limiting such review to new appropriations would undermine the law's purpose if a subsequent transfer or change of use could occur without considering the public interest.[106] Montana has integrated water-quality concerns into water rights decisions by requiring that a proposed change of use must not adversely affect the water quality of another appropriator or interfere with the ability of a discharge permit holder to satisfy effluent limitations in the discharge permit.[107] Oregon now allows changes of use away from land where the use was originally authorized if it will benefit endangered species,[108] and it allows new uses for public benefits without requiring that plans be amended.[109]

The "public trust doctrine," as a forceful common-law basis for infusing public-interest concerns into water decisions—even without an express state statutory or constitutional requirement—became a major topic of discussion in the 1980s after it was applied to force reevaluation of long-standing rights in California's Mono Lake litigation.[110] The concept is that the state holds water in trust for the people and cannot convey it for private uses without considering the effect on public uses such as fish and wildlife and recreation. The case that gave the doctrine life in the field of water allocation was concluded in 1994 with an amendment of Los Angeles's water rights to maintain instream flows and restore channels in tributaries to Mono Lake.[111]

It appears that the ideals expressed in the Mono Lake decision requiring respect for public values in water will remain a part of California water law, thus influencing administrative practices and statutory interpretation. For instance, an appellate court has found that the State Water Resources Control Board must condition use of water rights on release of sufficient water from a dam licensed by the board to restore and maintain fisheries as they existed prior to the diversion of water.[112] Some commentators decried the California court's mandate to revisit established water rights in the *Mono Lake* case as the greatest threat ever to the prior appropriation doctrine, whereas others hailed it as necessary to protect public resources.[113] Whatever fears there may have been about the doctrine's potential to undermine western water law, other states have not embraced the public trust doctrine in water rights born in the *Mono Lake* case.

States did virtually nothing during the 1990s to improve public-interest review of water decisions. Indeed some states took steps backward. In Idaho the supreme court pulled back from its earlier announcement[114] that the public trust doctrine applied to waters of that state. The court did not disavow the doctrine entirely but held that it does not apply to the massive Snake River basin adjudication.[115] The Idaho legislature then went further, prohibiting application of the public trust doctrine in water appropriations and other matters related to water rights.[116]

Colorado is the only state that has implemented no requirement for public-interest considerations or protections. The state constitution states that water is "declared to be property of the public, and [is dedicated] to the use of the people."[117] The Colorado Supreme Court nevertheless refused to require water courts deciding whether a large new water right is for a "beneficial use" to accept or consider evidence of impacts of the proposed water uses on the environment, local economies, farming, rural communities, or other public values.[118]

Planning

Because water law traditionally has encouraged the independent appropriator to make key development decisions and because most of the water in the West is tied to uses or physical structures designed for specific purposes, changing demands require comprehensive planning. Indeed many of the goals for water reform—efficiency, conjunctive use of groundwater, protecting instream flows, respecting the public interest—seek to reverse conditions created by incentives embedded in the old system.[119]

Problematic incentives are numerous. Subsidies to development projects and a policy of allowing water rights to be appropriated and used free of charge have created wasteful and environmentally destructive projects. The rule of capture, when applied to groundwater, disconnected pumpers from responsibility for depletion of surface flows. Endowing water users with property rights rewarded depletion of streams and frustrated protection of flowing water and other public values. Moreover, states imposed legal obligations on water suppliers such as municipalities and districts to serve new demand, thus clearing the way for growth.[120] Growth was further facilitated by a special exception to the antispeculation, use-oriented ideal of prior appropriation. State courts and legislators sanctioned the early acquisition of water rights by growing cities—dubbed by some the "great and growing cities doctrine"—long before actual demand materialized.[121] The Colorado Supreme Court recently reiterated its license for a city to accumulate water rights to meet future growth projections in a case involving a city's acquisition of sufficient water for a nearly 500 percent population expansion.[122] Idaho, however, has authorized the Department of Water Resources to decide whether it is necessary to change water to a municipal use and for how long a municipal provider can "hold water rights to meet reasonably anticipated future need."[123]

During the 1980s a few states had planning processes that attempted to integrate water supply, quality, and some other issues such as environmental values and flood control.[124] Presumably, planning continued in those states, although it is not necessarily evident in a survey of legislation and court decisions. The majority of western states, which had no water planning programs or in which planning was limited to supply issues, did not make planning more comprehensive during the 1990s. The few changes in this area dealt instead with piecemeal planning to address specific problems such as groundwater, conservation, or drought.[125] Instead of leading to more integrated water decisions, this approach further divides consideration of related factors and effects that inhere in water.

Water planning remains segregated from land-use planning. Given the serious and widespread concern in the West over population growth and sprawl in urban areas,[126] it is surprising that water planning has not been considered in connection with land-use management efforts. California alone has made advances by including water conservation, evaluation of supply, and pollution control as required elements of land-use plans. In 1995 it passed a law requiring cities to determine that a firm water supply is in place before approving major new developments.[127] This determination does not necessarily connect with any state or regional land-use or water plans or the plans of any neighboring municipality. The courts have shown some willingness to enforce requirements of the California Environmental Quality Act that municipalities produce an EIR evaluating the effects of growth in water demand, including effects on an area from which water would be exported.[128]

It seems obvious that land-use patterns drive water demand. Because about half of all urban water demand in the West today is for watering lawns and gardens, land-use choices can have a dramatic effect on future water demand. By foregoing sprawling subdivisions in favor of denser development, population growth can be accommodated with far less water. Moreover, states might soften or head off the federal regulatory presence by instituting strong land-use planning and control programs. Federal air and water pollution controls, solid waste regulation, wetlands protection, and endangered species preservation laws all affect the way in which land is used. If states sought to ameliorate effects of land-use decisions on water quality, wetlands, and fish and wildlife habitat by considering these effects in planning and permitting development and growth, there would be less need for federal regulation.[129]

The failure of states to relate growth management to water planning is commensurate with the paucity of state-level land-use planning and management programs. Only Oregon and Hawaii have assumed any significant level of statewide land-use and growth-management control.[130] Even they do not integrate water-resource planning in the state-mandated planning responsibilities. Most states have abdicated to local governments, resulting in widespread public dissatisfaction in the lack of effective, coordinated planning. The appealing assumption that local communities are in the best position to determine and regulate land-use patterns may be valid for relatively isolated and slow-growing municipalities. But the ideal gives way when intense economic pressure comes to bear on relatively unsophisticated or unprepared local governments, and it does not work in any event if some municipalities in a region have the will and foresight to control growth and their neighboring municipalities do not. Thus without statewide or regional programs, effective land-use planning and growth management are unlikely, as is the emergence of a framework for integrating water planning.

Transfers and Marketing

Free transferability of water rights facilitates movement of water to economic uses of higher value and promotes greater efficiency and conservation. These are goals touted by economists and water reformers with which western states apparently agree in principle. State activity in the area of transfers during the 1990s, however, tended to restrict rather than encourage water markets as proposed in the 1980s.

Because states are subject to political pressure to control transfers between basins, out of state, and from agricultural to urban uses, it is reasonable for them to attempt to ameliorate adverse effects of transfers. Thus states have passed a number of laws that impede or increase the cost of transfers. Oregon and Kansas have subjected major appropriations to a special level of review.[131] Colorado has focused on the effects of transfers out of agricultural uses; it now requires revegetation when farmland is dried up.[132] Oregon, Nevada, and Texas have placed controls on transbasin diversions.[133] Six states have tried to take control of interstate transfers without offending dormant commerce-clause limitations.[134]

The only significant state effort to promote intrastate transfers was based on subsidy, not liberalization of the market. Desperate to reduce dependence on imported water, in 1998 the California legislature appropriated $235 million to promote transfer of rights from agricultural to urban uses.[135] It would move water from the Imperial Irrigation District to San Diego through existing facilities of the Metropolitan Water District. Curiously, the transfer of low-value uses to high-value uses needed political rather than economic impetus; it still has not been consummated.

A development that ostensibly encourages water marketing among states on the Colorado River was the creation of the Arizona Water Banking Authority in 1996.[136] The law was motivated by Arizona's desire to capture presently unneeded Colorado River water to which it is legally entitled. Thus other states, primarily California, could not continue using the water surplus to meet current demands in excess of their entitlements. Rather than being motivated by market forces, the "banking" proposal was created by state action that required new taxes to subsidize the expensive pumping and transport of water from the river for storage underground in Arizona. The notion was that in dry years, Arizona could rely on this stored groundwater and market to Nevada and possibly California the right to use the quantity to which it is entitled directly from the river in those years. Acceptance of the scheme by the Colorado River basin states marks a significant departure from their former unyielding opposition to interstate water marketing.[137]

Outside-the-Box Approaches

The most important innovations in water policy in the 1990s could be called outside-the-box approaches because they are not part of the traditional institutional framework. Typically, they have been motivated by federal pressure and local initiative. Although sweeping state institutional reforms are theoretically possible, experience of the recent past suggests that the most promising advances in the foreseeable future will be such ad hoc responses to problems arising in specific geographic areas.

Macro-watershed initiatives. In several major watersheds specific responses have been made to issues not being addressed successfully by existing institutions. The most notable is CALFED, which involves representatives of agriculture, business, the environment, and cities in efforts to solve problems growing out of the so-called Bay-Delta dispute.[138] Originally seen as a problem of controlling water quality in the delta of the Sacramento and San Joaquin Rivers at San Francisco Bay, it implicated water use and protection throughout California. Environmental and supply problems and operations of water-diversion facilities were causing serious violations of the federal Clean Water Act and Endangered Species Act. State institutions were not equipped to deal comprehensively with the related problems of water quality, watershed protection, ecosystem restoration, water-use efficiency, water transfers, and facilities for water storage and conveyance and for flood control. In the past these issues had been dealt with piecemeal by state legislation, agencies, and courts. When state institutions failed to produce effective solutions, however, CALFED was created. The participants are now engaged in statewide water planning to ensure more reliable water supplies and improved water quality for the environment, cities, and farms.

On the Colorado River, the operations of Glen Canyon Dam are now subject to recurring review by an "adaptive management work group."[139] A group of states, Indian tribes, power purchasers, recreational users, federal agencies, and environmental organizations is charged with monitoring the effects of policies and plans for operating Glen Canyon Dam. Releases of water from the dam have been primarily for power generation in the past, but this caused serious environmental and social problems. The group evaluates the consequences of dam operations for protection of Grand Canyon National Park, fulfillment of recreational and environmental purposes, recovery of endangered fish species, and other goals. It then recommends changes in the dam's operating regime as necessary to meet varied management objectives.

Another macro-basin effort is under way on the Platte River. After twenty years of conflict over the effects of water-development projects on endangered species in the central Platte River, the states of Colorado, Nebraska, and Wyoming have signed a cooperative agreement with the U.S. Department of the Interior for a joint program of restoration and management of the river. The plan addresses endangered species and various collateral issues.[140]

In the case of the Truckee River–Pyramid Lake issue in Nevada, the results of long and complex negotiations among diverse interests were embodied in federal legislation.[141] It provided for resolution of multiple issues ranging from tribal water rights to interstate allocation of waters between California and Nevada. Important wildlife issues were addressed, including water rights to benefit the Stillwater National Wildlife Refuge and implementation of fishery recovery plans for the imperiled cui-ui and the Lahontan cutthroat trout. The act also called for negotiation among all interested parties of an operating agreement for federal facilities on the Truckee River with increased efficiencies for the Newlands Reclamation Project.

Finally, representatives from Arizona, California, and Nevada have come together with various water and power agencies to form a regional partnership to develop the lower Colorado River basin Multi-Species Conservation Program.[142] The program is intended to comply with the Endangered Species Act (ESA) by conserving endangered species and their critical habitat in the 100-year floodplain of the Colorado River within the United States. Because species that become endangered may lie in the path of future development, the program will attempt to conserve eighty-eight mostly nonlisted species in the lower Colorado River basin. This will allow present development to proceed as part of an approved habitat conservation plan under provisions of the ESA that allow some incidental harm to endangered species if the plan provides overall benefits to habitat that otherwise might not be attained.[143]

Local watershed efforts. In addition to macro-watershed efforts, hundreds of local efforts involve people in neighborhoods, watersheds, and communities in solving water and other resource problems. Their approach is far different from the traditional approach to water problems. Instead of looking narrowly at specific projects or particular issues or problems, they attempt to include as many issues and interests as possible.

In watershed groups, people who have never dealt with one another may be drawn together by different concerns radiating from use of water. When a river runs through a town or community, for instance, some residents may be concerned about pollution caused by an upstream mine and seek ways to solve the problem, whereas others may be concerned because an endangered species issue inhibits the way in which the town can accommodate growth. Together, disparate interests become involved in developing multifaceted solutions, often employing a systems approach.[144]

The Animas River Stakeholder Group is typical of collaborative problem solving at the watershed level. Local residents came together over a common concern for the quality of the Animas River in southern Colorado after water-quality studies by the state Department of Health confirmed the extent of contamination of the river by cadmium, lead, and other metals leached from several abandoned mines. Encouraged by a federal Environmental Protection Agency (EPA) policy favoring watershed-based solutions to water-quality problems, the state agency hired a facilitator and helped get a U.S. Department of Energy grant to set up the stakeholder group. People were attracted to the group for reasons that varied from wanting to restore a fishery to avoiding Superfund listing and liability. Participants included two towns, a county, three state agencies, two federal agencies, a tribe, landowners, mining companies, environmental activists, and other concerned citizens. The group has undertaken monitoring, identifying pollution hot spots, prioritizing sources, and planning solutions to problems.[145]

The Natural Resources Law Center at the University of Colorado School of Law has been studying local watershed groups and has published a collection of profiles of many of them.[146] These groups are typically grassroots efforts created and operated from the ground up. Oregon[147] and Washington[148] have passed legislation that encourages and provides funding for such groups. Still, the use of local watershed groups remains largely an outside-the-box approach.

Dam removal. Breeching or tearing down dams is surely an outside-the-box approach. It is a revolutionary and controversial issue that was hardly even a topic of serious discussion until recently. Some experts urged "reoperation" (redefinition of the operating mode) of major dams and facilities, especially Bureau of Reclamation projects, to fulfill new environmental purposes and demands on water resources.[149] Indeed, reoperation is a widely applicable method of ameliorating the adverse effects of water development, but outright removal of some dams appears appropriate in many cases. Several years ago, as an option in relicensing proceedings, the Federal Energy

Regulatory Commission began considering removal of ("decommissioning") antiquated dams that were doing little good and much harm. The Edwards Dam on the Kennebec River in Maine was ordered removed to restore Atlantic salmon runs that had been impeded for almost 200 years. The Elwha Dam, which has blocked salmon spawning on the Elwha River in Washington, probably will be taken out in the near future.[150] American Rivers has reported that nearly 100 dams have been removed across the country.[151]

Surprisingly, removal of some of the nation's largest dams also is being seriously discussed. A current study assesses the consequences of breeching four major dams on the Snake River.[152] The plan would remove impediments to fish passage, thus improving the possibility of fulfilling Indian treaty rights to fishing and water. One northwestern governor has spoken in support of the controversial plan.[153]

Proposals also have been made to remove Glen Canyon Dam.[154] At first the idea was raised casually at a Sierra Club meeting. Some members of Congress called hearings on the idea, however, thus giving it unexpected attention and legitimacy. Since then the public has become more intrigued, and the proposal remains viable at least as a vehicle for studying the effects of the dam.[155]

General stream adjudications. At least four states have initiated general stream adjudications to determine water rights of users throughout an entire river basin. This is arguably not an outside-the-box approach, inasmuch as stream adjudications are created by state legislative action and occur in state courts or agencies. Nor is the idea of bringing a single lawsuit to determine all rights within an entire river system new. But there has been a resurgence of complex, multiparty litigation to determine water rights. Creating such special procedures and rules to govern a proceeding before a special-purpose court, master, or agency tends to remove policy making for entire watersheds from ordinary processes for several years while rights are determined.

Basinwide adjudications can be massive in scope. In Idaho the Snake River basin adjudication involves 174,459 claims and has been in active litigation since 1987. The Gila River adjudication in Arizona began in 1974 and seeks to resolve the claims of 24,000 water users.[156] In Oregon the Department of Water Resources began an adjudication process in 1990 that joined 25,000 claimants to determine federal and pre-1909 water rights in the Klamath basin.[157]

Revisiting water rights and claims in the context of entire watersheds and modern values can alter long-standing expectations, which in turn can lead to political reactions. If adjudications treat all the claims in a basin with the same level of concern for equity and efficiency, they may cast doubt on old rights. State legislatures have attempted to change procedures and alter the substantive rules that determine water rights of the parties where court rulings seem to disrupt settled expectations. For example, in Idaho newer, more efficient groundwater uses were in competition with older, less efficient surface-water uses. When the court in the Snake River basin adjudication applied a standard of reasonable efficiency, it disadvantaged some of the most senior users whose uses depended on old surface diversion works. The state legislature tried to change the rules of decision after the special master and court in the adjudication ruled in favor of the junior users.[158] Similarly, in the Gila River adjudication, revisions to the enabling legislation were mostly struck down after challenges by the United States and by tribes.[159] The attempted revisions changed the rules for findings of forfeiture and abandonment, as well as adverse possession and filing requirements for water-rights applications, and would have eliminated the availability of alternative water sources as a bar to surface-water-rights claims. The legislature intended that the new rules apply retroactively for the benefit of senior users in the adjudication, but the court found that the vested rights of the would-be junior users could not be changed without violating due process under the state constitution.

Indian water-rights settlements. Under venerable Supreme Court precedent, Indian tribes and the United States hold impliedly reserved rights to water in sufficient quantities to fulfill the purposes of their reservations.[160] These rights are superior to the rights of most water users who depend on rights to use water under state prior appropriation laws because they date from the time a federal or Indian reservation was established, not from first use. Unless specifically quantified for particular purposes, reserved water rights cast a cloud of uncertainty over state-created rights, which have a later origin.

Desire to quantify federal and Indian reserved water rights to provide greater certainty for other water users was the motivation for most states' general stream adjudications. The federal government and the tribes tried to keep actions that might quantify reserved water rights out of state courts by arguing that the McCarran amendment,[161] a federal law consenting to the adjudication of federal water-rights claims in state courts, does not apply to reserved rights. Although the legislation does not mention reserved rights, the courts uniformly ruled that Congress intended to cover these rights in passing the McCarran amendment.[162] The states then seized the opportunity to quantify federal and Indian reserved rights in their own courts. Most western states commenced some type of proceeding for this purpose. Even quasi-administrative adjudications have been held to qualify as proceedings under the statute.[163]

Proving claims to reserved rights requires that a court quantify future needs, not simply present usage. In the case of a reservation set aside for agricultural purposes, the court must find out how many practicably irrigable acres are on the reservation and then determine how much water is needed for those lands.[164] That this is enormously expensive and time-consuming was proved in the only McCarran amendment reserved rights case to be fully adjudicated.[165] Some experts had long urged that such claims should be settled rather than litigated, and states and the federal government embraced the idea.[166] Following a flurry of activity in the 1980s, when a dozen or more settlements were concluded, the Native American water-settlement process stagnated. The Clinton administration concluded only one settlement in addition to those that were in the works when it came into office in 1993.[167] Montana finally approved a settlement on the Rocky Boy's Reservation in the closing days of 1999.[168] Other settlements have been in negotiation and may move forward in the near future.[169]

Forces for Change: Federal Regulation and Local Action

Today, water decisions are sensitive to the public interest. The environmental, social, and economic efforts of major water-development proposals have become important factors in many decisions about water management. Local groups are finding ways based on more comprehensive consideration of consequences to solve problems caused by past decisions. Moreover, decisions today tend to include diverse interests. These changes have not resulted from reforms in state legal and institutional frameworks. Scattered among all of the western states are some interesting efforts at reform, but even if all were concentrated in a single state the record would fall short of the needs for reform that were identified in the 1980s. Most reforms in water policy in the 1990s were federally motivated or outside the box; they resulted mostly from intensified activity within federal programs and the growth of watershed or other place-specific efforts by citizens.

Influence of Federal Regulation

The two most powerful legal influences today on development and use of water are not state laws but the federal Endangered Species Act[170] and section 404 of the federal Clean Water Act.[171] Section 7 of the Endangered Species Act requires that an official who grants any federal permit or approval must consult with the U.S. Fish and Wildlife Service (or in some cases the National Marine Fisheries Service) to determine if the proposal would jeopardize the continued existence or habitat of a species listed by the government as threatened or endangered. Because the construction, alteration, or operation of virtually every major water facility—whether publicly or privately sponsored—requires some kind of federal permit and much of the undeveloped water in the West affects habitat, the ESA often affects plans for water management. If a water project would jeopardize an endangered species or its habitat, an alternative must be found. Moreover, under section 9 of the ESA, even nonfederal actions that result in habitat modification on private lands can constitute a prohibited "taking" of endangered species.[172]

Application of the ESA can potentially halt new development and uses of water. Consequently, parties seeking practical solutions that comply with the law have created many of the voluntary, collaborative processes at the macro- and local watershed levels.[173] These groups are attractive because they provide forums for development of more flexible plans than the alternatives sometimes offered by the U.S. Fish and Wildlife Service. Also, the plans can be designed to satisfy diverse interests. The Fish and Wildlife Service can propose "reasonable and prudent alternatives" that allow the federal government to approve actions that do some harm to endangered species.[174] The agencies also can approve "habitat conservation plans" that allow some incidental taking of endangered species by private parties who otherwise would be subject to severe sanctions.[175] Some of these plans, however, produce environmental benefits that could not be required by the government. By including anyone who might be able to raise a substantial legal objection in the process of developing reasonable and prudent alternatives or a habitat conservation plan, the action is likely to be susceptible to challenge.

Outside-the-box processes, including many of the local and macro-watershed initiatives being used to solve water problems in the West, were frequently motivated by fear of ESA enforcement. Colorado, Nebraska, and Wyoming were brought to the table over endangered species issues on the Platte River because the central Platte is habitat for whooping cranes and other endangered species.[176] In the Snake and Columbia Rivers, after seven species of salmon were listed as threatened or endangered,[177] the ESA began to drive the major changes in water management and hydropower generation throughout the Northwest. Local watershed groups have attempted to devise solutions to the complex problems of fisheries and hydropower facilities.[178] The Bay-Delta problems that led to the CALFED process arose under the Endangered Species Act and the Clean Water Act.[179] Colorado River management is heavily influenced by the need to protect endangered species of fish, thus limiting the scope and type of water use and the extent of power generation on the river and in the seven basin states. Nearly every stretch of the Colorado River has been designated as critical habitat for one or more species of endangered fishes.[180]

Section 404 of the federal Clean Water Act is also a significant limitation on water development. The law prohibits the dredging and filling of U.S. waters—including wetlands—without a permit. The U.S. Army Corps of Engineers administers the program and includes among activities requiring a permit the building of a dam.[181] The courts have validated the ability of the Corps (and of the EPA, which can veto permits granted by the Corps)[182] to deny section 404 permits for new dams and water facilities.[183] The EPA vetoed a permit for the proposed Two Forks Dam in Colorado, for instance, after it was disclosed that the project would flood miles of an extraordinary trout stream, eliminate elk habitat, and degrade water quality.[184] The EPA saw alternatives to the Two Forks project because Denver had not exhausted possibilities for water conservation, groundwater use, or exchanges of water and leasing of water rights. The EPA intervened in the absence of any state agency empowered under Colorado law to protect or even evaluate the public interest in the development of a water right. Because the section 404 permitting process is a federal action, it also triggers the need for assessing effects of a water-development project on endangered species.[185]

Faced with immediate and expanding threats of federal regulatory action that would limit their ability to allocate and develop water, the states seemed committed to improving their water laws in the 1980s. State reforms were expected to be effective enough to soften the effect of federal laws and supplant much of the need for federal presence. In fact, state actions have been piecemeal and reactive. Most of the limited reforms of state laws in the 1990s were motivated by the mandates of federal laws, especially the ESA and section 404. In some instances states did the minimum necessary to ensure compliance with federal laws. For instance, the need for reform of Texas groundwater law was increased by litigation that enforced the Endangered Species Act. The legislation that integrated management of surface and groundwater was specific to the Edwards Aquifer; true conjunctive management of groundwater still has not been achieved in the rest of Texas.

Because the pressure on California comes from palpable fears of water shortage, its reforms have gone beyond what was strictly necessary to comply with federal law. Still, far less would have happened in California without federal pressure.

Involvement of Multiple Interests in Local Decisions—An Unexpected Federal Role

Nearly all the outside-the-box approaches that characterize the most important water-policy reforms of the 1990s utilize inclusive processes. They generally include, at the watershed or community level, representatives of a variety of interests historically excluded from formal state processes. Several states have recognized the value of watershed-based decision making and supported processes at that level.[186]

Although watershed groups are used in the water-quality programs in at least ten states,[187] only Washington and Oregon have passed laws that encourage and assist these groups in addressing a variety of watershed issues. Water-quality issues dominate the attention of most but not all groups because of the EPA's watershed policy, which provides assistance and financial incentives and a way to avoid blanket federal regulation under the Clean Water Act.[188] The resulting shift in governance promotes local initiative and activism. Moreover, parties who had influenced decisions by elevating disputes to the courts sometimes use these new problem-solving institutions. Environmentalists, for instance, increasingly participate in these processes in place of litigation or after getting the attention of regulated parties by bringing a lawsuit. Nevertheless, federal financial support and the participation of federal agency personnel as much as any other factors account for the progress of many watershed groups.[189]

One would not have predicted that the federal government would foster local, collaborative processes. If the states failed or refused to respond to statutory requirements, the federal government could have assumed a stricter regulatory stance. Federal mandates, such as those of the ESA, are unequivocal. If the states did not change, it seemed in the late 1980s that the irresistible force of the federal command-and-control regime was headed for the immovable object of state water policy. Instead of meeting state resistance with greater force, however, the federal government emerged as more agile than the states by encouraging and participating in alternative processes. This is remarkable in that the federal agencies have unequivocal missions demanding that they force compliance with the law, whereas federal statutory support for collaborative processes is almost nonexistent. Moreover, the agencies face practical constraints and institutional disincentives to embrace multi-interest collaboration. These efforts can be long, expensive, and uncertain in outcome. They require agencies to commit limited resources and personnel to seemingly endless meetings instead of simply applying the law. Bureaucrats must be open to outsiders' suggestions for experimental approaches to compliance, and they may become involved in solving problems raised by others that go beyond the federal mandate. For instance, benefits to sport fishing, aesthetics, and tourism may be urged as part of a consensus solution to an endangered species problem.[190]

Today representatives of the Bureau of Land Management, the Environmental Protection Agency, the Fish and Wildlife Service, and the Forest Service are actively participating in collaborative processes. The Bureau of Reclamation also has changed to a water-management agency committed to using its facilities to accomplish a broad range of purposes beyond irrigation, power generation, and other uses historically associated with the agency; and it has involved public participation in its processes.[191]

There are two problems with the present state of affairs. One is a criticism by environmentalists, soundly based in law, that it is the job of federal agencies to enforce the laws rather than to be facilitators of compromise. When a federal agency plays the role of facilitator, it can blur the bright line of federal legal requirements. The federal government has unique missions in protecting species and habitats under federal law, protecting navigable waters, asserting Indian water rights and water rights for public lands, and dealing with international and interstate water allocation. These matters were the subjects of national legislation precisely because they were too important to be left to state governments or local interests.[192]

The other problem with ceding responsibility to federal agencies is the converse of the first. Federal programs inherently risk substituting national determinations of the public interest for state or local judgments, even on matters involving state or local importance and expertise. Just as some broad federal interests are best represented and driven by federal standards, some state and local interests need to be defined and dealt with at a political level closer to the physical resource.

THE FUTURE OF STATE WATER-POLICY REFORM: IT TAKES A CRISIS

Can states begin to assert greater leadership in areas where the federal government has become active in responding to public demands for change in water policy? If so, they could supplant much of the federal role, including decisions and functions often best understood by the parties closest to the affected ecosystems. This assumption was implicit in the agenda that the western governors set for themselves in the 1980s. Even more appropriate, states could envision water problems as part of their dominant and largely unaddressed problem of land use and growth management.

What is necessary to motivate the states to pursue such an agenda? It will take the force of a perceived crisis to move the western states to significant action in reforming water policy. Future reforms of state water law and policy will depend on the fortuity of crises and on states' preparedness to act when opportunities arise.

It may have been unrealistic to accept the rhetoric of the 1980s and therefore to expect that the states would embrace an agenda of water-policy reform without the pressure of a crisis. A series of four "Park City Workshops" led to tacit agreement on the Park City Principles by the parties who attended. The principles required state action.[193] But without conditions that approach crisis proportions, state reform of water policy is unlikely to happen in the near future. It is reasonable to expect, however, that the consequences of failing to confront demands that exceed supplies and problems of noncompliance with federal laws that could further constrain the use of existing supplies will eventually create a crisis sufficient to provoke action on water policy.

The success of water reform in meeting modern needs depends on readiness of the varied interests with a stake in improved water policy to take advantage of crisis situations when they arise. The stresses that will contribute to future water crises in the western United States—thus presenting opportunities for change in water policy—include demographic change, effects of present policies, and climate.

Demographic Change

The West is the fastest-growing, most urbanized region in the country (Chapter 13). Water policies embedded in the law were designed to promote the rapid settlement and development of the West. Today the dominant problem for the West is not how to promote growth but how to manage it. Nevertheless, few states have adopted effective land-use controls. Most states leave the task to local governments, and none has truly integrated water planning with land-use planning.

California and Texas show large population increases from birth and immigration. These states alone will account for about half of the nation's total growth from 1995 to 2025. Growth in the West will be concentrated in what some researchers have termed "urban archipelagos"—dense metropolitan areas surrounded by sparsely populated rural areas.[194] These urban areas, which will account for most of the West's expansion in water demand, are usually distant from water sources and are already confronting the limits of their supplies. Water sources now available in the West may be sufficient to accommodate growth in the foreseeable future, but only if water is moved from agricultural to urban uses (Chapter 13).[195] These new and changed water uses portend economic and social dislocation and may require the proliferation of new facilities—dams and pipelines—as well as further depletion of stream flows. Attempts to satisfy the demands of population growth consequently will create major environmental challenges and contribute to crises.

Effects of Current Policy

Water policy contributes to the likelihood of a crisis. This is true of old components of policy that are inflexible and fail to respond to broader values and changing conditions, but it is also true of changes in policy that demand new services from existing supplies. Inattention of the law to degradation and depletion of supplies creates stresses. For example, failure to respond to problems of aquifer depletion and contamination of supply can curtail production from present sources, creating conditions that are irremediable during the planning horizon for water managers. Net depletion of aquifers, especially where recharge occurs only over geologic time, can ruin the economies of entire regions, as is threatened in the several states overlying the Ogallala Aquifer.[196]

Pollution of surface or groundwater can also render some sources of water useless for the foreseeable future. Contaminated groundwater is especially difficult to rehabilitate. Another groundwater problem needing attention in many areas is saltwater intrusion. As pumping removes fresh water—particularly in coastal areas—seawater is drawn into the aquifer, destroying it as a freshwater source,[197] thus exacerbating water crises.

Even laws that respond to new public values can add to pressures on fixed supplies. Laws promoting efficient use, environmental protection, and maintenance of stream flows are intended to add balance to the system, but they too are water demands that can stress the system. The demand for additional and different services limits the amount of water that can be extracted and the ways and places in which water can be developed. The Endangered Species Act, Clean Water Act section 404, the Federal Power Act, and the Grand Canyon Protection Act contribute to conditions that may form a crisis. As water reform advances and environmental protections are implemented, shortages will occur in part because of these new demands.

Laws promoting efficiency are part of the agenda for legal reform, but their implementation also can cause stresses. The historical allocation of most water in the West to inefficient agricultural uses and the presence of barriers that block transfer of water to more efficient uses have motivated reform. In recent years market transfers have taken advantage of overly generous (or wasteful) appropriations for agriculture where water rights may not have been fully used. The resulting efficiency, however, removes a margin of safety from the system. For instance, the inefficiency of overstating rights and of applying excessive quantities of irrigation water that ultimately seeps back to a river has ameliorated environmental consequences of consumptive uses. As urban suppliers divert the same quantities of water but put it to more efficient use, they may decrease the water in streams by inefficient agricultural applications that incidentally maintained stream flows. When irrigation ditches are lined, water is saved and new crops can be grown, but wetlands sustained by seepage may dry up. Generally speaking, the most efficient application of water to consumptive uses tends to be at odds with nonconsumptive uses, such as fishing and boating. A legal reform agenda should include processes that consider these competing values.

Shortages may be more sharply and immediately felt by water users as the system becomes more efficient. This is especially true when transfers move water from one geographic area or economic sector to another, as would occur through a transbasin diversion or a transfer from agricultural to urban uses. These kinds of transfers tend to cause economic and social dislocations. The solution is not to perpetuate inefficiency but to account for the consequences more completely and thoughtfully in decision-making processes.

Climate

Climate can contribute to future water crises. Whereas demographic change and effects of current policy create stresses that set the stage for water crises, the perception of a crisis will almost always be triggered by a climatic event. Thus the occurrence of climatic events will be the most readily identifiable antecedent to a crisis of magnitude sufficient to attract the political attention needed for significant water-policy reform.

Three kinds of climatic effects potentially cause crises for water services: cyclical variations, severe drought, and long-term climate change. The first type results from the seasonal and annual cycles that cause precipitation, snowpack, and river flows to vary widely in the West. The degree of variation can be enormous, but the probable extent of such cyclical variations can be predicted based on historical records. Planners project the "safe yield" of water sources and reservoir storage based on long-term averages. Cyclical variations are unlikely to trigger crises unless they coincide with other stresses.

Severe droughts, like major floods, are inevitable but infrequent events and can cause significant challenges for water planning. To some extent these extreme events are included in the long-term statistical information used to project safe yield. Even so, droughts can be severe enough to create a crisis, thereby arousing public concern and political action.

Long-term climate change may exacerbate the effects of cyclical variations and severe droughts.[198] Identifying and predicting climate change is a developing field, and although scientists seem to agree that human activities are altering climate, significant uncertainty remains about the effects of climate change on regional water resources.[199] Long-term climate change generally has not been considered or reflected in water managers' plans or in the decision-making frameworks of institutions charged with shaping water policy. This is perhaps to be expected in light of uncertainty about the extent and nature of climate change. Even when different climate models predict the same range of expected changes in temperature, they may predict different kinds of effects. Moreover, current predictions may not be reliable for a specific region or watershed. The risks of overreacting or underreacting to the predictions are equally troubling. Although the evolving science of climate change can be perplexing, it is possible that scientists will develop realistic scenarios that can guide the planning and problem-solving processes.[200]

Episodes of actual or anticipated water shortage have always excited popular and political interest in water problems. The resulting attention to actual or anticipated supply problems opens a window of opportunity for water-policy reform. This does not mean a climate event or trend will effectively focus attention on the problem of anticipating or adapting to future problems of climate variability and change. In practice, legislatures may respond to a natural climate variation by initiating a politically popular project or program, such as building a dam. If the response addresses the climatic event at all, it may be only temporary, as was a water-marketing scheme implemented in California during the drought of the late 1980s.[201] Unlike other crisis-provoking stresses, effects of climate are not human-induced, reasonably predictable beyond normal cycles, or controllable by laws and institutions. These facts make climate variability a uniquely difficult issue for policy makers.

If policy makers consulted more directly and consistently with climate experts, they could improve their ability to account for climate variability, especially long-term climatic trends. In addition to assisting with data interpretation, these experts could help develop appropriate water policies. Climate experts should also be included in formulating the place-specific, outside-the-box efforts that anticipate rather than respond to crises. They could thereby test approaches to solving water problems and glean transferable experiences that might later be institutionalized.

The key to using crisis-driven opportunities for water reform is to engage people who can influence policy. This will be a difficult challenge for such projects. Proposals for reform threaten established interests. In the past, changes in state water law and policy were relatively modest and were designed to facilitate water allocation and development. Today's water decisions require input from a variety of previously unrepresented interests because of the public's panoply of interests in the environmental, economic, and social consequences of water use. Well-established interests with great influence in state politics, such as special districts and state agencies, correctly perceive that changing policies may threaten their influence. These traditional institutions appear willing to risk gradual obsolescence rather than suffer an immediate marginal loss of economic and political influence by opening the process to other interests, values, and people.

Between Crises: Outside-the-Box Approaches as Laboratories for Broader Reforms

If states develop sound water-policy–reform proposals in advance, the reforms can be put forward when a crisis arrives. Today's outside-the-box projects targeting particular local problems can be useful laboratories for developing and testing policies and approaches that could be applied more broadly. Localized processes can provide opportunities for refining methods for including multiple parties and dealing with multiple issues. The solutions developed in the outside-the-box approaches should also be cataloged for consideration by problem solvers in the future because the measures intended as a response to place-specific issues may be transferable to comparable situations elsewhere.

Outside-the-box efforts have proliferated because people with a stake in a particular local issue or problem perceived the potential for developing a viable solution and understood that the consequences of failure could be grave and imminent. Apart from more pervasive crises, these multi-interest, place-specific problem-solving processes will likely continue producing the most significant changes in water policy as they respond to the exigencies of localized problems.

States can build upon the local approach and formulate proposals for integrating some of the processes and solutions that emerged in the several outside-the-box approaches of the 1990s into established institutions for water governance. These place-specific projects have experimented with more open and inclusive public participation. They represent attempts to address a wide range of interconnected environmental and economic issues rather than focusing only on water allocation or development. Many involve integrated consideration of environmental preservation or restoration, habitat protection, demand management, land-use controls, and the use of markets and pricing.

Outside-the-box efforts also may be improved and may offer lessons for water-policy reform. Researchers should examine the kinds of solutions that emerge and their typical elements. In addition, researchers should consider who participates, what motivates participants, how problems are identified and framed, and how the effort is organized. Potential research topics include the importance of leadership; differences in effectiveness of consensus decision making and majority rule; the presence, absence, or type of federal participation and its relevance to the success or failure of the effort; and the influence of scientific and technical expertise on effectiveness.

Solutions to water problems must be based on access to sound technical and scientific information and judgment. Scientific experts, including hydrologists, biologists, agronomists, foresters, and meteorologists, can bring essential information and valuable insights to water-policy discussions. Yet these experts rarely have participated in formulating policy or even advising those who create policy. In the past their involvement in water policy has been limited to narrow issues or to the evaluation of proposals already developed by policy makers.

Climate experts in particular have been consulted infrequently in water-policy formulation. Moreover, they have taken little initiative to relate their data and research to actual problems or policies, even though climate variability is the stress most likely to trigger crises offering policy-reform opportunities. Given the endemic uncertainty in climate models and predictions, scientists are especially needed to interpret and assist in the rational application of the best data available concerning long-term trends (Chapter 16). Failing to consider and react appropriately could undermine the value of new information. Climate experts should be involved or at least consulted in most aspects of water policy and planning, from local and macro-watershed efforts to formulation of broader policies. They should assist in preparing for the physical manifestations of climate variations and in predicting coincidence of stresses that could create the crisis-driven opportunity for reform.

Greater involvement of scientific experts in policy making requires that scientists recognize the value of applying their work to real-world problems. Because the immediate material and professional rewards of doing so are often small, scientists may not find it attractive to commit to long-term involvement in solving the problems of planners or managers. Experts may, however, be attracted to working with specific groups if their contributions can be extended to broader policy reform.

CONCLUSION

The issues in western water today are far more complex than in the days when the West's policies were being formed. Initially, water could be developed without considering the effects on anyone or anything but holders of water rights. Policies allowed users to do as they pleased because state water laws and institutions facilitated and mediated productive use of water as a means of expanding the economy of the West. Later the essential policy goal was to find and develop new supplies. Financial support for structures to transport or store water became centralized, often at the federal level. The federal government focused almost entirely on engineering feasibility and not on the economic, environmental, or social impacts of its projects.

Problems resulting from unregulated appropriation and subsidized development are now widely acknowledged. The problems include inefficient use, damage to natural systems, social and economic impacts of moving water from agricultural to urban uses, and pollution caused by return flows and depletions. Now public preferences for conserving resources for both economic and ecological reasons confront the old, inflexible water laws and policies. Dams and canals have tied water to specific places and uses, and vested property rights to use of water have made changes more costly.

As federal laws asserted national environmental goals and as federal funds became scarce, the national government retreated from major water-development subsidies. Eventually, the best sites for dams were used up, and most proposals for developing major new supplies encountered environmental conflicts. Federal agencies became more active in enforcing environmental laws that thwarted both public and private attempts at water development. Meanwhile, state laws changed little to adapt to the new situation of the West.

Large demographic changes are occurring in the West because of unprecedented population growth and concentration of human settlements in urban areas (Chapter 13). Urban water demands are expanding, although much of the West's water is still committed legally to agricultural uses. At the same time, we understand the limits of nature better than ever. The public increasingly appreciates the value of protecting ecosystems and of conserving natural amenities for human enjoyment, and strong political constituencies oppose environmentally damaging construction or operation of water facilities. Nearly all practicably developable supplies are fully committed or unavailable because of the costs that would be incurred or the environmental damage that would be caused by developing them. Instead of building new dams and canals, water suppliers must stretch existing supplies through demand management by moving water from old to new uses. These phenomena have incited advocacy for policy reform. State officials in the 1980s embraced an agenda that included integrating water planning with other interests, using market forces to enable efficient water transfers, allowing greater public participation, regulating to protect stream flows needed for fish and wildlife and recreation and to prevent pollution, and satisfying the public interest in water decisions. During the 1990s, however, the western states made little progress on the water reform agenda. Although existing state legal and institutional frameworks endure virtually unchanged, changes did occur largely because the federal government pressed its regulatory requirements and local interests responded by seeking place-specific solutions. Thus water reform along the lines proposed by state leaders in the 1980s advanced in the 1990s in spite of the states' inaction. Proponents of reform, ranging from environmentalists to urban suppliers to commercial recreational users to scientists, are not part of a unified political movement. Consequently, measurable progress in dealing with the inadequacies of water policy has tended to occur through outside-the-box efforts.

The continuing need for creative institutional responses was emphasized in the 1996 report of the Western Water Policy Review Advisory Commission, which called for improved governance of water through "integration of federal programs with state, tribal, and local efforts."[202] Although it is possible to reinvigorate the state role in water policy, it is unrealistic to expect that pervasive policy development will proceed rationally and at an even pace. The light of political attention shines on water issues only episodically, typically in times of stress, when a crisis develops in the availability or condition of resources. Producing carefully considered, well-informed solutions in a crisis atmosphere is awkward and unpromising. Yet in the midst of a perceived crisis, states can seize opportunities for making incremental changes by institutionalizing the reforms pioneered by groups already involved in outside-the-box efforts. States can assimilate these experiences as models for alleviating broader problems and reforming outdated state laws and institutions.[203]

Developing wise policies is a long-term, continuous process. Opportunities to implement broad changes in water policy, however, are likely to arise only episodically; therefore thoughtful and workable policies need to be ready when opportunities arise to institutionalize change. Outside-the-box approaches such as those that emerged in the 1990s can be valuable laboratories for cultivating the elements of broadly applicable institutional reforms. States should recognize this opportunity and formulate integrated water policies and new institutions relevant to the issues of the twenty-first century. If they do not, western water policy will continue to change—but it will change in spite of rather than because of states' efforts.

ACKNOWLEDGMENTS

The author is grateful for the excellent research assistance of Niccole Sacco, class of 2000, University of Colorado School of Law. This chapter benefited from advice, materials, or comments from Hope Babcock, Meg Caldwell, John Firor, Douglas S. Kenney, Lawrence J. MacDonnell, Dale Pontius, and Barton Thompson. A version of this chapter was published as an article at 20 *Stanford Environmental Law Journal* 3 (2001).

NOTES

1. Samuel C. Wiel, Water Rights in the United States 88 (3d ed. 1911).
2. *See* Sarah F. Bates, David H. Getches, Lawrence J. MacDonnell, and Charles Wilkinson, Searching Out the Headwaters 31–33 (1993).
3. *See* Desert Land Act of 1877, 43 U.S.C. § 321; Mining Act of 1866, 43 U.S.C. § 661.
4. *California-Oregon Power Co. v. Beaver Portland Cement Co.*, 295 U.S. 142 (1935); *see also Andrus v. Charlestone Stone Products Co.*, 436 U.S. 604, 614 (1978).
5. *United States v. Rio Grande Irrigation Co.*, 174 U.S. 690 (1899).
6. *See, e.g., California-Oregon Power Co.*, 295 U.S. at 155; *United States v. Appalachian Electric Power Co.*, 311 U.S. 377, 298 (1940); *California v. United States*, 438 U.S. 645, 662 (1978).
7. *Winters v. United States*, 207 U.S. 564 (1908); *Arizona v. California*, 373 U.S. 546 (1963).
8. *See* Charles T. DuMars and A. Dan Tarlock, *New Challenges to State Water Allocation Sovereignty*, 29 Nat. Resources J. 331, 332 (1989).
9. 43 U.S.C. § 383 (Reclamation Act). *See also* 16 U.S.C. § 821 (Federal Power Act); 33 U.S.C. § 1251(g) (Clean Water Act).
10. *Arizona v. California*, 373 U.S. 546 (1963); *Ivanhoe Irrigation Dist. v. McCracken*, 357 U.S. 275 (1958); *City of Fresno v. California*, 372 U.S. 627 (1963); *California v. Federal Energy Regulatory Commission*, 495 U.S. 490 (1990). *But see California v. United States*, 438 U.S. 645 (1978) (legislative history of the Reclamation Act of 1902 indicates Congress intended to defer to state water law not in direct conflict with the act).
11. *E.g., Riverside Irrigation Dist. v. Andrews*, 758 F.2d 508 (10th Cir. 1985).
12. On the deficiencies of early record keeping, which led to creation of state agencies, *see* McIntire, *The Disparity Between State Water Rights Records and Actual Use Patterns—"I Wonder Where the Water Went,"* 5 Land and Water L. Rev. 23 (1970).
13. *See* Charles F. Wilkinson, Crossing the Next Meridian: Land, Water, and the Future of the West 240 (1992) (describing western water agencies as "captured agencies").

14. *E.g., Young & Norton v. Hinderlider,* 110 P. 1045 (N.M. 1910) ("public welfare" includes consideration of costs of projects for protection of investors as well as consideration of public health and safety); *Tanner v. Bacon,* 136 P.2d 957 (Utah 1943) ("public welfare" requirement allows state to prefer a junior applicant for water rights with the "more beneficial use" over a senior applicant); *Shokal v. Dunn,* 707 P.2d 441 (Idaho 1985) (state engineer can consider "any locally important factor," such as water quality, fish and game, health, and economics, as part of the "local public interest").
15. *See* Susanne Hoffman-Dooley, *Determining What Is in the Public Welfare in Water Appropriations and Transfers: The Intel Example,* 36 Nat. Resources J. 103 (1996). *See* Charles F. Wilkinson, Crossing the Next Meridian: Land, Water, and the Future of the West 240 (1992).
16. *See* John D. Leshy, *Special Water Districts—The Historical Background, in* Special Water Districts: Challenge for the Future 14 (James N. Corbridge, ed. 1983).
17. Wells A. Hutchins, Water Rights Laws in the Nineteen Western States 160 (1971) (completed by Harold H. Ellis and J. Peter DeBraal).
18. *See generally* Special Water Districts: Challenge for the Future (James N. Corbridge, ed. 1983).
19. *See* John R. Mather, Water Resources: Distribution, Use and Management 294–297 (1984); Western Water Policy Review Advisory Commission, Water in the West: Challenge for the Next Century at 2–9 (1998).
20. *See* Robert G. Dunbar, Forging New Rights in Western Waters 34–45 (1983).
21. *See* D. Monte Pascoe, *Plans and Studies: The Recent Quest for a Utopia in the Utilization of Colorado's Water Resources,* 55 U. Colo. L. Rev. 391, 398 (1984).
22. David H. Getches, *Water Planning: Untapped Opportunity for the Western States,* 9 J. Energy L. and Pol'y. 1 (1988).
23. Water Resources Planning Act of 1965, 42 U.S.C. § 1962–1962d-18.
24. Ludwik A. Teclaff, *Evolution of the River Basin Concept in National and International Water Law,* 36 Nat. Resources J. 359, 371 (1996).
25. National Environmental Policy Act, 42 U.S.C. §§ 4331–4344 (1969).
26. United States Water Resource Council, *Economic and Environmental Principles and Guidelines for Water and Related Land Resources Implementation Studies* (March 10, 1983).
27. *See* Rodney T. Smith, Troubled Waters: Financing Water in the West 2 (1984); D. Craig Bell, Jo S. Clark, Julia Doermann, and Norman K. Johnson, *Retooling Western Water Management: The Park City Principles,* 31 Land and Water L. Rev. 303 (1996).
28. Richard W. Wahl, Markets for Federal Water: Subsidies, Property Rights, and the Bureau of Reclamation 189 (1989).
29. Charles F. Wilkinson, Crossing the Next Meridian: Land, Water, and the Future of the West 259 (1992).
30. Western Water Policy Review Advisory Commission, *supra* note 19 at 2-9.
31. For example, impermeable clays create poor soil drainage in a district along Arizona's Gila River, causing saline groundwater to destroy crops. The Bureau of Reclamation undertook multiple, expensive, and ultimately ineffective projects to try to remedy the situation. *See* David Getches, *Colorado River Governance: Sharing*

Federal Authority as an Incentive to Create a New Institution, 68 U. Colo. L. Rev. 573, 587 (1997). *See also* Marc Reisner, Cadillac Desert: The American West and Its Disappearing Water 7, 481 (1986). Also, the Kesterson Project in California's Central Valley caused leaching of naturally occurring selenium from irrigated farmlands that was drained from fields into the Kesterson Wildlife Refuge. The water was toxic to migratory birds in the refuge, inciting costly, not totally effective responses. Laura H. Kosloff, Comment, *Tragedy at Kesterson Reservoir: Death of a Wildlife Refuge Illustrates Failings of Water Law,* 15 Environmental Law Reporter 10386, December 1985.

32. *See* Bates et al., *supra* note 2 at 36.
33. *See* Western Water Policy Review Advisory Commission, *supra* note 19 at 2-9.
34. *Ibid.* at 2-12–2-13.
35. Elizabeth Losos, Justin Hayes, Ali Phillips, David Wilcove, and Carolyn Alkire, *Taxpayer-Subsidized Resource Extraction Harms Species,* 45 BioScience 446, 448 (1995). The researchers examined the impacts of a variety of resource extraction activities—including grazing, mining, logging, recreation, and water development—on endangered species. Water development was found to affect the greatest number of species.
36. *See* Marc Reisner, Cadillac Desert 293–300 (1986); Bates et al., *supra* note 2 at 43–47.
37. *See* Clean Water Act, 33 U.S.C. §§ 1251–1387; Clean Air Act, 42 U.S.C. §§ 7401–7671q; Federal Land Policy and Management Act, 43 U.S.C. §§ 1701–1782 (1976); National Forest Management Act, 16 U.S.C. §§ 1600–1614 (1976); Endangered Species Act, 16 U.S.C. §§ 1531–1544 (1973).
38. *See* Alexandra M. Shafer, *The Reclamation Reform Act of 1982: Reform or Replacement?* 45 U. Pitt. L. Rev. 647, 666 (1984).
39. Richard D. Lamm and Michael McCarthy, The Angry West 188–189 (1982).
40. *See* Smith, *supra* note 27 at 143. Although the Reagan administration, under the banner of "New Federalism," promised to remedy the trend toward greater federal control, it actually shifted burdens to the states for cost sharing for construction of water-resources projects and reduced the federal share of grants for wastewater treatment facilities, with little devolution of control to the states. *See* Peter Rogers, America's Water: Federal Roles and Responsibilities 179–180 (1993).
41. Although the government is involved in fewer major water projects than in the past, controversy continues over its role in projects considered to be environmentally destructive and wasteful. *See* Jeff Stein, Peter Moreno, David Conrad, and Steve Ellis, Troubled Waters: Congress, the Corps of Engineers, and Wasteful Water Projects (2000).
42. *See, e.g.*, Hoover Commission (Commission of Reorganization of the Executive Branch of the Government), Department of the Interior, H.R. Doc. No. 122, 81st Cong. (1949); Hoover Commission, Report on Water Resources and Power, H.R. Doc. No. 208, 84th Cong. (1955); President's Advisory Committee on Water Resources Policy, H.R. Doc. No. 315, 84th Cong. (1956); Senate Select Committee on National Water Resources (Kerr Committee), S. Rep. No. 29, 87th Cong. (1961); National Water Commission, Water Policies for the Future: Final Report to the President and to the Congress of the United States (1973); National Acad-

emy of Sciences, Water and Choice in the Colorado Basin: An Example of Alternatives in Water Management (1968).

43. *See, e.g.*, Bates et al., *supra* note 2; Report of the Long's Peak Working Group on National Water Policy, America's Waters: A New Era of Sustainability (December 1992); Water Quality 2000, a National Water Agenda for the 21st Century: Final Report (November 1992); Zach Willey and Tom Graff, *Federal Water Policy in the United States—An Agenda for Economic and Environmental Reform,* 13 Colum. J. Envtl. L. 325 (1988); Water Science and Technology Board of the National Academy of Sciences, National Research Council, Water Transfers in the West: Efficiency, Equity, and the Environment (1992); Marc Reisner and Sarah Bates, Overtapped Oasis: Reform or Revolution for Western Water (1990).
44. *See* America's Waters: A New Era of Sustainability, Report of the Long's Peak Working Group on National Water Policy, 24 Envtl. L. 125, 128 (1994).
45. *See* Water Efficiency: Opportunities for Action, Report of the Western Governors' Association Water Efficiency Working Group, Western Governors' Association Resolution 86-011 (Appendix A), signed by Governor George Deukmejian, California; Governor Richard Lamm, Colorado; Governor George Ariyoshi, Hawaii; Governor Ted Schwinden, Montana; Governor Bob Kerrey, Nebraska; Governor Toney Anaya, New Mexico; Governor George Sinner, North Dakota; Governor Vic Atiyeh, Oregon; Governor William Janklow, South Dakota; Governor Norman Bangerter, Utah, and Governor Booth Gardner, Washington.
46. Bruce Driver, Western Water: Tuning the System, Report to the Western Governors' Association from the Water Efficiency Task Force (July 7, 1986).
47. Water Efficiency: Opportunities for Action, Report to the Western Governors from the Western Governors' Association Water Efficiency Working Group (July 6, 1987).
48. For an important article on the need to integrate water-resources planning and local land-use planning and management, *see* A. Dan Tarlock and Sarah B. Van de Wetering, *Growth Management and Western Water Law From Urban Oases to Archipelagos,* 5 Hastings W.-N.W. J. Envtl. L. and Pol'y 163 (1999). *See also* Kevin M. O'Brien and Barbara Markham, *Tale of Two Coasts: How Two States Link Water and Land Use Planning,* 11 Nat. Resources and Env't 3 (1996); David H. Getches, *From Ashkabad, to Wellton-Mohawk, to Los Angeles: The Drought in Water Policy,* 64 U. Colo. L. Rev. 523 (1993).
49. An early modification of the doctrine altered the most basic requirement of prior appropriation—that water needs to be diverted before a legal right exists—and allowed a later diversion to relate back to the time the intent to appropriate was formed. The purpose was to provide security for investments in water development. *See* David H. Getches, *Water Law in a Nutshell 101* (3rd ed., 1997). When a single state's laws proved inadequate to deal with development and use of waterways that crossed state lines, the interstate compact was introduced. *See* Jerome C. Muys, Interstate Water Compacts: The Interstate Compact and Federal-Interstate Compact, National Water Commission, Legal Study No. 14 (1971). All states now have laws providing for administratively issued permits to use water, except Colorado, which uses specialized water courts to perform a similar function. *See* A. Dan Tarlock, James N. Corbridge, and David H. Getches, Water

Resource Management: A Casebook in Law and Public Policy 292–303 (5th ed. 2002).

Some changes in the basic state water laws in the West responded to demands for change that were not primarily development oriented—for example, laws to protect instream flows. *See* A. Dan Tarlock, *Appropriation for Instream Flow Maintenance: A Progress Report on "New" Public Western Water Rights,* 1978 Utah L. Rev. 211 (1978).

50. *See* Gregory J. Hobbs Jr., *Ecological Integrity, New Western Myth: A Critique of the Long's Peak Report,* 24 Envtl. L. 157 (1994); Gregory J. Hobbs Jr., *The Reluctant Marriage: The Next Generation (A Response to Charles Wilkinson),* 21 Envtl. L. 1087 (1991).
51. *See* Kenneth J. Burke, Esq., Dr. Ronald G. Cummings, and Jerome C. Muys, Esq., Interstate Allocation and Management of Nontributary Groundwater: A Discussion Paper Prepared for the Western Governors' Association 129; United States Advisory Commission on Intergovernmental Relations, Coordinating Water Resources in the Federal System: The Groundwater–Surface Water Connection 27–49 (October 1991).
52. *Public Utility District No. 1 v. Washington Department of Ecology,* 511 U.S. 700 (1994).
53. *See California v. United States, supra* note 6.
54. *See* Getches, *supra* note 22 at 25.
55. *See* Western Water Policy Review Advisory Commission, *supra* note 19 at 3-22. *See also* Bonnie G. Colby, Mark A. McGinnis, Ken A. Rait, and Richard W. Wahl, Transferring Water Rights in the Western States—A Comparison of Policies and Procedures, U. Colo. Nat. Resources L. Center Occasional Paper Series (February 1989).
56. The methodology for compiling the data in this section was as follows. All western state statutory enactments and appellate court decisions for the period that apparently dealt with water law were compiled. This was checked against three sources that periodically report on recent developments in water law. *See* American Bar Association, *Natural Resources, Energy, and Environmental Law* (University of Tulsa College of Law); *Water Strategist (Analysis of Water Marketing, Finance, Legislation and Litigation)* (Stratecom, Inc., Claremont, Calif.); *Rocky Mountain Mineral Law Foundation Water Law Newsletter* (Denver, Colo.). These secondary sources were consulted to (1) discover omissions in research into statutory law (e.g., an uncodified appropriation of funds in connection with interstate water marketing), and (2) have the benefit of the reporters' expertise in evaluating the significance of the legislation. Then I judged which legal developments should properly be considered state law reforms. This process surely has resulted in the omission of some developments others would have included. The data include some legal developments from 1989, and therefore the study period is longer than ten years. It seemed preferable to overcompensate for the fact that research was concluded before the end of 1999 than to use fewer than ten years of data.
57. David Getches, *Water Use Efficiency: The Value of Water in the West,* 8 Pub. Land L. Rev. 1, 12 (1987); A. Dan Tarlock, *Appropriation for Instream Flow Maintenance: A Progress Report on "New" Public Western Water Rights, supra* note 49;

Steven J. Shupe, *Waste in Western Water Law: A Blueprint for Change,* 61 Or. L. Rev. 483 (1982).

58. 1991 Kan. Sess. Laws, 292 § 4 (amending Kan. Stat. Ann. 82a-732 (1989) (water-conservation plans required for use of existing water rights).
59. Tex. Water Code Ann. § 16.051(a) (West 1998) (state water plan for water conservation).
60. *State of Washington v. Grimes,* 852 P.2d 1044 (Wash. 1993) (quantity of adjudicated rights limited by "reasonable efficiency").
61. Mont. Code Ann. §§ 85-2-103(14), 1-402(2)(e), and 1-419 (1991) (salvaged water can be used by appropriators who conserve).
62. 1996 Wash. Laws 320 § 11(2).
63. 1993 Or. Laws ch. 641 (West) (amending Or. Rev. Stat. §§ 537.470, 537.480, 537.485, 537.490, 537.495, and 540.510).
64. 1998 N.M. Laws ch. 37 (H.B. 460) (amending N.M. Stat. Ann. § 72-5-28 (1998) (enacted to protect conserved water rights from forfeiture as a result of non-use).
65. Ariz. Rev. Stat. Ann. §§ 45-311 to 319 (West 1992); Colo. Rev. Stat. § 9-1.3-102 (1991) (low-flow plumbing requirements).
66. Nev. Rev. Stat. § 628.
67. 1993 Water Recycling Act, 1991 Cal. Legis. Serv. 187 (West) (amends Cal. Water Code §§ 13050, 13241 and adds ch. 7.5 [commencing with § 13575]) (use of new water is waste if reclaimed water available); A.B. 541, 1997–1998 Reg. Sess. (Cal. 1997) (amending Cal. Water Code § 13271 and adding Cal. Water Code §§ 13529, 13529.2, 13529.4) (treated recycled water exempt from the definition of "sewage," allowing it to be used to supplement water supplies); Cal. Legis. Serv. 733 (West 1994) (amending Cal. Water Code §§ 13575, 13576 and adding Cal. Water Code §§ 13579, 13580, 13581, 13582) (water suppliers must identify potential recycled water uses within their service areas); 1993 Cal. Legis. Serv. 980 (West) (amending Cal. Water Code §§ 13552.4, 13552.8 (reclaimed water use must be considered in planning).
68. 1992 Cal. Legis. Serv. 226 (West) (adding Cal. Water Code § 14875) (DWR can adopt gray-water standards for residential systems).
69. Cal. Legis. Serv. 980 (West) (amending Cal. Water Code § 13554.2 and adding §§ 13552.2, 13552.6) (using potable water).
70. A.B. 609, 1997–1998 Reg. Sess. (Cal. 1998) (amending Cal. Water Code §§ 13,575–76, 13,579–81) (authorizes feasibility study of providing recycled water for groundwater replenishment); 1998 Cal. Legis. Serv. ch. 164 (A.B. 609) (West) (amending §§ 13575, 13576, 13579, 13580, 13581 of the Water Code) (allows agreements between groundwater replenishment agencies and recycled water wholesalers for the use of recycled water for groundwater replenishment).
71. Leg. Bill 108, 94th Leg., 2d Sess. (Neb. 1996) (amending various water statutes, specifically Neb. Rev. Stat. § 46-656) (groundwater hydrologically connected to surface water to be conjunctively administered if use of that water is affecting or is likely to affect surface-water supplies).
72. Act of May 30, 1993, 73rd Leg., R.S. ch. 626 (mandating management and control of groundwater in Edwards Aquifer, departing from absolute ownership rule).

73. *Friendswood Dev. Co. v. Smith-Southwest Indus., Inc.*, 576 S.W.2d 75 (Tex. 1978) (announced a prospective rule imposing liability for negligent pumping that causes subsidence).
74. *Sierra Club v. San Antonio,* No. MO-96-CA-097, *slip op.* (W.D. Tex. August 23, 1996).
75. *See* Todd H. Votteler, *The Little Fish That Roared: The Endangered Species Act, State Groundwater Law, and Private Property Rights Collide Over the Texas Edwards Aquifer,* 28 Envtl. L. 845 (1998).
76. Tex. Water Code Ann. § 11-151 (West 1998) (requiring water-rights permits for new and existing wells from Authority to be based on consideration of the effects of extracting groundwater on surface uses; protects spring flows needed for endangered species).
77. *Barshop v. Medina County Underground Water Conservation District,* 925 S.W.2d 618 (Tex. 1997) (upholding statute against constitutional challenge).
78. *Sipriano v. Great Spring Waters of America, Inc.*, 1 S.W.3d 75 (Tex. 1999) (landowner has right under the "rule of capture" to pump the groundwater under owned land without regard to impacts on others; decision whether to make changes in this principle left to the legislature).
79. Tex. Water Code Ann. § 11.151 (West 1998).
80. 1997 Nev. Stat. 572 (1997).
81. 1999 Nev. Stat. 534.120 (1999) (A.B. 408).
82. Idaho Code § 42-5201 *et seq.* (Supp. 1999) (adding ch. 52 to provide for Ground Water Districts).
83. Idaho Code § 42-4201A (Supp. 1994) (recognizing the appropriation and storage of water underground to be a beneficial use of water). Some states have administratively provided for storage of recharged waters, e.g., 2 Colo. Code Regs. § 402-11 (1995) (allowing evacuated aquifer space to be recharged and used as underground storage reservoirs, upon approval of state engineer).
84. N.M. Rev. Stat. § 72-5A-1 to 72-5A-17 (1999) (H.B. 2162) (permit program for groundwater storage and recovery). *Cf.* Ariz. Rev. Stat. Ann. § 45-2401 *et seq.* (Arizona water banking legislation [*infra* notes 136–137] depends on groundwater storage).
85. *See* David Gillilan and Thomas Brown, Instream Flow Protection: Seeking a Balance in Western Water Use (1997).
86. *In re Application A-16642,* 463 N.W.2d 59 (Neb. 1990) (new statute upheld).
87. The state engineer has said, however, he will recognize rights to flow in a public agency if a proper application is received. *See* Natural Resources Law Center, University of Colorado, *Instream Flow Protection in the West* 331–356 (Lawrence J. MacDonnell et al., eds. 1989).
88. Utah Code Ann. § 73-3-3 (Supp. 1992) (Division of Parks [as well as Wildlife] can change existing water rights to benefit fish, recreation, and environmental preservation).
89. Mont. Code Ann. §§ 85-2-102, 85-2-402, 85-2-404 (1989) (experimental leasing allowed on five streams); Mont. Code Ann. § 85-2-408 (1995) (opportunities for temporary changes or leases of existing water rights for instream flow uses increased).

90. *Hubbard v. Washington Department of Ecology*, 936 P.2d 27 (Wash. Ct. App. 1997).
91. Idaho Code § 42-1501 to 42-1505 (1996).
92. On Indian treaty rights to water, *see* Snake River Basin Adjudication, Judgment Granting Motions for Summary Judgment (Nez Perce Consolidated Subcase 03-10022) (11/10/99).
93. *See Carson-Truckee Water Conservancy Dist. v. Clark*, 741 F.2d 257 (9th Cir. 1984), *cert. denied*, 470 U.S. 1083 (1985) (ESA obligations empowered secretary to release water from federal project for endangered fish protection).
94. 1999 Idaho Acts ch. 364 (S.B. 1151).
95. In re *General Adjudication of All Rights to Use Water in the Big Horn River System*, 835 P.2d 273 (Wyo. 1992).
96. Colo. Rev. Stat. § 37-92-102(3).
97. 1994 Colo. Sess. Laws 766 (amending Colo. Rev. Stat. § 37-92-102[3] [Supp. 1994]) (state may not acquire conditional rights for instream flows); Colo. Rev. Stat. § 37-92-102(3) (1990 and Supp. 1994) (to relinquish instream flow rights, state board must hold administrative hearing; enacted in reaction to *Aspen Wilderness Workshop, Inc. v. Colorado Water Conservation Board*, 901 P.2d 1251 [Colo. 1995], holding that power to appropriate a minimum stream flow imposes a fiduciary duty between the state board and the people of the state).
98. *City of Thornton v. City of Fort Collins*, 830 P.2d 915 (Colo. 1992) (appropriation is for diversion and not instream flows if structures for "use" of water right are built at each end of stretch of river to be protected).
99. *In the Matter of the Application for Water Rights of the United States*, District Court, Water Division No. 6, Colorado, Case No. W-85 (March 1, 1985) (Dinosaur National Monument); *In the Matter of the Amended Application of the United States for Reserved Water Rights in the Platte River, District Court, Water Division No. 1*, Colorado, Case No. W-8439-76 (February 12, 1993) (National Forest lands).
100. *U.S. v. City of Challis*, 988 P.2d 1199 (Idaho 1999); *In re SRBA*, Case No. 39376, 1999 WL 778325 (Idaho, October 1, 1999).
101. The governor publicly declared that "the lives and livelihoods of tens of thousands of Idahoans could be affected." Given the rather limited opportunities for water development upstream of wilderness areas, this statement and the moratorium appear to be tactical moves to attract publicity and influence the state supreme court in the state's appeal of the decision. The political reaction against the decision recognizing federal reserved rights for wilderness areas led to the justice who wrote the opinion being voted out of office. Mark Warbis, *Power of GOP Machine, Water Rights, Drive Election Victories*, Associated Press Newswires, Boise, Idaho, May 24, 2000.
102. Douglas Grant, *Public Interest Review in Water Allocation and Transfer in the West: Recognition of Public Values*, 1987 Ariz. St. L. J. 681.
103. *Ibid.* at 147. *See generally* Norman K. Johnson and Charles T. DuMars, *A Survey of the Evolution of Western Water Law in Response to Changing Economic and Public Interest Demands*, 29 Nat. Resources J. 347 (1989).
104. *E.g.*, Wash. Rev. Code § 43.21C.030 (West 1998); Cal. Pub. Res. Code §§ 21000–21781 (West 1999). *See County of Inyo v. City of Los Angeles*, 139 Cal. Rptr. 396 (3d Dist. Ct. App. 1977) (EIR for proposed increased groundwater extraction by Los

Angeles from Inyo County should consider the alternative of a water-conservation program).

105. *See Galante Vineyards v. Monterey Peninsula Water Management Dist.*, 71 Cal. Rptr. 2d 1 (6th Dist. Ct. App. 1997).

106. *Bonham v. Morgan*, 788 P.2d 497 (Utah 1989) (change of use application subject to public welfare standard, although requirement was explicit only as to new applications); *Hardy v. Higginson*, 849 P.2d 946 (Idaho 1993) (although not explicitly required by statute, local public-interest standard applies to application to amend a water-right permit). Idaho Code § 42-211 (Supp. 1993); 1993 Mont. Laws 244, 445 (amending Mont. Code Ann. § 85-2-319) (changes must be in conformity with water-quality protection goals).

107. Mont. Code Ann. § 85-2-402 (1999).

108. 1999 Or. Laws 611 (amending Or. Rev. Stat. § 537.2.11).

109. 1999 Or. Laws 703 (amending Or. Rev. Stat. § 536.295).

110. *National Audubon Society v. Superior Court*, 658 P.2d 709 (Cal. 1983).

111. California State Water Resources Control Board, Decision 1631 (September 28, 1994) (Los Angeles's water rights amended to require water releases and limits on exports as necessary to provide instream flows and a channel restoration program to protect fish and other public trust resources).

112. *California Trout, Inc. v. Superior Court*, 266 Cal. Rptr. 788 (3d Dist. Ct. App. 1990).

113. *See* various perspectives expressed in Richard J. Lazarus, *Changing Conceptions of Property and Sovereignty in Natural Resources: Questioning the Public Trust Doctrine*, 71 Iowa L. Rev. 631 (1986); Charles F. Wilkinson, *Symposium on the Public Trust and the Waters of the American West: Yesterday, Today and Tomorrow*, 19 Envtl. L. 425 (1989); Timothy J. Conway, *National Audubon Society v. Superior Court: The Expanding Public Trust Doctrine*, 14 Envtl. L. 617 (1984); Michael C. Blumm and Thea Schwartz, *Mono Lake and the Evolving Public Trust in Western Water*, 37 Ariz. L. Rev. 701 (1995).

114. *See Kootenai Envtl. Alliance v. Panhandle Yacht Club, Inc.*, 671 P.2d 1085 (Idaho 1983).

115. *Idaho Conservation League, Inc., v. State of Idaho*, 911 P.2d 748 (Idaho 1995).

116. Idaho Code §§ 58-1201 to 1203 (1996) (doctrine solely a limit on state's alienation of title to beds of navigable waters).

117. Colo. Const., art. XVI, sec. 5.

118. *Aspen Wilderness Workshop, Inc. v. Colorado Water Conservation Board*, 901 P.2d 1251 (Colo. 1995).

119. *See* Bates et al., *supra* note 2 at 128–151.

120. *See* A. Dan Tarlock, *Western Water Law, Global Warming, and Growth Limitations*, 24 Loy. L. Rev. 979, 1010–1011 (1991).

121. *E.g., Denver v. Sheriff*, 96 P.2d 836 (Colo. 1939); *see* Janis E. Carpenter, *Water for Growing Communities: Refining Tradition in the Pacific Northwest*, 27 Envtl. L. 127, 128 (1997).

122. *Thornton v. Bijou Irrigation Co.*, 926 P.2d 1 (Colo. 1996).

123. Idaho Code § 42-202B. The section also requires a determination that no significant effect on agriculture will result and says that determination of future need should exclude demand in areas with conflicting land-use plans. *See* §§ 42-202B(5) and (6).

124. Kansas, Montana, Oregon, and Washington have the most comprehensive planning processes. *See* Getches, *supra* note 22.
125. Drought planning was the subject of Mont. Code Ann. § 2-15-3308 (1991); Tex. Water Code Ann. § 16.051(a), §§ 11.1271, 11.1272, 11.134 (West 1998); Or. Rev. Stat. § 536.720 (1998); Wash. Rev. Code Ann. § 43.83B.410 (West 1999).
126. *See* Brad Knickerbocker, *USA, What City Dwellers Really Want,* Christian Science Monitor, February 16, 2000 (announcing results of opinion polls declaring sprawl and growth as top concerns of the public).
127. Cal. Gov't Code § 65302 (West 1999).
128. Cal. Water Code §§ 10910–10914 (West 1999).
129. *See Stanislaus Natural Heritage Project v. County of Stanislaus,* 55 Cal. Rptr. 625 (5th Dist. Ct. App. 1996).
130. *See* Donald G. Hagman and Julian Conrad Juergensmeyer, Urban Planning and Land Development Control Law §§ 9.1 to 9.9 (2d ed. 1986); Haw. Rev. Stat. §§ 205-1 to 205-18 (1998); Or. Rev. Stat. §§ 197.005 to 197.860 (1998). Outside the West, Vermont has one of the strongest state land-use planning programs. *See* 10 Vt. Stat. Ann. §§ 6001–6108; Richard O. Brooks, Toward Community Sustainability: Vermont's Act 250 (1997).
131. Or. Rev. Stat. § 540.510 (1989) (appropriations >50 cfs subject to legislative approval); Kan. Stat. Ann. §§ 82a-1501 to 1506 (1989) (special procedures apply to transfers >2000 AF/year or > 35 miles).
132. Colo. Rev. Stat. §§ 37-92-305(4.5) and 103(10.4) (Cum. Supp. 1992) (revegetation required for changes out of agriculture).
133. Or. Rev. Stat. §§ 537.801, 537.803, 537.805, 537.809, 537.810, 537.830 (1989); Nev. Rev. Stat. §§ 533.345, 533.530 (1995); Tex. Water Code Ann. § 11.085 (West 1997) (impacts and alternatives must be assessed).
134. Alaska Stat. § 46.15.035 (Michie 1992) (also fee for export); Ariz. Rev. Stat. Ann. §§ 45-291 to 45-294 (West 1989) (director of water resources considers effects on water conservation, availability of alternative sources, current water supply, and future demands); Idaho Code § 42-401 (1990) (standards for exports generally similar to requirements for in-state appropriations); Nev. Rev. Stat. § 533.520 (1991) (similar to Idaho statute); Okla. Stat. Ann. tit. 82, § 1085.2(2) (West 1990) (legislative approval required); Utah Code Ann. §§ 73-3a-101 to -109 (Supp. 1991) (exports enabled on conditions designed to protect state interests).
135. California legislature appropriated $235 million to promote a trade between Imperial Irrigation District and Metropolitan Water District of Southern California for the benefit of San Diego. S.B. 1765, 1997–1998 Reg. Sess. (Cal. 1998), Cal. Water Code § 12.560–65 (West 1998). This reportedly would be the largest water transfer agreement in the history of the United States. *See California: The Metropolitan Water District of Southern California Endorses Colorado River Water Transfer Agreement Between San Diego County Water Authority and Imperial Irrigation District,* Western Water Law and Policy Reporter, Vol. 2, No. 8, 191 (June 1998).
136. Ariz. Rev. Stat. Ann. § 45-2401, *et seq.* (1996). See also Ariz. Rev. Stat. §§ 45-291–294 (permitting interstate transport of water).
137. The secretary of the interior has proposed rules to facilitate the Arizona plan. *See* 64 Fed. Reg. 68373 (December 1999). These rules are far more limited than rules

allowing interstate water marketing that Interior proposed several years earlier. A similar but more comprehensive approach to banking and marketing had also been proposed earlier by California, but it met hostile opposition from Arizona and the other basin states. *Ibid.* at 610.

138. *See* Elizabeth Ann Rieke, *The Bay-Delta Accord: A Stride Toward Sustainability,* 67 U. Colo. L. Rev. 341 (1996); A. Dan Tarlock, *Federalism Without Preemption: A Case Study in Bioregionalism,* 27 Pacific L. J. 1629 (1996).
139. Warren T. Coleman, *Legal Barriers to the Restoration of Aquatic Systems and the Utilization of Adaptive Management,* 23 Vt. L. Rev. 177, 187 (1998).
140. *See* J. David Aiken, *Balancing Endangered Species Protection and Irrigation Water Rights: The Platte River Cooperative Agreement,* 3 Great Plains Nat. Resources J. 119, 144 (1999).
141. Truckee-Carson–Pyramid Lake Water Rights Settlement Act of 1990, Pub. L. No. 101-618, 104 Stat. 3289 (1990); A. Dan Tarlock, *The Creation of New Risk Sharing Water Entitlement Regimes: The Case of the Truckee-Carson Settlement,* 25 Ecology L. Q. 674 (1999).
142. *See* Robert Wigington and Dale Pontius, *Toward Range-Wide Integration of Recovery Implementation Program for the Endangered Fishes of the Colorado River* in The Colorado River Workshop: Issues, Ideas, and Directions 63 (Grand Canyon Trust, February 1996).
143. *See* 16 U.S.C. 1539(a). *See also* Department of the Interior, Habitat Conservation Planning Handbook: November 1996; 61 Fed. Reg. 63854 (1996).
144. *See generally* William E. Taylor and Mark Gerath, *The Watershed Protection Approach: Is the Promise About to Be Realized?* 11 Nat. Resources and Env't 16 (1996).
145. Douglas S. Kenney, Report to the Western Water Policy Review Advisory Commission, Water in the West: Challenge for the Next Century 12–16 (1998).
146. *See* Natural Resources Law Center, University of Colorado, The Watershed Source Book: Watershed-Based Solutions to Natural Resource Problems (1996).
147. Or. Rev. Stat. § 541.384 (1998) (Watershed Management Act requiring Water Resources Department to cooperate with other state and federal agencies, local watershed councils, nonprofits, and citizens' groups to administer watershed management programs); 1995 Or. Laws 404, § 4 (allocating funds from state lottery revenue to Governor's Watershed Enhancement Board).
148. 1998 Wash. Laws ch. 247 (adding Wash. Rev. Code Ann. § 90.82.060 [1998]) (watershed planning act that provides for local governments, Indian tribes, and local citizens to prepare and adopt watershed plans through a consensus process).
149. *See* L. MacDonnell, *Managing Reclamation Facilities for Ecosystem Benefits,* 67 U. Colo. L. Rev. 197, 200 (1996).
150. *See A New Era for Hydropower,* American Rivers, Winter 1998 at 3. See also John McPhee, *Farewell to the Nineteenth Century,* The New Yorker, September 27, 1999, 44.
151. Patrick Graham, *Idea of Draining Lake Makes Waves (A Movement to Empty Lake Powell Is Gaining Ground, Albeit Slowly, Despite Harsh Effects),* Portland Oregonian, March 7, 1999, A-21; American Rivers, Friends of the Earth, Trout Unlimited, *Dam Removal Success Stories: Restoring Rivers Through Selective Removal of Dams That Don't Make Sense* <www.amrivers.org/successstories.html>.

152. Michael C. Blumm, Laird J. Lucas, Don B. Miller, Daniel J. Rohlf, and Glen H. Spain, *Saving Snake River Water and Salmon Simultaneously: The Biological, Economic, and Legal Case for Breaching the Lower Snake River Dams, Lowering John Day Reservoir, and Restoring Natural River Flows,* 28 Envtl. L. 997, 1004 (1998).
153. *See* Brent Hunsberger, *Kitzhaber Calls for Breaching Four Dams,* Portland Oregonian (February 19, 2000), 1.
154. *See* Scott K. Miller, *Undamming Glen Canyon: Lunacy, Rationality, or Prophecy?,* 19 Stan. Envtl. L. J. 121 (2000).
155. *See* Oversight Hearing on the Sierra Club's Proposal to Drain Lake Powell or Reduce Its Water Storage Capability: Hearings Before the Subcommittees on National Parks and Public Lands, and Water and Power Resources of the House of Representatives, 105th Cong. (1997); Ed Marston, *Sierra Club Moves to Fortify Its "Drain Lake Powell" Campaign,* High Country News, October 15, 1997.
156. *See* John E. Thorson, *State Watershed Adjudications: Approaches and Alternatives,* 42 Rocky Mtn. Min. L. Inst. 22-1, 22-39 (1996).
157. *United States v. Oregon,* 44 F.3d 758, 762 (1994); *see also* Reed D. Benson, *Maintaining the Status Quo: Protecting Established Water Uses in the Pacific Northwest, Despite the Rules of Prior Appropriation,* 28 Envtl. L. 881, 902 (1998).
158. *Idaho v. United States,* 912 P.2d 614 (1995) (partially overturning Act of April 12, 1994, chs. 454–455, 1994 Idaho Sess. Laws 1443–91 [1994 Act], codified at Idaho Code §§ 42-1401 to 1428 [Supp. 1994]). *See also* A. Lynn Krogh, *Water Right Adjudications in the Western States: Procedures, Constitutionality, Problems and Solutions,* 30 Land and Water L. Rev. 9, 39 (1995); Robert E. Bakes, *The Snake River Basin Adjudication: From the Beginning to the Present,* 38 Feb. Advocate (Idaho) 10 (1995).
159. *San Carlos Apache Tribe v. Superior Court,* 972 P.2d 179 (Ariz. 1999) (constitutional and McCarran amendment challenges resulted in setting aside most attempted legislative alterations of general stream adjudication law, H.B. 2276).
160. *Winters v. United States,* 207 U.S. 564 (1908).
161. 43 U.S.C. § 666.
162. *Colorado River Water Conservation Dist. v. United States,* 424 U.S. 800 (1976); *Arizona v. San Carlos Apache Tribe,* 463 U.S. 545 (1983).
163. The United States and Klamath Tribe unsuccessfully challenged application of the McCarran amendment to quasi-administrative proceedings in *United States v. Oregon,* 44 F.3d 758 (9th Cir. 1994), *cert. denied,* 516 U.S. 943 (1995).
164. *See Arizona v. California,* 373 U.S. 546 (1963).
165. *In re General Adjudication of All Rights to Use Water in the Big Horn River System,* 753 P.2d 76 (Wyo.), *affirmed sub nom. Wyoming v. United States,* 492 U.S. (1989). *See also* Teno Roncalio, *The Big Horns of a Dilemma,* in Indian Water in the New West 209, 211 (Thomas R. McGuire et al., eds., 1993).
166. *See* Susan D. Brienza, *Wet Water v. Paper Rights: Indian and Non-Indian Negotiated Settlements and Their Effects,* 11 Stan. Envtl. L. J. 151 (1992); Joseph R. Membrino, *A Federal Perspective,* in Indian Water in the New West 57, 59 (Thomas R. McGuire et al., eds., 1993); John E. Thorson, *Proceedings of the Symposium on Settlement of Indian Water Rights Claims,* 22 Envtl L. 1009 (1990).
167. *E.g.,* Yavapai-Prescott Indian Water Rights Settlement Act of 1994, Pub. L. No. 130-434, 108 Stat. 4526 (1994); Central Utah Project Completion Act, Pub. L. No.

101-575, 106 Stat. 4611 (1992). States continued to ratify settlements essentially concluded before the Clinton administration. *E.g.*, Mont. Code Ann. § 85-20-301 (1991) (ratifying Northern Cheyenne–Montana Compact); Act of April 2, 1991, Idaho Sess. Laws 547, ch. 228, § 1 (1991).

168. Chippewa Cree Tribe of the Rocky Boy's Reservation Indian Reserved Water Rights Settlement and Water Supply Enhancement Act of 1999, P.L. 106-163, December 9, 1999, 113 Stat. 1778. *See* Barbara A. Cosens, *The 1997 Water Rights Settlement Between the State of Montana and the Chippewa Cree Tribe of the Rocky Boy's Reservation: The Role of Community and of the Trustee,* 16 UCLA J. Envtl. L. and Pol'y 255, 257 (1997–1998).
169. The terms of a settlement on the Warm Springs Reservation have been agreed upon. *See* Jonathan Brinckman, *Water Rights Pact Comes Peacefully,* Portland Oregonian, November 18, 1997, E07.
170. Endangered Species Act, 16 U.S.C. §§ 1531–1544 (1973).
171. Clean Water Act, 33 U.S.C. §§ 1251–1387.
172. *E.g., Babbitt v. Sweet Home Chapter of Communities for a Great Oregon,* 515 U.S. 687 (1995).
173. Lee P. Breckenridge, *Nonprofit Environmental Organizations and the Restructuring of Institutions for Ecosystem Management,* 25 Ecology L. Q. 692, 697 (1999).
174. 50 C.F.R. § 402.02.
175. 16 U.S.C. § 1539(a); *see also* Department of the Interior, Handbook, *supra* note 143; Frederico M. Cheever, *An Introduction to the Prohibition Against Takings in Section 9 of the Endangered Species Act of 1973: Learning to Live With a Powerful Species Preservation Law,* 62 U. Colo. L. Rev. 109, 169 (1991).
176. Aiken, *supra* note 140.
177. *See generally* The Northwest Salmon Crisis: A Documentary History (Joseph Cone and Sandy Ridlington, eds., 1996).
178. *See* Michael C. Blumm and Greg D. Corbin, *Salmon and the Endangered Species Act: Lessons From the Columbia Basin,* 74 Wash. L. Rev. 519 (1999).
179. *See* Rieke, *supra* note 138 at 356.
180. James H. Bolin Jr., *Of Razorbacks and Reservoirs: The Endangered Species Act's Protection of Endangered Colorado River Basin Fish,* 11 Pace Envtl. L. Rev. 35 (1993).
181. 33 U.S.C. § 1344.
182. 33 U.S.C. § 1344(c).
183. *Riverside Irrigation Dist. v. Andrews,* 758 F.2d 508 (10th Cir. 1985); *Sierra Club v. Sigler,* 695 F.2d 957 (5th Cir. 1983).
184. *See* Daniel F. Luecke, *Two Forks: The Rise and Fall of a Dam,* 14 Nat. Resources and Env't. 24 (1999).
185. Endangered Species Act, 16 U.S.C. § 1536(2). If the action is a major federal action having a significant effect on the human environment, it also triggers the requirement of the National Environmental Policy Act for preparation of an environmental impact statement. National Environmental Policy Act, 42 U.S.C. § 4332(2)(c). *See also Riverside Irrigation District v. Andrews, supra* note 183.
186. Or. Rev. Stat. § 541.345 (1997); 1998 Wash. Laws ch. 247 (adding Wash. Rev. Code Ann. § 90.82.060 (1998)).

187. Natural Resources Law Center, University of Colorado School of Law, *The State Role in Western Watershed Initiatives* 37–54 (Research Report RR-18, 1998).
188. U.S. Environmental Protection Agency, *Watershed Framework Approach,* U.S. Environmental Protection Agency, Office of Water, EPA 840-S-96-001, June 1996.
189. *See* Kenney, *supra* note 145 at 61–63 (finding that "the Federal Government plays a significant and essential role in the effective functioning of most watershed initiatives").
190. *See* Nancy Perkins Spyke, *Public Participation in Environmental Decisionmaking at the New Millennium: Structuring New Spheres of Public Influence,* 26 B.C. Envtl. Aff. L. Rev. 263 (1999).
191. Reed D. Benson, *Whose Water Is It? Private Rights and Public Authority Over Reclamation Project Water,* 16 Va. Envtl. L. J. 363 (1997).
192. Reed D. Benson, *Recommendations for an Environmentally Sound Federal Policy on Western Water,* 17 Stan. Envtl. L. J. 247 (1998).
193. *See* Bell et al., *supra* note 27.
194. *See* Western Water Policy Review Advisory Commission, *supra* note 19 at 2-14–2-18.
195. *See* Lawrence J. MacDonnell and Teresa A. Rice, *Moving Agricultural Water to Cities: The Search for Smarter Approaches,* 2 Hastings W.-N.W. J. Envtl. L. and Pol'y 27 (1994).
196. *See* Sandra Postel, Pillar of Sand: Can the Irrigation Miracle Last? 77 (1999).
197. James Wilson, Groundwater: A Non-Technical Guide, Academy of Natural Sciences 48, 50 (1982).
198. *See generally* John Firor, The Changing Atmosphere: A Global Challenge (1990); Kenneth D. Frederick and Peter H. Gleick, Water and Global Climate Change: Potential Impacts on U.S. Water Resources (1999); Nigel W. Arnell, *Climate Change and Global Water Resources,* 9 Global Envtl. Change S31 (1999); William W. Kellogg, *Human Impact on Climate: The Evolution of an Awareness, in* Societal Responses to Regional Climatic Change: Forecasting by Analogy (Michael H. Glantz, ed., 1988); William E. Riebsame, *Adjusting Water Resource Management to Climate Change,* 13 Climatic Change 69 (1988); Ernest T. Smerdon, *Impact of Global Change on Water Resources,* 9 Ariz. J. Int'l and Comp. L. 155 (1992).
199. *See* Martin Parry, Cynthia Rosenzweig, Ana Iglesias, Günther Fischer, and Matthew Livermore, *Climate Change and World Food Security: A New Assessment,* 9 Global Envtl. Change S1 (1999); *The Science of Global Change: Intergovernmental Panel on Climate Change,* 9 Ariz. J. Int'l and Comp. L. 9 (1992).
200. *Cf.* Richard W. Katz, *Statistics of Climate Change: Implications for Scenario Development, in* Societal Responses to Regional Climatic Change: Forecasting by Analogy (Michael H. Glantz, ed., 1988).
201. *See* Brian E. Gray, *The Market and the Community: Lessons From California's Drought Water Bank,* 1 Hastings W.-N.W. J. Envtl. L. & Pol'y 17 (1994).
202. *See* Western Water Policy Review Advisory Comm'n, *supra* note 19 at xxxi.
203. *See* Larry Morandi, Rethinking Western Water Policy: Assessing the Limits of Legislation (1994) (concluding, based on several case studies, that the legislative process can be valuable in resolving differences and promoting better water use and other progressive goals).

FIFTEEN

Economic and Institutional Strategies for Adapting to Water-Resource Effects of Climate Change

JOHN LOOMIS
JESSICA KOTEEN
BRIAN HURD

This chapter presents an overview and synthesis of policy tools that state and federal agencies can use to assist water managers in adapting to climatically driven variability in water demand and supply. Some of these policy tools can also help address the current conflicts in water demand, supply, and valuation between traditional water uses and newer, growing uses.

INCREASING CONFLICTS BETWEEN COMPETING WATER USES

Off-stream water use in the United States increased over tenfold during the twentieth century in response to population and economic growth (Brown 1999). Growth is expected to continue in the western United States, and domestic and commercial demands for water probably will continue to grow as well (Chapter 13). As domestic demand for water increases, additional sources must be acquired. Even with increased conservation and water-use efficiency, population growth will outstrip municipal allocations in a majority of the West's urban growth areas. In many of these regions, surface supplies are fully subscribed, and growing urban demands will require transfers of water from other users or regions. What does this mean for aquatic ecosystems?

Table 15.1—Trends of estimated water use in the United States, 1950–1995 (based on Solley et al. 1998, table 31).

Year	1950	1955	1960	1965	1970	1975	1980	1985	1990	1995
Population (millions)	151	164	179	194	206	216	230	242	252	267
					Off-stream uses (10^9 gpd)					
Public supply[1]	14.0	17.0	21.0	24.0	27.0	29.0	34.0	36.5	38.5	40.2
Rural domestic and livestock	3.60	3.60	3.60	4.00	4.50	4.90	5.60	7.79	7.89	8.89
Irrigation	89	110	110	120	130	140	150	137	137	134
Industrial[2]	77	111	138	176	217	245	255	217	225	219
Total	184	242	272	324	379	419	445	398	408	402

[1] Includes municipal and commercial uses.

[2] Includes thermoelectric power production (cooling).

Stream flows have dropped as scientists have demonstrated the need of specific stream flows for support of aquatic life and the maintenance of river channels. Also, rising incomes and urbanization have intensified demand for recreation and water quality that depend on flow (Gillian and Brown 1997). These issues have increased the complexity of water management and thus have raised the desirability of integrated water management and the usefulness of quasi-market institutions at the watershed level. Only by these approaches can a flexible system emerge that accounts for the linkages between upstream and downstream users, allocates the impact of minimum flow constraints efficiently across all users, and generates market signals that tend to internalize conservation strategies.

Demands for Water

The adequacy of water supply depends on the relationship of water availability to water use (Brown 1999). A great many competing uses exist, including off-stream uses for municipal, agricultural, and industrial purposes or in-stream uses such as hydropower, recreation, and environmental purposes. Policy tools that can better allow these differing demands to be met will be the focus of the second half of this chapter.

Trends in Water Use

Growth in total U.S. water withdrawals between 1950 and 1990 consistently outpaced population growth (Table 15.1). The change from 1950 to 1990 (which has slowed since then) corresponds to an annual increase of 1.2 percent per year for population and 2.4 percent per year for withdrawal (Brown 1999). As shown by Table 15.1, this growth in withdrawals resulted largely from increases in irrigation and thermoelectric cooling, which together account for 83 percent of off-stream water uses in the United States. In the western United States irrigated agriculture alone accounts for 80–90 percent of withdrawals in some states (Gibbons 1986).

Withdrawals for municipal use consistently increased between 1950 and 1995. Although population growth explains much of the trend, an increase in per capita use brought about by growth in technology and the conversion of older or rural households to complete plumbing was also important. In the last few decades, increased use of dishwashers, washing machines, swimming pools, and lawn sprinkler systems has added demand (Brown 1999). To explain the future outlook of withdrawals for municipal uses, recent studies have mainly concentrated on population growth and rising per capita income. It seems likely that municipal water demand will continue to increase because of the upward trend in population, especially in the western United States. If increases in water prices and awareness of the need for water conservation and implementation of water-conservation programs have a larger effect than rising income, however, we could expect to see a leveling or even a decrease in per capita demand. Reductions in per capita water use will be needed in the western United States in part to offset the increase in population and to allow for balancing of existing and new water uses while minimizing the need to transfer water from other uses and regions.

Industrial and commercial withdrawals showed growth until 1980 (Table 15.1), but between 1980 and 1995, withdrawals for industrial uses dropped sharply. New technologies that require less water, improved plant efficiencies, water recycling, higher energy prices, and changes in laws and regulations have resulted in decreased industrial water use (Solley et al. 1998). In the future for the industrial sector we should see a decline in total water use and use per unit of production, but change probably will occur at a slower rate than since 1980. Industrial water demand is not fixed; it has been responsive to external influences.

Rural use of water is for domestic purposes and livestock. The steady increase in rural use has been caused mainly by an increase in withdrawals for livestock. It is useful to consider rural and irrigation water demands together. Water demand for irrigation rose at a steady rate until 1980, when it began to decline. Between 1950 and 1980, irrigation systems expanded, but higher energy prices in the 1970s and a large drawdown of groundwater levels in some areas increased the cost of irrigation water. In the 1980s, improved application techniques increased competition for water, and a downturn in the farm economy reduced demands for irrigation water (Solley et al. 1998). The future may bring greater pressure to reduce water use for irrigation. Studies have shown that in some areas it is more valuable and beneficial to keep a greater portion of water in streams for purposes of aesthetics, recreation, support of fish and wildlife, and maintenance of water quality than it is to preserve irrigation uses (Ward 1987; Loomis 1987, 1994). Thus the future may bring greater competition between irrigation and other uses.

EFFECTS OF CLIMATE CHANGE ON DEMAND FOR AND SUPPLY OF WATER

Changes in the spatial and temporal distributions of runoff can have profound social and economic consequences (Hurd et al. 1999a). Anticipated greenhouse warming may significantly affect precipitation, evapotranspiration, stream flow, and water availability. Hydrologic changes that may accompany global climate change are expected to have a variety of effects on human settlements and ecological systems. The effects include changes in water supply and quality for drinking and for irrigation, recreational, commercial, and industrial uses; stream flows that support aquatic ecosystems, recreation, hydropower, navigation, and wastewater assimilation may change. In addition, changes may occur in the extent and productivity of wetlands with consequences for fish, wildlife, and wastewater assimilation as well as frequency and severity of floods. Regions where water resources are stressed under current climatic conditions are likely to be the most vulnerable to changes in climate and in extreme events (Hurd et al. 1999b). With the exception of higher temperatures and greater rates of evapotranspiration, however, changes in precipitation and runoff are uncertain—particularly at the regional scale, which is of greatest interest to local water planners, utilities, and government agencies (Chapter 6).

Frederick and Gleick (1999) describe the effects of climate change on the United States as shown by two general circulation models—the Hadley and the Canadian. Estimates of change in runoff from these models are similar in some cases and vastly different in others. For example, the Canadian model suggests that average runoff will decline everywhere except California by 2030; a decline of more than 20 percent is estimated in two-thirds of the regions. In contrast, the Hadley model suggests that runoff will increase in most basins; much of the nation's arid and semiarid regions would have considerably more water, reducing problems of water scarcity but perhaps increasing flooding. For 2090 the Hadley model shows much of the nation with even wetter conditions, whereas the Canadian model suggests further drying in the East but an increase in moisture over much of the West. Frederick and Gleick conclude that the effects of greenhouse warming on future water supplies are uncertain (cf. Chapter 6).

In the Colorado River basin, Hurd et al. (1999a) have measured the economic effects of hypothetical changes in climate and precipitation. With a small change in temperature and a large change in precipitation, runoff increases, which leads to an increase in economic benefits. With large in-

Table 15.2—Effects of climate change on welfare in the Colorado River basin for various possible combinations of change in mean temperature (°C) and annual precipitation (% increase or decrease) (from Hurd et al. 1999a; 1994 dollars).

Change in Simulated Benefits (+) and Costs (–) by Sector								
Change °C	2.5	5.0	1.5	2.5	5.0	5.0	1.5	2.5
Change % precipitation	+15	+15	+7	+7	+7	+0	–10	–10
Change in benefits, 10^6 $/year								
Agriculture	29	–1	13	–2	–40	–112	–99	–127
Municipal and industrial	3	–4	1	–3	–20	–263	–76	–332
Hydropower	159	–112	40	–66	–300	–436	–401	–468
Recreation	1.0	–1.0	0.2	–0.5	–7.0	–18.0	–16.0	–28.0
Salinity damage	102	–57	31	–34	–202	–301	–291	–336

creases in temperature combined with small increases in precipitation, however, runoff decreases, which in turn decreases economic well-being. These findings have implications not only for water supplies but also for hydropower generation and salinity. In years when reservoirs are relatively full (e.g., 1984), increased spring flows create less hydropower value than if flows develop more gradually or can be stored to meet summer air-conditioning demand or demands for peaking power. Table 15.2 shows the economic changes for agriculture, hydropower, salinity, municipalities and industries, and recreation under a variety of climate-change scenarios. Hydropower losses could be 30 to 66 percent of total damages, as compared with agriculture at only about 7–10 percent of damages. The second-largest source of damage is increased salinity, which reduces crop production and may affect municipal and industrial water systems. For the central case climate scenario (+2.5°C, +7% precipitation), total annual costs are estimated to be about $105 million. Much of the total damages is a result of the currently inflexible water allocation rules associated with the Colorado Compact, prior appropriation doctrine, and similar measures, which protect the senior water rights of irrigated agriculture. In the upper basin, irrigated agriculture is often the lowest-valued water use. If legal flexibility and economic incentives were provided to upper basin irrigators, the damages from climate change could be cut by nearly half. Specifically, total damages would be estimated at $65 million per year, a $40 million per year saving with legal flexibility and economic incentives to allow irrigators to sell water to higher-valued users downstream. The absolute magnitude of the cost savings would be even larger with more extreme climate change.

In the remainder of this chapter we describe several tools that would allow policy makers and managers to cope with increased variability and reduction in utility of water supplies. Institutional changes and economic approaches have been repeatedly called for but not detailed. For example, Schilling and Stakhiv (1998) conclude that new policies and institutions are the ultimate mechanisms for steering adaptive strategies and that climate impact assessments and evaluations of adaptive-management strategies often neglect basic economic principles about behavior and the role of incentives.

Although the basic institutional inadequacies concerning western water law and the implications of prior appropriation doctrine have long been understood, these institutions and policies have been frustratingly resistant to changes needed to reflect the nonagrarian economy of the new West (Chapter 14). Furthermore, although evolving institutions are a necessary step in the right direction, they are rarely elaborated or emphasized in articles dealing with effects of climate change on water-resource management. For example, a recent issue of *Water Resources Update* (issue 112, 1998, editors Schilling and Stakhiv) devoted to the topic of global climate change and water resources listed water pricing as a policy tool important in adaptation to climate change, but the authors devoted most of their attention to recommendations for change in water-resource infrastructure investments and operating rules. Prolonged severe droughts in California and elsewhere over the past decade have demonstrated, however, the effectiveness of state water banks (markets), federal provisions for water transfers, increasing block water rates, and drought-year water leasing in moving hundreds of thousands of acre feet of water very quickly from one use or area to another. Although market responses to drought have many times filled much of the shortfall between off-stream water supply and demand and might also do so for climate change, they have been less successful in meeting environmental demands for water. Much recent progress has been made in this area, however.

CHANGES IN CALIFORNIA WATER MANAGEMENT: A TEMPLATE

Hurd et al. (1999a) have shown that market-oriented adaptations that improve the efficiency of water use and facilitate water transfers from uses of lower to those of higher value can substantially mitigate adverse effects of water scarcity (see also Chapter 14). California provides examples through irrigation pricing reforms, water transfers, water markets, legal changes, and several other adaptations.

Irrigation Water Pricing Reform

Greater reliance on market institutions may decrease the inefficient use of water for irrigation and alleviate some of the demand that arises from water shortages. As long noted by many observers of western water, underpricing of irrigation water, reinforced by institutional barriers to markets, is rampant and is a significant stumbling block to more efficient use of water

(Moss 1998). The potential gains from even a 10 percent reduction in irrigation water use are substantial because irrigation accounts for nearly 90 percent of water use in the intermountain West (Gibbons 1986). A 10 percent reduction in consumptive use by agriculture represents a 90 percent increase in water availability for urban or environmental purposes. Thus a small reduction in agricultural water use releases large quantities of water to cities and environmental uses. Reductions in agricultural water use can be achieved in several ways but are most likely to occur through water pricing reforms and incentives for water transfers (either on an individual transactions basis or through a formal water market); the latter will be discussed in the next section.

The Central Valley Project Improvement Act (CVPIA) can serve as a template for updating other state and federal water projects in the face of climate change. CVPIA is part of the Reclamation Projects Authorization and Adjustment Act of 1992 (P.L. 102-575). As the title implies, this act is intended to improve the management and efficiency of linked water projects in California. The desirable features of CVPIA for improving water-use efficiency in the West include increasing block-rate pricing of irrigation water and allowing water transfers to other uses and small quantities of water for wetlands and rivers.

At the time of passage, water contract prices for many Central Valley Project (CVP) irrigation districts were as low as $3.50 per acre foot, whereas the cost of service (the operation and maintenance cost plus repayment of the investment cost at no interest) was $6–$22 per acre foot, and the full service cost (which includes the interest cost of the investment) was $12–$50 per acre foot (Bureau of Reclamation 1993a). Section 3405(d) of the act requires that all long-term water contracts (of three years or more) and renewed water contracts be based on an inverted block-rate structure. The rate structure specifies the exact contract rate for the first 80 percent of the contract, a rate halfway between the contract and full-cost rate for the next 10 percent, and the full rate for the remaining 10 percent of water. The U.S. Bureau of Reclamation (USBR) is now applying these rates at the wholesale level in its contracts with water districts (Bureau of Reclamation 1993b).

Numerous studies have demonstrated gains in efficiency associated with an increasing block-rate water pricing schedule (Dinar et al. 1989; Wichelns 1991) caused by greater incentive for conservation of water and reduced agricultural drainage (Dinar and Letey 1991; Wichelns 1991). Of course, the magnitude of these gains depends on the level and rate of increase in the price with each block. In principle, steeply increasing block prices for higher levels of water use could result in substantial decreases in demand for irrigation water. Given the conservative nature of the increasing block price, the actual outcome may be only modest water conservation. Nonetheless, the precedent has been established and can be improved as the vagaries of climate change necessitate. The next step is to ensure that the irrigation districts pass on the USBR's increasing block-rate structure to farmers rather than just averaging water prices across the blocks. Further, an incentive mechanism (elaborated on in the following sections) is necessary to ensure that the conserved water is made available to other users (including other agricultural users). Otherwise, increased irrigation efficiency may actually reduce stream flows if such efficiency reduces return flows or results in expansion of irrigated acreage with the conserved water.

Water Transfers

Economists have long pressed for institutional changes that would allow for water transfers from government projects that would promote flexibility and efficiency in water allocation (Anderson 1983; Vaux and Howitt 1984; Howe et al. 1986; Frederick 1986; Saliba and Bush 1987). Water in the West is one of the few cases where water is allocated not by economic value but rather by who uses it first. The prior appropriation system for water allocation was established in the second half of the nineteenth century initially to promote mining, which needed to divert large amounts of water, and then later to encourage settlement of western lands. Imagine what the economy would be like today if wood and steel were allocated to the same uses as they were in the 1800s (e.g., buggies but not automobiles). The U.S. economic miracle is that resources flow to uses of higher value as consumer demand changes. Climate change may finally break our anachronistic restrictions on the freedom of water rights holders to seek the most valued uses for their water. If we can provide the same legal flexibility for water rights that we provide for other property, water will flow to new, more valuable uses (including environmental uses). In the end we could even have water voluntarily allocated to uses associated with today's nonagrarian economy.

The CVPIA was one of the first congressional acts to allow for transfers of federal water both out of the project area and to project purposes other than those originally authorized. Section 3405(a) states that "all individuals or districts who receive CVP water . . . are authorized to transfer all or a portion of the water subject to such contract to any other California water user, State or Federal agency, Indian tribe or private nonprofit organization for project purposes or any purpose recognized as beneficial under applicable State law." This clearly opens the door for transfer of water across geographic regions as well as across uses. Thus any number of different users can purchase water.

The precedent the CVPIA sets for other Bureau of Reclamation water projects throughout the West is important. With this act Congress signaled its desire that water transfers to other uses and users outside the project areas be allowed. Congress addressed environmental concerns via a surcharge to finance mitigation through either habitat management or water purchases. The act shows that changes can occur, even in federal projects.

Water transfers are not limited to California. During 1998, for example, formal water trades occurred in the western United States for municipal purposes (104), irrigation (24), public trusts (12), and other reasons (13) (Water Strategist 1999). Also, dozens of more informal transactions within water districts are not widely publicized. In addition, cities occasionally buy farms to acquire the associated water rights.

Water Markets

Water markets can be decentralized (buyers and sellers must locate each other, sometimes through use of brokers); moderately centralized, as in the case of computer bulletin boards used for posting water offers within a water district (e.g., California); or even state-run, centralized water banks. Water banks are public or private institutions that match buyers and sellers of water in a given geographic area. Markets emerge and become more formalized when the difference in value of water for different uses becomes large enough to pay brokers for matching buyers and sellers. Water markets facilitate the movement of water from uses of low value to those of high value. Even with the electronic information revolution, however, transaction costs can be high. The establishment of a highly visible water bank by a state agency reduces transaction costs, thereby facilitating exchanges.

California has had a successful experience in coping with drought by developing a state-run water bank. As the result of a severe drought in California, cities and farmers with permanent crops (e.g., orchards) faced substantial water-supply shortfalls. In 1991 Governor Wilson created the California Water Bank, which was designed to meet four essential needs: municipal and industrial uses, agricultural uses, protection of fish and wildlife, and carryover storage for 1992 (Department of Water Resources 1992; Loomis 1992). The rules of operation were simple. The state would buy water at $125 an acre foot and sell it for $175 an acre foot, and the buyer would pay delivery costs. Of the $50 spread between purchase and sale price, $45 was to cover losses in transit from the source to the buyer. The remaining $5 covered contract negotiation and administrative costs. In part, the purchase price represents foregone income from fallowing different crops and the crops' consumptive water use.

Legal Changes

Functioning of the California Water Bank was facilitated through a standardized water-purchase contract. Also, two potential legal concerns were resolved with two California State Assembly bills. Assembly Bill Number 9 gave water suppliers explicit authority to enter into contracts with the California Water Bank for transfer of water outside the service area of the water supplier; Assembly Bill Number 10 assured farmers and other water suppliers that their transfer of water in 1991 would not affect their water rights (Department of Water Resources 1992). Bill Number 10 was intended to reassure farmers that they would not lose their property rights in water if they leased their water to the Water Bank. In addition, the Department of Water Resources (DWR) coordinated water transfers with the State Water Resources Control Board and the U.S. Bureau of Reclamation.

Sources of Water

About 50 percent of the water for the California Water Bank came from fallowing corn, wheat, pasture, sugar beets, and other crops (Department of Water Resources 1992). Aerial photographs were used to monitor compliance with crop-fallowing contracts (Department of Water Resources 1992). About one-third of the water came from farmers who were able to substitute groundwater for surface water, thus releasing surface flows to the Water Bank. Sellers were required to meter the groundwater that they pumped. The DWR also increased its well monitoring from semiannually to monthly (Department of Water Resources 1992). Water owned by water agencies or irrigation districts that was stored in reservoirs accounted for the final 17 percent of water sold to the Water Bank (Department of Water Resources 1992). Because the program was strictly voluntary, only farmers whose water was worth less than $125 per acre foot participated. In a matter of weeks, 351 contracts were finalized through the Water Bank for a total of 830,000 acre feet. This rapid response shows that markets could be of great value in dealing with water shortages caused by climate change.

By any standard, 830,000 acre feet is a substantial amount of water; in fact, it is more water than the city of Los Angeles uses in a normal year (Los Angeles Department of Water and Power 1985). At the same time it reflects less than 5 percent of agriculture's water use in a normal year. As Howitt (1991) notes, this amount of water would have required a new dam costing about $3 billion. Yet the water was provided at little cost to the public and at relatively low cost to most farming communities. For example, no rural communities were economically devastated by the Water Bank (Howitt 1991). In the case of northern California, water-receiving industries included Silicon Valley computer chip manufacturers and San Francisco bakeries (M. Lennihan, personal communication, November 1992). Idling pasture in the Central Valley to provide water for computer manufacturing no doubt increased California's net income during the drought.

Water Purchasers

Major purchasers of water from the California Water Bank were (1) the Metropolitan Water District of Southern California, which bought 215,000 acre feet on behalf of several cities, including Los Angeles; (2) the Kern County Water Agency and Westlands Water District in the southern San Joaquin Valley, which used much of the water for agriculture (in total, about 20% of all water purchased went to agriculture; S. Macaulay, unpublished data); and (3) the city of San Francisco, which bought 50,000 acre feet (S. Macaulay, unpublished data). Several other Bay-area urban water districts purchased water as well. The delivered cost of this water was high. In the Bay area the addition of transport cost to the purchase price raised delivered cost to $400 an acre foot (M. Lennihan, personal communication, November 1992).

Dealing With Environmental Concerns

Water (50,000 acre feet) was provided for environmental purposes (primarily wetlands) as separate sales at substantially lower prices. In one such sale, California Fish and Game (CFG) obtained 28,000 acre feet of water for much less than the California Water Bank price as a condition of the sale of 130,000 acre feet at the $125-per-acre-foot price by the Yuba County Water Agency (Bob Aldridge, personal communication). The water was allocated to wetlands at four state wildlife management areas (Department of Water Resources 1992).

The DWR personnel operating the Water Bank also sought input from the CFG regarding Water Bank releases. The DWR recognized that altering the location and timing of historical flows would affect fisheries. By coordinating the timing of water releases from non–Water Bank sources and pumping from the San Francisco and Sacramento Delta with Water Bank deliveries, effects of drought on the Sacramento River and Delta fisheries were minimized (Department of Water Resources 1992).

Future water banks might be structured similarly to oil markets in that they would have a firm futures price and a spot market. That is, farmers and cities could contract months in advance for options to take delivery of water. Additionally, contracts for delivery of water in the months ahead could be signed during the winter; for example, July water deliveries could be signed in January at prices that, given the available information on snowpack and reservoir storage, would clear the market. A buyer might ultimately pay a price that later seemed too high or too low, depending on the amount of precipitation. Uncertainty for both seller and buyer would be reduced with such preseason contracts. A spot market would allow nearly instantaneous sales and purchases of water between willing buyers and sellers at any time.

The prices in this market would fluctuate widely depending on weather and temperature.

The California Water Bank, founded in 1991, was an important demonstration of the power of a centralized market and adequate incentives to move a large amount of water from agricultural uses of low value to alternate urban and agricultural uses of high value at a time of water scarcity.

MANAGEMENT THROUGH FLEXIBLE AGREEMENTS

Drought-Year Water Leasing

During drought years, many water rights cannot be fulfilled because of the prior appropriation doctrine, which requires that holders of senior water rights get 100 percent of their water before holders of junior water rights get any. Because senior water rights are held mainly for agricultural purposes, streams are dry by the end of the summer. Outright purchases of senior water rights for drought protection, however, may impose unwanted costs on the economies of local communities that rely on irrigated agriculture (Michelsen and Young 1993).

A new approach that could allow agricultural holders of senior water rights to retain their rights but have the option to lease their rights for other uses during droughts is through the use of water-supply option contracts. The exercise of the option would transfer water to the highest-valued uses when needed while preserving water for agriculture during normal water-supply situations (Michelsen and Young 1993).

Conditions Necessary for Establishing Water-Supply Option Contracts

Several conditions favor the establishment of water-supply option markets (Michelson and Young 1993): (1) The water supply must provide sufficient water for use of the option in drought years and also be able in average years to supply the lower-valued use. (2) Property rights must be definable and transferable. This can be difficult in some areas of the western United States where use of irrigation water is based on estimates from crude proxies such as diversion time or acres. (3) Agricultural operations must be capable of temporary suspension (usually this is possible only for annual crops). (4) Both buyer and seller must have realistic knowledge of the value of water and alternative water-supply costs. (5) The probability and severity of drought must be known within acceptable limits of risk for both parties. (6) Transaction costs and costs of transporting the water to the purchaser's point of intake must be less than the costs of the purchaser's next-most-costly water-supply alternative.

Peak-Load Water Pricing in Urban Uses

Encouraging more efficient urban use of water through pricing is a mechanism for managing water during drought. It comes from the idea of marginal-cost pricing, according to which water managers must charge prices based on the seasonal marginal cost rather than annual average prices of water. Thus the price of water would increase in summer, thereby rationing available supplies to the highest-value uses. This pricing approach has been adopted in Los Angeles (Hanemann 1993).

Climate-induced changes in water supply may result in shortages when demand is high, and the marginal cost of water will rise under a marginal-cost pricing system (Baumann et al. 1998). Thus during drought the cost of water goes up. This is an example of peak-load pricing. When discussing peak-load pricing the water industry usually uses the terms *seasonal pricing* and *seasonal rates.* Peak-load pricing may be a valuable tool for water managers who must deal with climate variability and climate change.

CONCLUSIONS

Changes in climate and, consequently, in water supply and demand are very likely over the next 50 to 100 years. Although confidence is high that temperatures and potential evapotranspiration rates will rise, confidence is lower with regard to specific regional changes in precipitation and runoff. Although the exact magnitude of change is uncertain, managers can best serve their customers by improving the flexibility of their water systems to respond to a wide range of changes in water supply and demand, as might be induced by climate change. Institutional and pricing reforms as well as water markets would facilitate management flexibility and could mitigate damages potentially caused by climate change more effectively than construction of new, more efficient reservoirs.

REFERENCES

Anderson, T., ed. 1983. *Water Rights: Scarce Resource Allocation, Bureaucracy and the Environment.* Ballinger, Cambridge, MA.

Baumann, D. D., J. J. Boland, and W. M. Hanemann. 1998. *Urban Water Demand Management and Planning.* McGraw Hill, Washington, DC.

Brown, T. C. 1999. *Past and Future Water Use in the United States.* Rocky Mountain Research Station, USDA Forest Service, Fort Collins, CO.

Bureau of Reclamation. 1993a. *1993 Irrigation Water Rates: Central Valley Project, California.* Mid-Pacific Region, Bureau of Reclamation, Sacramento, CA.

———. 1993b. *Interim Guidelines for the Implementation of Required Payments and Water Rates and Crediting of Revenues.* Central Valley Project Restoration Fund, Sacramento, CA.

Department of Water Resources. 1992. *The 1991 Drought Water Bank.* Department of Water Resources, Sacramento, CA.

Dinar, A., K. Knapp, and J. Letey. 1989. Irrigation water pricing policies to reduce and finance subsurface drainage disposal. *Agricultural Water Management* 16: 155–171.

Dinar, A., and J. Letey. 1991. Agricultural water marketing, allocative efficiency and drainage reduction. *Journal of Environmental Economics and Management* 20: 210–223.

Frederick, K., ed. 1986. *Scarce Water and Institutional Change.* Resources for the Future, Washington, DC.

Frederick, K. D., and P. H. Gleick. 1999. *Water and Global Climate Change: Potential Impacts on the U.S. Water Resources.* Pew Center on Global Climate Change, Arlington, VA.

Gibbons, D. C. 1986. *The Economic Value of Water.* Resources for the Future, Washington, DC.

Gillian, D. M., and T. C. Brown. 1997. *Instream Flow Protection: Seeking a Balance in Western Water Use.* Island, Washington, DC.

Hanemann, M. W. 1993. Designing new water rates for Los Angeles. *Water Resources Update* 92: 11–21.

Howe, C., D. Schurmeier, and W. Shaw. 1986. Innovative approaches to water allocation: The potential for water markets. *Water Resources Research* 22: 439–445.

Howitt, R. 1991. Editorial. *UC/AIC Quarterly* 5(3): 2. Agricultural Issues Center, University of California, Davis.

Hurd, B., M. Callaway, J. Smith, and P. Kirshen. 1999a. Economic Effects of Climate Change on U.S. Water Resources. In *The Economic Impacts of Climate Change on the U.S. Economy.* R. Mendelsohn and J. Neumann, eds. Cambridge University Press, Cambridge, pp. 133–177.

Hurd, B., N. Leary, R. Jones, and J. B. Smith. 1999b. Relative regional vulnerability of water resources to climate change. *Journal of the American Water Resources Association* (December) 35(6): 1399–1410.

Loomis, J. 1987. Balancing public trust resources of Mono Lake and Los Angeles water rights: An economic approach. *Water Resources Research* 23: 1449–1456.

———. 1992. The 1991 State of California Water Bank: Water marketing takes a quantum leap. *Rivers* 3: 129–134.

———. 1994. Water transfer and major environmental provisions of the Central Valley Project Improvement Act: A preliminary economic evaluation. *Water Resources Research* 30: 1865–1871.

Los Angeles Department of Water and Power. 1985. *Urban Water Management Plan.* Los Angeles Department of Water and Power, Los Angeles.

Michelsen, A. M., and R. A. Young. 1993. Optioning agricultural water rights for urban water supplies during drought. *American Journal of Agricultural Economics* 75: 1010–1020.

Moss, R. H. 1998. Water and the challenge of linked environmental changes. *Water Resources Update* 112: 6–9.

Saliba, B. C., and D. Bush. 1987. *Water Markets in Theory and Practice.* Westview, Boulder.

Schilling, K., and E. Stakhiv. 1998. Global change and water resource management. *Water Resources Update* 112: 1–5.

Solley, W., R. Pierce, and H. Perlman. 1998. *Estimated Use of Water in the United States in 1995.* U.S. Geological Survey Circular 1200.

U.S. Congress. 1992. Reclamation Projects Authorization and Adjustment Act of 1992. H.R. 429, Washington, DC.

Vaux, H., Jr., and R. Howitt. 1984. Managing water scarcity: An evaluation of interregional transfers. *Water Resources Research* 20: 785–792.

Ward, F. 1987. Economics of water allocation to instream uses in a fully appropriated river basin: Evidence from a New Mexico wild river. *Water Resources Research* 23: 381–392.

Water Strategist. 1999. *1998 Annual Transaction Review.* Stratecom, Claremont, CA.

Wichelns, D. 1991. Increasing Block-Rate Prices for Irrigation Water Motivate Drain Water Reduction. In *The Economics and Management of Water and Drainage in Agriculture.* A. Dinar and D. Zilberman, eds. Kluwer Academic Publishers, Boston, pp. 275–294.

Wilson, P. 1991. Executive Order W-3-91. Executive Department of California, Sacramento.

SIXTEEN

Climate Variability: Social, Policy, and Institutional Issues

KATHLEEN MILLER
STEVEN GLOSS

BACKGROUND

A great deal has been learned about climate and its variability in the West over the past several decades (Chapter 1). We have extended the period of hydrologic records from a few decades to well over a century (Chapter 3). This longer period of record has contributed to our appreciation of the nature and extent of climate variability and its effects on water resources. Furthermore, substantial scientific and technical advances are increasing our understanding of the mechanisms and processes (local, regional, and global) underlying variation in climate. Scientific advances are contributing to our ability to predict future variability of climate. The usefulness to society of this additional knowledge will depend, however, upon our ability and willingness to take advantage of new information. This chapter summarizes the conclusions of a working group that focused on policy and institutional issues as they relate to the effects of climate variability on the socially valued services and attributes of water resources in the West.

The West shows great heterogeneity in (1) the problems caused by climate variability, (2) current flexibility, (3) institutional constraints, (4) interests and perceptions of participants, (5) resources available to them, and (6) policy options. The time frame for adaptation to climate variation or risk is also relevant. In general, fewer options are available for coping with risks posed by climate variations when time horizons are short and resources limited. Managing the effects of a drought in a small watershed with limited storage capacity entails fewer options than those of a major municipal area with several sources of water. Although the policy options in the first situation may be very limited, they may include some productive use of climate information and forecasts.

Water Resources and Hydrologic Variability in the West

The West includes mountainous areas that receive considerable precipitation, as well as arid or semiarid grasslands and deserts (Chapter 2). Most of the runoff of the West originates in the mountains, whereas many of the human uses of water—including irrigated agriculture and municipal use—are located in drier areas of lower elevation. Montane precipitation occurs mostly as snow, which produces high flows in spurts (Chapter 5). Thus water storage and management are important considerations in the allocation and use of water in the West. Water typically is transported through natural river channels and artificial conveyance facilities and may involve transbasin diversions.

Although runoff follows seasonal patterns, it shows a great deal of intraseasonal and interannual variability, which can be quantified by the coefficient of variation (CV, ratio of standard deviation to mean, expressed as percentage) of annual unregulated stream flow. Much of the West falls within the highest category for this measure of natural variability (67% CV; Figure 16.1).

Historical, Policy, and Institutional Framework

Waters of the West traditionally have been viewed as working resources. The first major influx of settlers into the region brought the Mormons, who built their communities around cooperative irrigation projects. A second wave of settlement was driven by a mining boom in the mid-nineteenth century. Following the lead of mining districts in California, Colorado and surrounding states adopted the prior-appropriation rule of water law, which promoted the development and private use of water (see Chapter 14). Because much of the region is too dry for profitable rain-fed agriculture, further settlement and economic development depended upon the extension of irrigation. In the late nineteenth century numerous private and local community irrigation projects were constructed throughout the region (e.g., Chapter 11). As the direct flows were quickly appropriated, latecomers with junior water rights found it necessary to construct water-storage facilities in order to obtain reliable water supplies during the critical summer low-flow period.

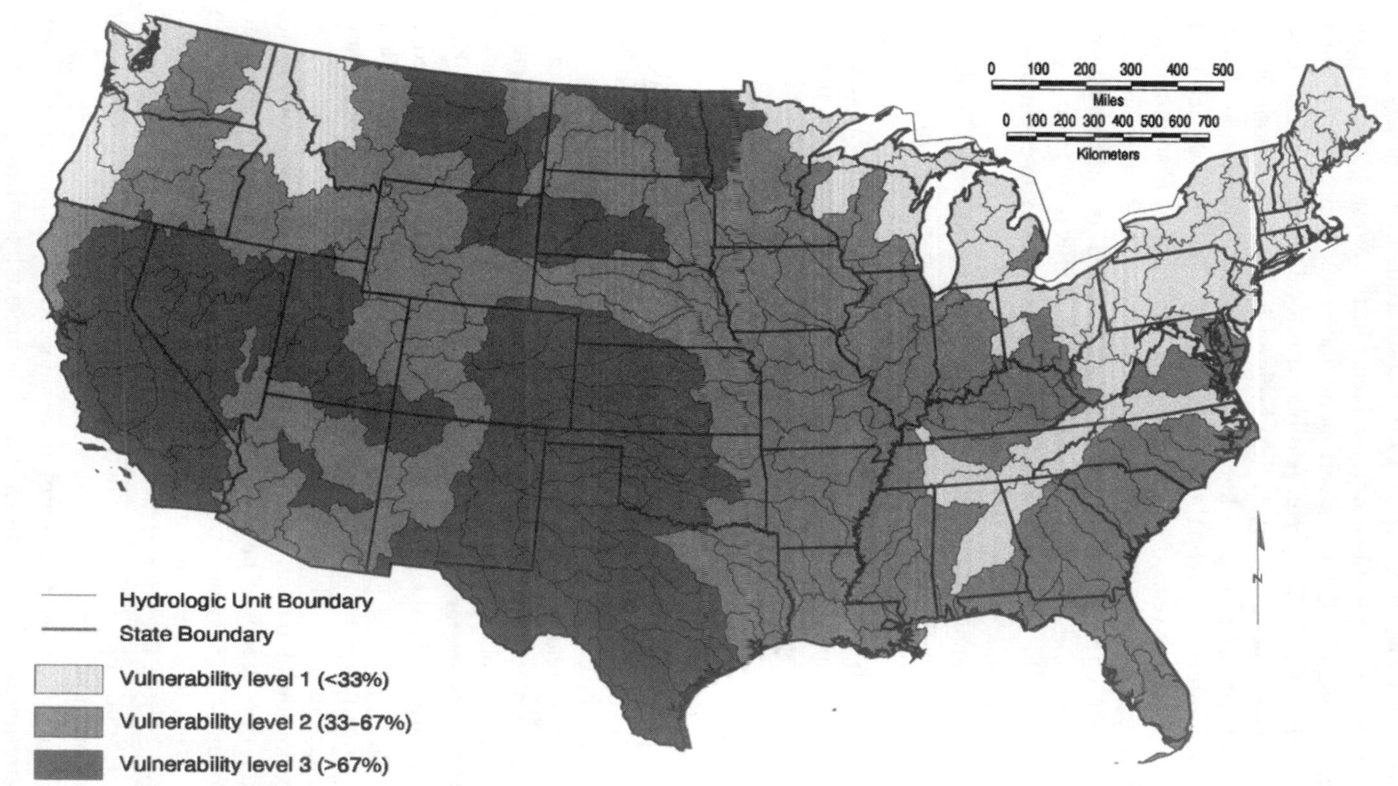

16.1 Coefficient of variation, expressed as percentage, for natural stream flow in the contiguous forty-eight states. Coefficients of variation were computed as the ratio of the standard deviation of annual unregulated stream flow to the unregulated mean annual stream flow. SOURCE: Stratus Consulting, Inc. (d:/ginwater/gis/projects/vogelmaps.aml).

The federal Reclamation Act of 1902 and subsequent federal acts greatly accelerated the development of water projects in the West, particularly on the largest rivers. The federal government undertook projects on a much larger scale than would have been feasible with local capital. In most cases the federal government substantially subsidized irrigation development by basing repayment obligations on liberal calculations of ability to pay and by providing zero-interest loans on those obligations (Rucker and Fishback 1983; NRC 1996). In recent years the federal role has shifted away from promotion of irrigation toward greater emphasis on environmental protection. Even the U.S. Bureau of Reclamation (USBR) has redefined its mission. Among other things, one of the goals of the "reinvented" USBR is to "promote the sustainable use of the water and associated land resources in an environmentally sensitive manner" (NRC 1996: 112).

The relative importance of federal, as opposed to private and local, irrigation projects varies considerably across the states of the West. In Colorado and Utah, for example, substantial nonfederal irrigation development began before the creation of the USBR. Of the approximately 13 million acre feet of water applied annually to irrigated farms in Colorado, only 8 percent comes from USBR sources (Table 16.1). In Utah and Nevada the USBR accounts for 9 percent of irrigation water use. Federal projects are more prominent in Idaho (44%), Arizona (36%), and New Mexico (21%). Irrigation accounts for 90 percent of consumptive water use in the West.

Groundwater pumping from the High Plains Aquifer and from alluvial aquifers in the Platte and Arkansas basins increased rapidly after World War II. The period of greatest expansion extended from the 1950s through the 1970s. Falling groundwater levels and higher pumping costs over parts of the High Plains Aquifer and water-quality problems in portions of the Central Platte alluvial aquifer have led to some reductions in irrigated acreage and water use and to greater public control over groundwater use (NRC 1996). Individual farmers in the High Plains Aquifer region have responded to rising pumping costs by increasing irrigation efficiency through use of low-pressure irrigation systems, conservation tillage, and crops with low water demand, such as wheat (Easterling 1996).

Irrigation development, impoundments, and management of water for production of hydropower have profoundly altered the hydrology and ecology of many of the region's watersheds. Water withdrawals cause certain stream reaches to dry up during the summer. In other cases storage releases and recharge from irrigation returns sustain summer flows in areas such as the lower South Platte River, where historical summer flows were lower than they are now.

In recent years, western states have adopted streamflow policies that allow water rights to be established for environmental, aesthetic, and recre-

Table 16.1—Annual irrigation water withdrawals and consumptive use (1990; Solley et al. 1993).

State	*Withdrawals (million acre feet)*	*Surface Water—Bureau of Reclamation (%)*	*Surface Water—Private Suppliers (%)*	*Groundwater—All Suppliers (%)*	*Irrigation Use (million acre feet)*	*Irrigation Use (%)*
California	31.3	20	42	38	21.8	93
Texas	9.5	5	30	66	8.0	79
Idaho	20.9	44	21	35	6.8	99
Colorado	13.0	8	70	22	5.6	94
Kansas	4.7	2	3	95	4.5	92
Nebraska	6.8	13	15	71	4.4	93
Arkansas	5.9	0	18	82	4.4	94
Arizona	5.9	36	25	39	4.0	82
Oregon	7.7	25	67	8	3.4	95
Washington	6.8	70	17	12	2.9	92
Wyoming	8.0	18	79	3	2.9	95
Florida	4.2	0	48	52	2.8	79
Montana	10.1	11	88	1	2.2	93
Utah	4.0	9	77	14	2.2	87
New Mexico	3.4	21	33	46	2.0	86
Nevada	3.2	9	60	31	1.6	86
Mississippi	2.1	0	7	93	1.5	74
Louisiana	0.8	0	36	64	0.7	39
Georgia	0.5	0	10	60	0.5	54
Oklahoma	0.7	6	12	82	0.4	58
All other states	3.9	6	45	49	3.0	25
United States	153.0	20	43	37	85.4	81

ational purposes (MacDonnell and Rice 1993). In most cases such rights may be held only by public agencies (Chapter 14). In addition, they are usually of such junior status that they provide little water during extreme drought (Wilkinson 1989). Some states, however, allow private entities to purchase senior water rights and donate them for environmental use (Colby 1993).

The allocation of water in all states of the West depends on the ownership of water rights and on the contracts and operating rules governing federal and other public water projects. Initial allocations can be modified by market transactions, but the cost of transferring water or water rights through markets varies considerably from state to state (Chapter 15). Most states in the West allow willing parties to exchange water, to transfer water rights permanently, or to buy and sell water on a short-term basis. This can enable uses of high value to continue during drought. Even so, some legal impediments exist to moving water between different types of uses. Nebraska, for example, has not allowed agricultural water rights to be transferred to another type of use and has only recently considered legislation to allow such changes.

Water banking is a relatively new concept in the West that may provide increased flexibility. The term *water bank* has been applied to two types of arrangements. Groundwater storage banks use surface water to recharge an aquifer, which then is used as a source of water when surface water is less abundant. Another use of the term refers to a formal mechanism that facilitates voluntary changes of the use of water under existing rights (Chapter 15). Idaho, for example, uses the latter type of water bank to facilitate one-season transfers of water stored in reservoirs on the Snake, Boise, and Payette Rivers (MacDonnell et al. 1994). Groundwater recharge is practiced in the lower South Platte basin in Colorado. Arizona has passed legislation that facilitates groundwater banking. Kansas is developing a water-banking program that may allow short-term water transfers while keeping ownership of the water right unchanged and may allow groundwater rights to go unused in exchange for future pumping credits. In addition to formal water transfers, municipalities and other water-right holders often make informal, voluntary arrangements to provide water for environmental purposes or to reduce flood risks when doing so is not too costly.

The allocation of surface water between states is governed in many instances by interstate compacts, but these agreements have been subject to costly litigation. Interstate lawsuits are in progress for both the Platte and Arkansas River systems.

Native American water rights remain an unresolved and important issue for water allocation in the Interior West (Chapter 14). Under the legal doctrine established by the 1908 *Winters* case (*Winters v. United States* [207 U.S. 564 (1908)]), tribes have reserved rights to substantial amounts of water (Western Water Policy Review Advisory Commission 1998). The vast majority of these claims have never been developed for the benefit of the tribes, nor have they been clearly quantified. In many cases nontribal water users have fully appropriated and used the sources of water that potentially would satisfy tribal rights. Tribal efforts to protect and develop their water rights have therefore encountered resistance from other water users, and during drought, some tribes have encountered difficulties in enforcing their water rights. In addition to substantial litigation (approximately sixty cases were pending as of 1995), about twenty current negotiation efforts are aimed at achieving settlement of tribal water-rights claims. The few successful settlements to date have often relied on federal funding to provide new water for tribal uses.

Systems of water administration vary by state. In Colorado the state constitution, adopted in 1876, proclaims that "the right to divert the unappropriated waters of any natural stream to beneficial uses shall never be denied" (Article XVI). This proclamation, related provisions in the state constitution, and subsequent legislation and court cases established a legal framework under which the waters of Colorado's streams can be appropriated to extinction and under which there are many more claims to the state's streams than can be simultaneously satisfied during low-flow periods. As a result, the state has developed a system of administration by state and division engineers who are responsible for ensuring that water rights are fulfilled in order of priority and that Colorado's water uses do not infringe on the state's obligations to neighboring states under the nine interstate compacts to which Colorado is a party (Getches 1987; Moses 1987).

The state uses a system of water courts to record the priorities and other features of new water rights and of those altered by transfers to new uses. Although the water court system has been criticized as unnecessarily costly (Martz and Raley 1987), it has the advantage of maintaining an accurate and up-to-date record of adjudication for all water rights. Short-term water leases and exchange agreements can be approved by the division engineer and do not require approval of the water court, although longer-term water leases could be challenged in the water courts, which may inhibit the development of water banks in Colorado. Colorado has experienced difficulties in sufficiently controlling groundwater development to prevent damage to downstream surface-water rights and compact obligations. This failure was the major trigger for a protracted and costly dispute with Kansas over the Arkansas Compact (*Western Water Law and Policy Reporter* 1996).

Other states in the Interior West approach water administration differently. State water agencies in several states have broad authority to prevent waste and to approve, condition, or deny permits for new uses as warranted to serve the public interest. Environmental values and preservation of stream flows are among the considerations these states use in determining consistency of a water use with the public interest (Tarlock 1989).

Water Management: Storage

The West contains several thousand dams. They range from massive, federally built structures such as Hoover Dam on the Colorado River to small irrigation impoundments (Graf 1999). In the South Platte basin alone there are almost 1,000 reservoirs. Most of these are small irrigation reservoirs located in Colorado (Bennett 1994). Owners of such reservoirs typically fill them once annually and release water as they need it. In some cases ditch companies engage in coordinated management of multiple water rights and reservoirs (Howe et al. 1980; Freeman and Wilkins-Wells 1999). In the continental United States, the Great Plains, Rocky Mountains, and Southwest have the highest ratios of reservoir capacity to mean annual runoff (Graf 1999). Many subbasins within these regions have a high ratio of mean annual water withdrawal to total reservoir storage (Stratus Consulting 1999). The latter measure indicates vulnerability to climate variability. In many parts of the Interior West, average annual water withdrawals from all sources equal or exceed 85 percent of the unregulated average annual stream flow (Figure 16.2).

Over the past quarter century, the federal government has assumed increasing authority over environmental protection, including maintenance of water quality. A prominent example is the Endangered Species Act, which requires the development of recovery plans for listed species.

SOCIAL AND INSTITUTIONAL RESPONSES TO WATER-RESOURCES ISSUES

Drought Management

Current management of climate variability focuses on historically documented extremes. For example, the Northern Colorado Water Conservancy District (NCWCD) uses the 1950s drought as a standard in its planning. Other water districts and ditch companies may use similar critical periods in their planning. If improved climate forecasts could provide reliable estimates of the probability of flows exceeding or falling short of those in the critical period, water managers could better adjust operations to take advantage of abundant supplies or to reduce disruption arising from short supplies. Individual irrigators can respond to anticipated water shortages by planting crops that use less water, fallowing a portion of their acreage, retiring marginal lands, or practicing conservation tillage. In places where short-term water transfers are feasible, farmers with orchards, vines, or other crops of high value may purchase water from neighbors who would otherwise use the water on lower-value crops.

Some past responses to droughts contributed to subsequent water-management problems. For example, many Colorado irrigators in the Arkansas River basin responded to the 1950s drought by drilling wells. Pumping then depleted the flow of the river at the state line, in violation of the Arkansas River Compact. Colorado's delay in restricting such pumping caused a substantial water debt to accrue, which the state must now repay to Kansas. Efforts to reduce the debt and comply with the compact currently restrict pumping to such an extent that water tables are rising on the Colorado side of the border, causing increased soil salinity that is damaging Colorado crops.

Severe, Sustained Droughts

Many water systems are designed to withstand the effects of a single-season drought, but droughts persisting two or more years may pose significant problems. The Highlands Ranch area south of Denver provides an example.

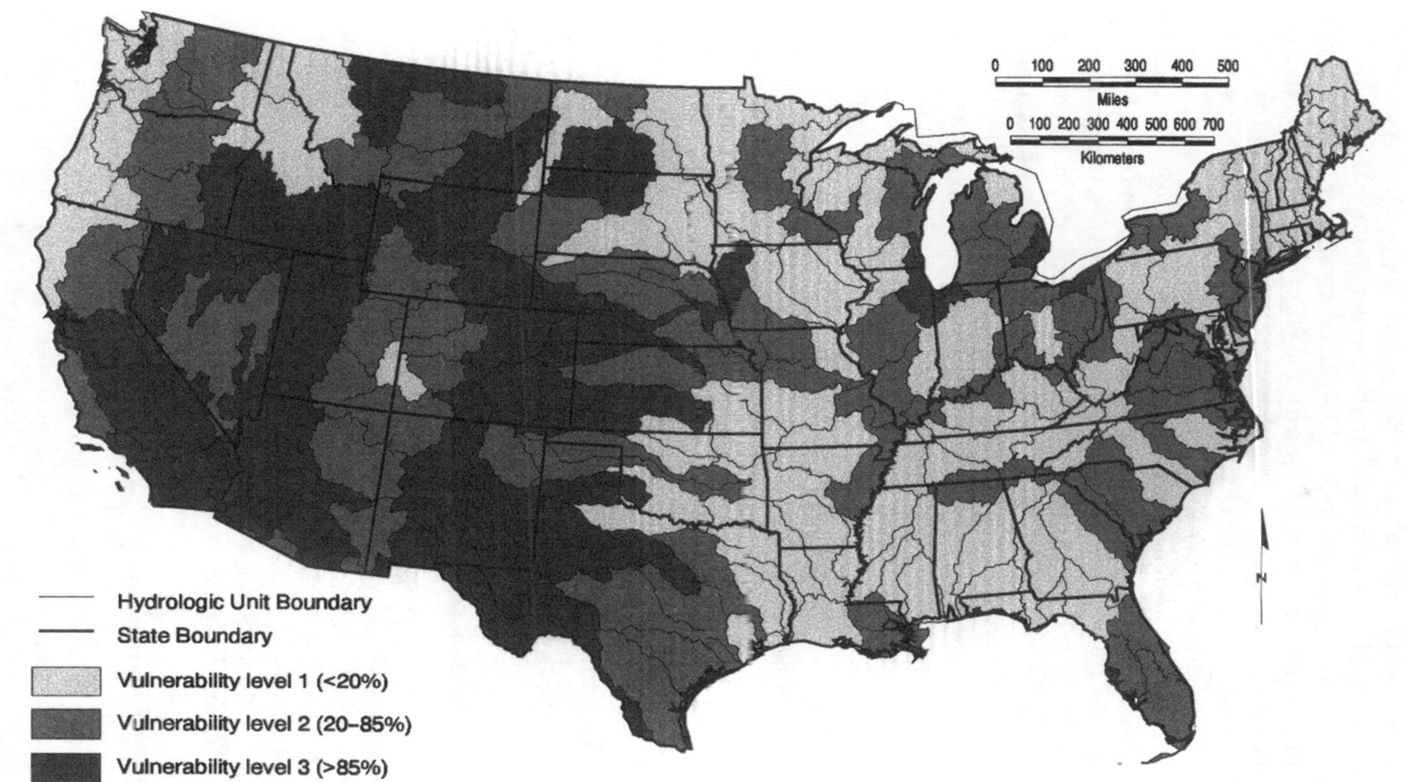

16.2 Proportion of stream flow, expressed as percentage, withdrawn for use. Ratios on which the map is based were taken from records of total annual surface and groundwater withdrawals in 1990 and are expressed as a ratio to the unregulated mean annual stream flow. SOURCE: Stratus Consulting, Inc. (d:/ginwater/gis/projects/vogelmaps.aml).

New homes depend on an aquifer for their water supply. Development has been permitted on the assumption that the aquifer has a 100-year life. A severe, sustained drought, however, could significantly shorten the life of the aquifer by diminishing recharge. At present, the effects of drought on the life of the aquifer are not understood. Thus a better understanding of the sensitivity of the aquifer, coupled with better climate forecasts, might be useful in setting development and water-management policies.

A study of severe, sustained drought on the Colorado River (Lord et al. 1995) highlighted the role of federal water in that system and the results of implementing the law of the river (composite of current laws, compacts, and treaties). Economic and other impacts were most severe in the upper basin and on hydropower, environmental, and recreational uses. Providing institutional flexibility, however, was predicted to result in substantial mitigation of the impacts experienced under the strict law of the river scenario. Although inefficient pricing of federal water has encouraged substantial application of water to uses of very low economic value, these uses can be cut back with little economic impact. This and the high storage capacity of the federal reservoirs provide a great deal of drought resilience. It is difficult to predict the consequences of a severe, sustained drought for the Colorado system, however, because the Secretary of the Interior has discretion over operation of the federal reservoirs and water deliveries on federal projects, particularly in the lower basin.

OPPORTUNITIES FOR IMPROVED PLANNING

Overallocation and Adjustment to Social and Demographic Change

Where water is very tightly allocated, as in Colorado and much of the West, there is little buffer to absorb the impacts of drought. During droughts, and in many watersheds during ordinary summer low-flow periods, some water rights must be shut off to satisfy the claims of downstream senior users. In other states, where streams have been closed to new appropriations, the buffer may be greater. As flows and storage have become more fully allocated or even overallocated, the need for planning has increased. Shortages are likely to become both more frequent and contentious as demands continue to grow and change.

The West is undergoing great demographic shifts (Chapter 13; Case and Alward 1997) that have led to transfer of water from agricultural to municipal uses. In many cases municipalities have acquired water rights in advance of need and are currently renting them back to agricultural users. This practice may provide a drought buffer for urban use of water. Better climate forecasts could facilitate more efficient planning for seasonal use of such water rights. For example, if a dry season is forecast, a city might be able to cancel a rental agreement before crops have been planted.

Recreation, which has also grown in importance, creates demand for maintenance of sufficient stream flows to support fishing, rafting, and other recreational activities. Increased interest in environmental restoration for its own sake has also come with the West's trend toward urbanization. Agriculture, however, remains the dominant water use throughout the Interior West. In some cases public agencies have used short-term purchases of water from agriculture to provide water for environmental purposes (Simon 1998). In addition to such short-term transfers, water has been permanently dedicated to environmental restoration. An example is the Oregon Water Trust, which acquires complete rights or rights to conserved water for maintenance of stream flows. If environmental water purchases become more common, seasonal climate forecasts might be useful in optimizing use of money for water purchases.

Possible responses to future drought episodes may resemble those identified in a study by Howe et al. (1980) of the 1976–1978 drought in Colorado. Many of the rural entities that supply water for irrigation in eastern Colorado and nearby towns cooperated in various ways to use water efficiently. For example, some holders of senior rights agreed not to call for their water, and some water users pooled their supplies in a single reservoir to reduce evaporation. The study also noted that enthusiasm for cooperative actions tended to wane during the second year of the drought. Other examples exist of voluntary cooperation to share water, reduce risk of flood damage, or sustain flows for environmental purposes. Denver, for example, has released water to push the annual peak flow of the South Platte to a pattern more favorable for fish. There is no assurance, however, that such cooperation could be obtained in a dry year.

Surplus Water and Natural Hazards; New Forecasting Tools

Flood damage may occur either during spring runoff or as a result of flash floods such as the one that flooded Fort Collins, Colorado, in 1998. Snowpack measurements provide estimates of the total volume of spring runoff, but both the timing and magnitude of the peak are sensitive to temperature and to rain falling on snowpack. Reservoir managers and state water authorities use short-term weather forecasts to predict the timing and magnitude of such peak flows, and the state engineer or other water authorities may request changes in reservoir release rates to reduce the risk of flood or failure of dams. Improved forecast accuracy and distribution should allow management decisions to be made further in advance of dangerous situations. For flash floods, communities might enhance preparedness in response to a forecast of an increased probability of such events. For example, they might increase expenditures for maintenance of storm drainage systems. Presently, however, it is difficult to establish any clear link to a climate forecast, and appropriate land-use planning may still be the best defense against flood damage.

Water users traditionally have used storage capacity and institutional arrangements such as the doctrine of prior appropriation to protect themselves from uncertainty caused by climate variation. New forecasting capability may offer other options for coping with variable water supplies.

Climate prediction is based on the statistical distributions of weather variables, which respond to slowly evolving components of the climate system. For example, slowly evolving ocean temperatures in the case of the El Niño–Southern Oscillation can provide boundary conditions for climate forecasts generated by numerical models of the coupled ocean-atmosphere system. The ability to forecast sea-surface temperature (SST) in the Tropical Pacific has been greatly enhanced by the improved quality of data on ocean and atmospheric conditions provided by the Tropical Atmosphere Ocean Array of instruments in that region (NRC 1999). The climate variables of most interest for water-resource management in the West, however, are temperature and precipitation. These variables are somewhat related to tropical SSTs but are much more difficult to predict. Improvements in the skill of climate forecasts relevant to water management in this region are likely to be modest (Chapter 2).

The procedure for making climate forecasts from forecasts of Tropical Pacific SST is to run forecast models many times with differing initial atmospheric and oceanic conditions, thus accounting for imperfect knowledge of initial conditions. The resulting ensemble of forecasts is used to produce a probability distribution of conditions that could occur (NRC 1999). Forecast maps displaying the probabilities that seasonal temperatures and precipitation will be above, near, or below normal are now routinely produced and distributed over the Internet by the International Research Institute for Climate Prediction (source: http://iri.columbia.edu/forecast/net_asmt/). These forecast products display a change in the odds of unusually wet or dry conditions; they are useful in risk management but are not a source of certainty.

CAPACITY TO RESPOND TO VARIABILITY: USE OF NEW INFORMATION

Our ability to use new forecast information will depend upon the extent to which decisions made by multiple actors at various levels add up to a rational allocation of water use and risk across the full array of water users and other affected interests. If avenues for communication and coordination are blocked, climate variation will be more damaging, and forecast information will be less useful.

Coordination and Efficiency

The West's population growth and the diversification of social values with respect to water resources present new challenges for managing the effects of climate variability. The amount of flexibility in infrastructure and institutions will determine how effectively we will be able to balance competing interests and avoid costly disputes as we respond to the impacts of climatic variations. Our ability to deal effectively with variability will depend on the time and effort devoted to consideration of these issues ahead of time. Many potential solutions are already known and should be considered for implementation in the near term.

Improvements in water-use efficiency have been demonstrated in the agricultural, industrial, and municipal sectors in many places around the world (Postel 1992). Many industrial improvements are cost-effective over very short time periods. Retrofitting and establishment of new building codes by municipalities and demand management by utilities are proving useful in the municipal sector (Michelsen et al. 1998). Agricultural efficiency is improved through a variety of cost-effective conservation measures that may involve transfers with municipalities or result in improvement of agricultural productivity.

Water Markets and Inclusion of Underrepresented Groups

Water markets offer a means of shifting water use. Often market-based transfers involve shifting water use from agriculture to another use. Because of the dominance of agriculture in western water use, transfer of a relatively small percentage of that water can satisfy demands in other sectors while providing beneficial capital and conservation incentives to agriculture (Chapter 13). Although water markets have potentially adverse impacts, their use should be encouraged as a means of increasing flexibility in meeting water needs.

The increasing urbanization and resettlement of the West have resulted in a populace only incidentally involved in water use at the individual level. This situation is much changed from a few decades ago. Likewise, increasing population has brought a new set of social challenges. Changing social values and needs require identification of mechanisms to include potentially underrepresented groups in the discussion, implementation, and administration of water-resources policy.

Flexibility and Vulnerability; Conjunctive Management

Institutions at all levels should examine their flexibility and potential vulnerability for various kinds of climate variation. Where vulnerability may be high or uncertainty large, policy, institutional, or infrastructural remedies should be sought. Hypothetical climate scenarios played out over a variety of spatial and temporal timescales may be useful in developing appropriate alternatives for coordination among institutions.

Groundwater is conjunctively managed with surface water in most alluvial aquifers in the West, but groundwater–surface-water interactions under conditions of climatic variability probably are not considered on any systematic basis. The role of altered recharge, not just traditional groundwater mining, under scenarios of increased climate variability should be considered.

Use of Data in Planning; Management and Dissemination of Data

New and improved data as well as climate and weather forecasting information are now available and should be integrated with the planning process. Institutional changes may be necessary to integrate the capabilities of scientific data generation with planning for water allocation and use. Each management sector has a unique culture and distinctive needs, consideration of which will best enable incorporation of new information in its planning processes.

The information age has increased the efficiency and speed with which data can be shared, as well as the formats in which it is available. Nevertheless, the increased availability of data and information has not erased the need for traditional means of information sharing. Coupled with the existing need to make climate data and forecasting most useful to the client is a need to recognize the breadth of information needs and user capabilities in the water-resources area.

Linkages Between Water Quantity and Water Quality; Education

Floods or droughts can affect water quality either directly or indirectly. Abnormally wet periods may increase transport or leaching of salts from soils as well as substances intended for slow infiltration or dissipation to the atmosphere. Thus unanticipated infiltration to groundwater tables or subsurface transport of natural elements or xenobiotic chemicals to surface waters may be unintended consequences of floods or abnormally wet conditions. Drought may reduce stream flow and water levels in lakes or reservoirs to below-normal levels, thus increasing the concentration of elements—including wastes or chemicals—in these waters. Planning that allows flexibility in discharge regulations or application rates of chemicals and fertilizers during periods of abnormal precipitation or runoff may help alleviate certain water-quality problems.

Education can be a foundation for enhanced public understanding and involvement in planning for and response to the impacts of climate variability on water resources. A low level of knowledge and interest on the part of the general public currently exists regarding climate variability and water resources in the West. There is some potential for increased public involvement in water-policy matters and support of contingency planning if the public is better informed. Educational efforts should be aimed at providing both a basic understanding of the way climate affects water resources and a good grasp of water supply and allocation policy at various scales.

Public education will likely lead to education of legislators and policy makers. Policies promoting sustainable use of water resources and flexibility in institutional arrangements for water allocation will ultimately be beneficial to society as a whole and supportive of the region's economic well-being. A crisis management approach is less likely to yield satisfactory results, as recognized recently by the Committee on Sustainable Water Supplies for the Middle East—an area facing these issues in such a mode (NAS 1999).

Education has and will remain fundamentally a local responsibility. Therefore organizations such as watershed councils and conservancy districts could assume a major role in developing a grassroots constituency that recognizes the need for planning (Chapter 11). Leadership from these organizations and their members would demonstrate concern for sustainable use of water resources. Although materials and expertise to support these efforts would likely need to be developed initially through state or federal programs, organizations such as these and the cooperative extension service should have primary responsibility for implementing programs at the local level.

One way to enhance acceptance of local educational programs by both the public and policy makers would be to develop local demonstration projects that are well articulated with state, regional, and federal agencies. These could take the form of physical models, computer gaming simulations, or even board games—any of which could be supported by video, web sites, and public presentations. Some water conservancy districts are exploring the use of modeling to demonstrate the implications of different climate scenarios.

Education should convey an understanding of the history and legal basis of western water rights. Relatively few people understand these issues. Many are new to the West, where allocation of water resources is a unique enterprise that needs to be understood in light of its historical and cultural importance in the region.

Roles of the Individual and of Government

People often fear change and are willing to tolerate inefficiency if they get certainty in return. When considering reforms, we must recognize this phenomenon. For example, the Western Water Policy Review Advisory Commission report (1998) is disliked by the agricultural community, which views it as an attack on agricultural water use. The reforms that have a future in this region must be crafted within the prior appropriation system, which entails, for example, purchase of water rights at fair market value for environmental purposes.

Programs that recognize the fundamental importance of the individual water-right holder in the West, in the context of larger historical institutions such as ditch and irrigation companies or municipalities, and that then align these historical patterns with emerging features of modern water demands will move toward sustainability.

Water rights are mostly administered under state law in the West, even when federal storage projects are involved. Federal projects often are coordinated in larger river basins through compacts or court decrees wherein the interests of the states or their legal rights are enforced. Currently, relatively few regional organizations deal with planning for the use or allocation of water across state boundaries or within river basins. To deal effectively with increased variability in water resources that may result from climate change, it will be important to improve coordination among states, regional organizations, and the federal government. Encouraging planning efforts at these three levels will reduce hardship, bring increased recognition of the need to address water-resources availability, and reduce the potential for conflict and litigation.

At present, there is a lack of trust among the federal, state, and local governments and water users. This inhibits cooperation and the protection of environmental water uses. Exchange of information is needed, but it requires building trust.

Planning for climate variability should probably occur as much as possible at the local and private-sector levels. It seems likely, however, that educational and incentive or regulatory programs as well as financial support will need to come at least in part from state and federal agencies, as explained next.

Because the federal government has been the primary sponsor of research on climate variability, federal agencies should demonstrate that their activities and programs address potential impacts of climate variability on water resources. This could first be done through administrative procedures or executive orders requiring that annual budgets and work plans demonstrate attention to climate variability. A more pervasive approach would require that federal agencies ensure that climate variability be systematically considered through the National Environmental Policy Act and through the issuance of federal permits under Section 404 of the Clean Water Act or Federal Energy Regulatory Commission relicensing procedures.

The federal government should provide money for model programs and technical assistance to state and local governments for incorporation of climate variability into their programs. Policies that promote the development of mitigation measures could help reduce the economic, social, and environmental impacts of climate variability and reduce the need for future government intervention. The U.S. Congress, through its Office of Technology Assessment (1993), recommends such options to make the country more resilient to potential climate variability. According to one study, mitigation for climate variability could save two dollars for every dollar invested (*Natu-*

ral Hazards Observer 1996). Various types of federal assistance also could be made contingent on effective state or local planning and implementation of mitigation measures.

States can be the liaison among individuals, water agencies, municipalities, the federal government, and other organizations. Their primacy in administration of water rights and the doctrine of prior appropriation could make their engagement in climate-variability issues both logical and important.

Progressive state governments in the West will want to demonstrate to their citizens and to future residents and businesses that they have plans and policies in place to prevent excessive adverse impacts from climatic variations. Although many western states have drought management plans, most have not developed strong mitigation measures. Colorado, for example, was one of the first western states to develop a comprehensive plan (Wilhite 1997), but it still lacks strong mitigation measures. Legislatures (perhaps in partnership with the federal government) should authorize programs that encourage incorporation of climate-variability planning at the state, regional, and local levels.

States also should provide support for public education regarding water resources and climate variability. An informed citizenry is likely to be more supportive of the need for programs dealing with the issue of climate variability.

SUSTAINABILITY

Water has perhaps more potential than any other resource to be managed in a sustainable manner. Demands, however, are dynamic over time, and the shifting demographic patterns and social values of the West illustrate the need for planning and flexibility in policies and institutions.

Natural ecosystem processes provide beneficial services that are typically either irreplaceable or costly to replace (NAS 1999). Water use and development in the West should be mindful of these ecosystem services.

The opportunity for sustainable development and use of water resources under the influence of climate variability is at hand, but planning and implementation will be required. We have numerous examples, notably with drought and floods, of managing by response rather than by planning. Lessons from the past combined with unprecedented levels of understanding give us opportunities for a more socially responsible collective behavior.

REFERENCES

Bennett, L. L. 1994. The Economics of Interstate River Compacts: Efficiency, Compliance, and Climate Change. Ph.D. dissertation, Department of Economics, University of Colorado, Boulder. Published as NCAR Cooperative Thesis No. 151. University of Colorado and National Center for Atmospheric Research, Boulder.

Case, P., and G. Alward. 1997. *Patterns of Demographic, Economic and Value Change in the Western United States.* Report to the Western Water Policy Review Advisory Commission, Denver.

Colby, B. G. 1993. Benefits, Costs and Water Acquisition Strategies: Economic Considerations in Instream Flow Protection. In *Instream Flow Protection in the West,* rev. ed. L. J. MacDonnell and T. Rice, eds. Natural Resources Law Center, University of Colorado School of Law, Boulder, pp. 87–101.

Easterling, W. E. 1996. Adapting North American agriculture to climate change in review. *Agricultural and Forest Meteorology* 80: 1–53.

Freeman, D. M., and J. Wilkins-Wells. 1999. Water exchanges in Colorado: Opportunities and constraints for the future. *Colorado Water: Newsletter of the Water Center of Colorado State University* 8: 10–11.

Getches, D. H. 1987. Meeting Colorado's Water Requirements: An Overview of the Issues. In *Tradition, Innovation, and Conflict: Perspectives on Colorado Water Law.* L. J. MacDonnell, ed. Natural Resources Law Center, University of Colorado, Boulder, pp. 1–24.

Graf, W. L. 1999. Dam nation: A geographic census of American dams and their large-scale hydrologic impacts. *Water Resources Research* 35: 1305–1311.

Howe, C. W., P. K. Alexander, S. Sertner, J. A. Goldberg, and H. P. Studer. 1980. *Drought Induced Problems and Responses of Small-Town and Rural Water Entities in Colorado: The 1976–78 Drought.* U.S. Department of the Interior, Office of Water Research and Technology. Project A-045-Colo, Completion Report 95. Colorado Water Resources Research Institute, Colorado State University, Fort Collins.

Lord, W. B., J. F. Booker, D. H. Getches, B. L. Harding, D. S. Kenney, and R. A. Young. 1995. Managing the Colorado River in a severe sustained drought: An evaluation of institutional options. *Water Resources Bulletin* 31: 939–944.

MacDonnell, L. J., C. W. Howe, K. A. Miller, T. Rice, and S. F. Bates. 1994. *Water Banking in the West.* Natural Resources Law Center, University of Colorado School of Law, Boulder.

MacDonnell, L. J., and T. Rice, eds. 1993. *Instream Flow Protection in the West,* rev. ed. Natural Resources Law Center, University of Colorado, Boulder.

Martz, C. O., and B. W. Raley. 1987. Administering Colorado's Water: A Critique of the Present Approach. In *Tradition, Innovation, and Conflict: Perspectives on Colorado Water Law.* L. J. MacDonnell, ed. Natural Resources Law Center, University of Colorado, Boulder, pp. 41–61.

Michelsen, A. M., J. T. McGuckin, and D. M. Stumpf. 1998. *Effectiveness of Residential Water Conservation Price and Nonprice Programs.* AWWA Research Foundation and American Water Works Association, Denver.

Moses, R. J. 1987. The Historical Development of Colorado Water Law. In *Tradition, Innovation, and Conflict: Perspectives on Colorado Water Law.* L. J. MacDonnell, ed. Natural Resources Law Center, University of Colorado, Boulder, pp. 25–39.

National Academy of Sciences (NAS). 1999. *Water for the Future: the West Bank and Gaza Strip, Israel, and Jordan.* National Academy Press, Washington, DC.

National Research Council (NRC). 1996. *Committee on the Future of Irrigation in the Face of Competing Demands: A New Era for Irrigation.* National Academy Press, Washington, DC.

———. 1999. *Making Climate Forecasts Matter: Panel on the Human Dimensions of Seasonal-to-Interannual Climate Variability*. P. C. Stern and W. Easterling, eds. National Academy Press, Washington, DC.

Natural Hazards Observer. 1996. Witt announces FEMA mitigation plan. *Natural Hazards Observer* 20: 10. Natural Hazards Information and Applications Center, University of Colorado, Boulder.

Postel, S. 1992. *Last Oasis: Facing Water Scarcity*. W. W. Norton, New York.

Rucker, R. R., and P. V. Fishback. 1983. The Federal Reclamation Program: An Analysis of Rent-Seeking Behavior. In *Water Rights: Scarce Resource Allocation, Bureaucracy, and the Environment*. T. Anderson, ed. Ballinger, Cambridge, MA, pp. 45–81.

Simon, B. 1998. Federal acquisition of water through voluntary transactions for environmental purposes. *Contemporary Economic Policy* 16: 422–432.

Solley, W. B., R. R. Pierce, and H. A. Perlman. 1993. Estimated use of water in the United States in 1990. *U.S. Geological Survey Circular 1081*, 76 pp.

Stratus Consulting. 1999. *Water and Climate Change: A National Assessment of Regional Vulnerability*. Revised Draft Report (May). Work Assignment 121, Contract 68-W6-0055 for the U.S. Environmental Protection Agency, Washington, DC.

Tarlock, A. D. 1989. *Law of Water Rights and Resources*. Clark Boardman, New York.

U.S. Congress, Office of Technology Assessment. 1993. *Preparing for an Uncertain Climate*. Volume 1, OTA-O-567. U.S. Government Printing Office, Washington, DC.

Western Water Law and Policy Reporter. 1996. Kansas v. Colorado: Status report on the Arkansas River Compact litigation. *Western Water Law and Policy Reporter* 1: 14–15.

Western Water Policy Review Advisory Commission. 1998. *Water in the West: Challenge for the Next Century*. National Technical Information Service, Springfield, VA.

Wilhite, D. 1997. *Improving Drought Management in the West*. Report to the Western Water Policy Review Advisory Commission. National Drought Mitigation Center, University of Nebraska, Lincoln.

Wilkinson, C. F. 1989. Aldo Leopold and western water law: Thinking perpendicular to the prior appropriation doctrine. *Land and Water Law Review* 24: 1–38.

PART FIVE

Analyzing the Analysts

Patricia Nelson Limerick, director of the Center for the American West at the University of Colorado and well-known author and speaker, has the last word in this book. Although trained as a historian rather than a scientist, Limerick is a proven shrewd observer of scientists, especially as they grapple with issues that bear directly on public policy and the welfare of society. Thus the steering committee of the symposium on which this volume is based asked her to sit through the oral proceedings and give the participants benefit of her observations and conclusions at the end. This she did, to the wide appreciation of the participants.

SEVENTEEN

Western Water Resources and "Climate of Opinion" Variables

Patricia Nelson Limerick

The knowledge that one is going to have the last word at a conference dealing with a subject as complicated as this one gets the neurotransmitters stirred up in a very memorable way. As a western American historian, I am a fascinated observer of the role of experts in natural-resource management, and so the assignment to listen closely and offer concluding remarks at this conference was an honor and a pleasure. On this occasion and others like it, I steer by the hope that commentary from the humanities can offer natural scientists an opportunity to step away from the detail and specificity of their inquiries and think about their situation in this historical moment. Reflections on the social, political, and cultural contexts in which experts work can be amusing, orienting, and maybe even necessary.

The naturalist John Muir found one route to exhilaration by climbing high in a tree in the midst of a wild Sierra Nevada rainstorm and holding to the tree while it whipped around in the wind. This kind of experience is not everyone's idea of a good time. But like it or not, those whose studies have bearing on society's use of water do their work in a position rather similar to Muir's preferred way of observing a storm. Hydrologists, limnologists, climatologists, engineers, and biologists specializing in aquatic species often have to hold—and hold hard—on to their perches while fierce winds of advocacy, desire, ambition, anger, resentment, and partisanship whip them around.

Now it is unlikely that a humanities-based observer can do much to calm those winds. But knowing more about the pattern of these forces, knowing from which directions the next gusts are likely to materialize, can make life in an arena of contestation a little more manageable. So I made it my task at the conference to keep an eye out for the most consequential trends, patterns, and issues in the social, cultural, and political contexts in which western water experts work. I have adopted the term *signal,* used often at the conference, which I understand to mean a distinguishable pattern discovered in a body of evidence. This exercise—identifying signals in the "climate of opinion" governing western water use—necessarily overstretches the license of the historian, extending the domain of discussion from the past tense into the present and even into the future. But the desire to make these comments useful to practitioners overrules timidity. Since a number of these signals are at present more potential than actual, like any predictions, they open the door to the possibility that the party making the forecast may be proven entirely, even hilariously, wrong. In truth, having a sense of both humor and humility is the absolute prerequisite for this line of work, which may help to explain its limited popularity among humanists!

THE TOP NINE SIGNALS FOR THE WATER-ORIENTED TO WATCH

1. Western Farming and Ranching May Lose Social Capital and Political Power.

Through the course of American history runs a deep and persistent stream of sentiment, often called "Jeffersonian agrarianism." Farmers are close to the soil; they are rendered more virtuous by that association with the earth. Because they can feed themselves, they cannot be bullied and overpowered in the way that dependent urban workers could be. Added to that, of course, is the cowboy phenomenon in which handsome men on even more handsome horses came to represent the toughness and independence of the national spirit.

These sentiments and images had an enormous impact on western water development. In the most smug forms of hindsight, an art form practiced with striking devotion by some contemporary environmentalists, the idea of turning arid western land into cropland was inherently improbable and unreasonable. In truth, federal funding of reclamation meant that eastern, midwestern, and southern farmers were subsidizing the provision of the water that would make it possible for western farmers to compete with them.

Changes in the livestock business, meanwhile, made the growing of winter feed crops like alfalfa into a substantial use of western water.

Over the last century, the percentage of farmers in the American population has been declining steadily. Sentiment in support of farming has not necessarily declined proportionately; many people whose own agrarian activities do not extend beyond a couple of fussed-over tomato plants still feel warmly about farmers. More important, many senators and congresspeople still have a pretty soft spot for agriculture in their hearts and legislative programs. In truth, small farms and ranches are cultural resources in themselves; in many western places the few remaining farmers and ranchers are living ties to an important past. And enthusiasts for open space have more and more come to realize that the maintaining of farms and ranches, whatever they may do to transform ecosystems, can be essential to keeping views open and horizons clear.

So a considerable amount of sentiment defends the generous allocation of water to agriculture in the West. But sentiment and emotion, although remarkably persistent, can also be remarkably changeable and fickle. Thus even though much of the current water infrastructure exists to supply water to agriculture, there is no guarantee or even likelihood that this arrangement will continue into the distant future. Hardheaded economic calculations may well bring more voters and politicians to the point of view that cities, suburbs, and industries might be the more appropriate users of the water than farms and ranches. Thus it becomes part of the job description for the scientist or engineer specializing in western water to keep a close eye on the fortunes of the idea known as Jeffersonian agrarianism, which throughout its history has provided a prime case study that abstract beliefs, situated in the human mind, have an enormous power to shape material reality.

2. Opponents to Growth May Discover a Way to Keep People Out (and Whether It Will Be Fair and Equitable Is a Whole Other Topic).

For a historian to whom the nineteenth century seems immediate and real, the recent shift in attitudes toward population growth can leave one breathless. Through the nineteenth century and into the twentieth, it was the ardent desire of most western settlers and residents to *have more people come to join them.* Manless lands called for landless men, several thousand promoters proclaimed. Land companies wanted purchasers; railroads wanted passengers and people producing crops or cattle or minerals in need of shipping; merchants wanted customers; politicians wanted voters; new settlers wrote persuasively to get relatives and friends to follow them. The booster was the representative figure, and a brighter future meant a better-populated future. Of course, in some situations like mining rushes, arriving late and finding too many people already in place, with the most profitable sites already claimed, could be a disappointment. But the assumption that population growth would mean a comparable increase in well-being had many, many believers. Water engineers provided the crucial infrastructure to support this habit of mind and of behavior.

At the turn of the millennium, the Interior West is the fastest-growing region in the country, and this state of affairs is no longer an occasion of universal happiness and celebration. Many westerners are deeply worried about the costs and burdens of growth, and the plans put forward to control growth cover a wide range of strategies and justifications. And yet the ethical position of people who would shut the door after their own quite recent arrival is a wobbly and precarious one; if the presence of too many people is the problem, then each person is himself a contributor to the dilemma, despite many vigorous efforts to claim an exemption as a true resident living in the proper spirit with the place. Worldwide, definitions of bearable population density vary enormously; one trip to Tokyo, London, or even New York City can call into question the idea that the West has become badly overcrowded. Moreover, this country has a long tradition of respecting a right to free mobility; occasions like slavery, Indian removal, or the relocation of Japanese Americans during World War II, in which people were told where they could and could not live, are not in the nation's finest tradition. Thus a fair and equitable form of prohibiting or even severely limiting new arrivals to an area is not an easy plan to design.

Growth control may prove to be mostly a matter of rhetoric: a free-flowing stream of words and declarations running parallel to an almost equally free-flowing, continuing stream of migration into the region. But then again, a movement that now has so many champions seems likely to have a significant impact, making calculations of residential growth—with accompanying water demand—misleading. If the flow of in-migration continues, then the clash between agricultural uses and residential uses of water will certainly escalate. Every morning's newspaper in the West offers another set of clues as to the degree to which antigrowth sentiment will produce results. Surely a prolonged drought, a pattern familiar from this region's past, would be the most consequential method of growth control imaginable.

3. Enthusiasm for the Reversal of Time and "Progress" May Grow in Power.

Reading the boosters of the nineteenth century as they exulted in the growth of population is now an exotic experience, proof of the often-used phrase that "the past is another country." In this instance and in quite a number of others, a striking number of Americans have taken what was once understood to be "progress" and recast it as affliction. Population growth, predator extermination, irrigated farming, mining, and dam building—many activities nineteenth-century westerners saw as progress—find themselves with a dramatically changed valuation. Activities that once flew under the flag of the "improvement" of raw nature now register as the degradation of ecosystems, and a whole new business of restoration, repair, and rehabilitation has arisen. Once a sign of the human capacity to build a better world, a dam can now be cited as an example of how people throw their power around and make a mess of things. The movement to breach or remove dams offers only one example of this growing trend of perception. Many, if not most, western advocates of reversing the meaning of progress live in ways made possible by the existing western water infrastructure—drinking water, watering plants, and typing their activist declarations on computers powered by hydroelectric power. Humans, as many episodes in history besides this one show, dwell in a medium of irony as fish live in water.

4. Federal Agencies Have Their Work Cut Out for Them in Regaining Flexibility and Credibility; In That Cause, It Might Help If They Admitted That They Brought Some of Their Troubles on Themselves.

Popular feeling toward the federal government has never been anything but complicated. The Articles of Confederation set up a weak central government; the Constitution then added some punch to federal power but also hedged it with checks and balances. For this new central government, westward expansion was the equivalent of weight training: in conquering Indian people, surveying and distributing land, subsidizing railroads, setting up new territorial governments, and building dams, the federal government built its muscles. Not surprisingly, the West as a region has remarkably intense and mixed feelings about federal agencies. Just as teenagers are at once very dependent on their parents and very resentful of that dependence, so westerners both rely on and complain about the feds. Add to this the divisions in western opinion, between environmentalists and resource users, and federal employees find themselves in an increasingly tough situation.

And yet it is necessary to recognize that federal agencies made their own solid contribution to the problem. A stance of arrogance, a posture of expertise beyond challenge, an exercise of power that sometimes did indeed look a lot closer to autocracy than to consultative democracy: the federal agencies did help to make the bed they now occupy. For episodes of its history, the Bureau of Reclamation was riding high, well supplied with money and congressional support. Error—whether it consisted of mistaking a period of high water flow as "normal" (as in the basis for allocation of the Colorado River) or the underestimate of siltation or evaporation in reservoirs—has had consequences. Federal agencies now face quite an accounting for the outcome of an earlier confidence that government experts could figure out nature and direct it to human needs. And this accounting has to take place in a political era in which distrust of the federal government has been mounting.

Can the Bureau of Reclamation and other federal agencies working in the management of natural resources address the distrust directed at them? Has the passage of time rendered them unavoidably inflexible, set in the pursuit of missions that have lost the setting in time in which they once made (more or less!) sense? Is it imaginable that federal employees could make a conscious and thorough effort to regain flexibility and to speak to the public with an honesty that recognizes and deals with the cynicism many feel toward any and all public officials? Imaginable or not, moves in that direction had better happen anyway if expertise in water issues is going to be directed toward a more appropriate fit between long-term and short-term agendas.

5. The Decentralization of Resource Decision Making Could Be Either a Dream Seen Through a Soft-Focus Lens or a Virtuous and Practical Alternative. Watershed Coalitions May Become Collaborative and Symbiotic With Water Experts, or, Then Again, Public Participation May Drive Us All Batty Before It Gets a Chance to Redeem Us.

From the origins of the conservation movement, the project of centrally planned resource use moved into a difficult relationship with the theory and practice of democracy. Advocates of conservation, like Theodore Roosevelt and Gifford Pinchot, frequently declared that wise resource use was in the best interests of "the people," and yet experts would make the key decisions. The most recent response to this contradiction between democratic rhetoric and elite decision making has been a widespread movement to form watershed coalitions, groups of "stakeholders" representing different occupations and philosophical positions but sharing a loyalty to their places of residence. Enthusiasm for this movement can sometimes make the observer a little nervous: in the pursuit of shared understanding and common goals, will scientific expertise become just another point of view, an equal (but no more than equal) occupant of a "seat at the table" of decision making? No doubt scientific expertise has a number of episodes of hubris to do penance for; no doubt science expertise could profit from repeated reminders of the constraints that social and cultural conditions pose. But it does seem more than possible that the penalty for past arrogance could be set too high, to a point where skepticism toward expertise deprives watershed residents of knowledge that will, in fact, determine the success or failure of their undertakings. For the historian, each of these watershed coalitions is its own instructive exploration of the question, Can natural-resource management and democratic process work together in ways we haven't had a chance to imagine over the last century?

And perhaps most consequentially, there is a question of patience and efficiency. All through the field of resource management, federal and state agencies have mandated all sorts of processes, venues, and arenas for "public participation." Whatever its virtues in the restoration of democratic faith, public participation is time-consuming; hours, days, weeks, months, years can pass while the discussion continues, and feelings are shared, and objections are raised, and concerns are expressed, and alternatives are considered. Do we have the capacity for endurance we will need, sitting in our chairs and listening to stakeholder after stakeholder? Given that question of efficiency, what are the prospects that public participation will help us to arrive at wiser ways of water use and allocation?

6. Scientists Could Go a Lot Further in the Cause of Taking Control of Their Public Image and Paying More Attention to What They Say and How They Say It.

However long they end up sitting at public hearings, forums, meetings, and convocations, scientists are going to have to absorb a cut in self-esteem. Because of unforeseen consequences from various scientifically arrived-at prescriptions and recommendations, as well as from an entrenched muddle when it comes to laypeople's understanding of differences in scientific opinion, public respect for expertise is not on the rise. If science sits on a pedestal, it is a pedestal with a bad wobble at its base. Scientists who simply assert their authority and feel resentful when their judgments are not deferred to are in for a particularly rough time. Some admissions of modesty and humility, and even some recognitions of past error and overconfidence, will play an important part in the construction of a more limber and more lasting version of scientific authority.

Scientists working on western environmental issues have one great asset available to them, waiting for them to recognize its usefulness and take possession of it. In the nineteenth century the best writers about the American West were naturalists and scientists. The most vivid descriptions of landscapes, flora and fauna, local people, and the experience of travel came from the pens of scientists. Can western scientists repossess their heritage as the West's best communicators? Do they *want* to repossess this heritage; can they *afford to* in a professional world in which publication in scholarly journals, with the target audience of other specialists, matters so much to career advancement? And yet the West's literary heritage contains some extraordinary writing—by nonscientists as well as scientists—on blizzards, rainstorms, fierce winds, and floods. Why not put some of those powerful passages to work on behalf of public understanding of climate change, even if such writing might run afoul of the outside referees of an academic journal?

7. Climatologists Will Have an Important Role to Play in Challenging Urban Complacency About Water Supply, but They Will Have to Be Pretty Agile to Avoid Getting Shot as Messengers Carrying Bad News.

Underlying most projections for future water use and delivery in the West are assumptions about the reliability of "normal" rainfall. And yet the record of precipitation in the region discloses prolonged periods of drought. How can this longer-range history of variation in what constitutes "normal" rainfall come into play in public discussion and planning? If climatologists don't get this message across, it is hard to think who else could. And yet a reluctance to take up this mission would be perfectly understandable, since the message of long-term climate variability—if taken seriously—would be so unsettling to boosters, developers, agriculturalists, and hundreds and thousands of people living in western cities and suburbs, complacently and unthinkingly turning faucets on and off and assuming that this comfortable arrangement will go on in perpetuity. Given the peevishness likely to come the way of the messenger, who can blame a climatologist who chooses to confine his communications to an audience of fellow specialists?

Besides the prospect of anger in response to the "bad news" of drought presented by the idea of climate variability, the climatologist attempting to communicate with the public on this issue has to reckon with one of the principal vexations in specialist/layperson communication: the difficulty the general public has in understanding the workings of probability and uncertainty as scientists and water managers would like to have those concepts understood. In a nutshell, nonspecialists will try to find, in a statement about the probability of a future drought, the answer to the question, What will happen to my lawn (or garden or farm or swimming pool)? Given this widespread habit of mind, how can natural scientists and water engineers and managers best approach the "re-education" of the public audience so that they can hear the information on probability as the experts intended for it to be heard?

8. It Will Take *Very* Creative Thought and a Lot of Intelligent Foresight If Environmental Protection Goals Are Going to Survive a Future Drought.

For nineteenth-century and early-twentieth-century western boosters, the fact that water flowed down riverbeds and dissipated into the sea was the very definition of waste. Coming out of that heritage of thought, the idea of maintaining in-stream flows really is revolutionary and transformative. Of course, the luxury of sufficient rainfall makes it possible to discuss and act on this remarkable idea that some water finds its best "use" by staying in the river and supporting aquatic life. Remove that luxury of sufficient rainfall, and the urgency of users' demands would seem to return the goal of maintaining stream flow to its earlier status as the embodiment of wasted resources. If the goals of environmental protection are going to survive an era of scarcity, the supporters of those goals are going to have to be a lot more innovative and forceful in their justifications and persuasions than they are at present. It would be wonderful if the climatologists who talk about future changes in rainfall could anticipate this dilemma and (with appropriate modesty and humility, of course!) use their expertise to direct society toward the most farsighted understanding of what we would lose if environmental protection were entirely sacrificed to other uses.

9. It Was Once Possible to Talk About Our Visions and Hopes for Society *at the Same Time* That We Talked About Hydrographs, Acre Feet, E-T, Release Planning, and Storage, and It Could Be Possible to Do That Again.

No doubt the reclamation movement of a century ago pushed us down the road toward a snarl of unintended consequences and conflicting uses. And yet there is no mistaking the fact that *ideals* played a big part in powering that movement: hopes for extending opportunity to those who might otherwise miss out on it, dreams of an egalitarian society where a growing distance between the rich and the poor did not fracture society, yearnings for a government that could pool society's resources and direct them toward mutually beneficial ends.

In contemporary discussions of water resources, technicalities and specificity rule. As an observer I suspect that the experts, who deal so comfortably in the language and assumptions of "water work," are, more often than not, guided by their own ideals, hopes, dreams, and yearnings. What sort of society would they like to have the water infrastructure support? How can we work out some sort of match between our manipulation of western water resources and our visions of what the West should be and how westerners should live, with their natural setting and with each other? What are the dreams of scientists and engineers, and, in a nation that values freedom of expression, shouldn't we encourage scientists and engineers to tell us those dreams, to listen to ours, and to offer us guidance in how to get the best possible match between our use of natural resources and our hopes for a just, generous, and yet provident society?

About the Contributors

Tom Cech, Central Colorado Water Conservancy District, Greeley.

Martyn P. Clark, Cryospheric and Polar Processes Division, Cooperative Institute for Research in Environmental Sciences, University of Colorado, Boulder.

Randall M. Dole, NOAA-CIRES Climate Diagnostics Center, Boulder, Colorado.

Terrance J. Fulp, U.S. Bureau of Reclamation, Lower Colorado Region.

Luis Garcia, Department of Chemical and Bioresource Engineering, Colorado State University, Fort Collins.

David H. Getches, University of Colorado School of Law, Boulder.

Steven Gloss, University of Wyoming, Laramie.

Alan F. Hamlet, Department of Civil and Environmental Engineering and Joint Institute for the Study of Atmosphere and Ocean, Climate Impacts Group, University of Washington, Seattle.

Lauren E. Hay, U.S. Geological Survey, Denver Federal Center, Lakewood, Colorado.

Brian Hurd, Department of Agricultural Economics and Agricultural Business, New Mexico State University, Las Cruces.

Jessica Koteen, Sapphos Environmental Inc., Pasadena, California.

John Labadie, Department of Civil Engineering, Colorado State University, Fort Collins.

George H. Leavesley, U.S. Geological Survey, Denver Federal Center, Lakewood, Colorado.

Dennis P. Lettenmaier, Department of Civil and Environmental Engineering and Joint Institute for the Study of Atmosphere and Ocean, Climate Impacts Group, University of Washington, Seattle.

William M. Lewis Jr., Center for Limnology, Cooperative Institute for Research in Environmental Sciences, University of Colorado, Boulder.

Patricia Nelson Limerick, Department of History and Center of the American West, University of Colorado, Boulder.

John Loomis, Department of Agricultural and Resource Economics, Colorado State University, Fort Collins.

Dave Matthews, U.S. Bureau of Reclamation, Denver, Colorado.

Gregory J. McCabe, U.S. Geological Survey, Denver Federal Center, Lakewood, Colorado.

Kathleen Miller, University Center for Atmospheric Research, Boulder, Colorado.

Phil Pasteris, National Water and Climate Center, Natural Resources Conservation Service, U.S. Department of Agriculture, Portland, Oregon.

Kelly T. Redmond, Western Regional Climate Center, Desert Research Institute, Reno, Nevada.

Rene Reitsma, Department of Information Systems, University of St. Francis Xavier, Antigonish, Nova Scotia, Canada.

Mark C. Serreze, Cryospheric and Polar Processes Division, Cooperative Institute for Research in Environmental Sciences, University of Colorado, Boulder.

Eugene Stakhiv, Institute for Water Resources, U.S. Army Corps of Engineers, Alexandria, Virginia.

Kenneth M. Strzepek, Civil, Environmental, and Architectural Engineering, University of Colorado, Boulder.

William R. Travis, Department of Geography, University of Colorado, Boulder.

C. Booth Wallentine, Utah Farm Bureau Federation, Sandy.

Robert L. Wilby, Department of Geography, University of Derby, Derby, United Kingdom.

Connie A. Woodhouse, Institute of Arctic and Alpine Research, University of Colorado, Boulder.

David N. Yates, Civil, Environmental, and Architectural Engineering, University of Colorado, Boulder, and National Center for Atmospheric Research, Boulder, Colorado.

Index

Page numbers in italics indicate figures or tables; "n" indicates a note.